DRILLING DEEPER

A Reality Check on U.S. Government Forecasts
for a Lasting Tight Oil & Shale Gas Boom

By J. David Hughes

post carbon institute

> ## Visit **postcarbon.org/drilling-deeper**
> ## for more information and related resources.

About the Author

David Hughes is a geoscientist who has studied the energy resources of Canada for four decades, including 32 years with the Geological Survey of Canada as a scientist and research manager. He developed the National Coal Inventory to determine the availability and environmental constraints associated with Canada's coal resources. As Team Leader for Unconventional Gas on the Canadian Gas Potential Committee, he coordinated the publication of a comprehensive assessment of Canada's unconventional natural gas potential. Over the past decade, Hughes has researched, published, and lectured widely on global energy and sustainability issues in North America and internationally.

In 2011, Hughes authored a series of papers on the production potential and environmental impacts of U.S. natural gas. In early 2013, he authored *Drill, Baby, Drill: Can Unconventional Fuels Usher in a New Era of Energy Abundance?*, which took a far-ranging and painstakingly researched look at the prospects for various unconventional fuels to provide energy abundance for the United States in the 21st century. In late 2013 he authored *Drilling California: A Reality Check on the Monterey Shale*, which critically examined the U.S. Energy Information Administration's (EIA) estimates of technically recoverable tight oil in the Monterey Shale, which the EIA claimed constituted two-thirds of U.S. tight oil; the EIA subsequently wrote down its resource estimate for the Monterey by 96%. In early 2014 he authored *BC LNG: A Reality Check*, which examined the issues surrounding the proposed massive scale up of shale gas production in British Columbia for LNG export.

Hughes is president of Global Sustainability Research, a consultancy dedicated to research on energy and sustainability issues. He is also a board member of Physicians, Scientists & Engineers for Healthy Energy (PSE Healthy Energy) and is a Fellow of Post Carbon Institute. Hughes contributed to *Carbon Shift*, an anthology edited by Thomas Homer-Dixon on the twin issues of peak energy and climate change, and his work has been featured in *Nature, Canadian Business, Bloomberg, USA Today*, as well as other popular press, radio, and television.

About Post Carbon Institute

Post Carbon Institute's mission is to lead the transition to a more resilient, equitable, and sustainable world by providing individuals and communities with the resources needed to understand and respond to the interrelated economic, energy, and ecological crises of the 21st century.

Acknowledgements

The author would like to thank geoscientist David Dean for his insightful review and helpful comments from the perspective of a long-term industry insider. Asher Miller and Daniel Lerch provided in-depth reviews and many helpful comments and suggestions. Daniel Lerch also provided tireless editorial services. John Van Hoesen provided GIS services and prepared the maps for each play. The report also benefited from contributions and exchanges with many other colleagues on all aspects of energy—usually on a daily basis.

ABSTRACT

Drilling Deeper reviews the twelve shale plays that account for 82% of the tight oil production and 88% of the shale gas production in the U.S. Department of Energy's Energy Information Administration (EIA) reference case forecasts through 2040. It utilizes all available production data for the plays analyzed, and assesses historical production, well- and field-decline rates, available drilling locations, and well-quality trends for each play, as well as counties within plays. Projections of future production rates are then made based on forecast drilling rates (and, by implication, capital expenditures). Tight oil (shale oil) and shale gas production is found to be unsustainable in the medium- and longer-term at the rates forecast by the EIA, which are extremely optimistic.

This report finds that tight oil production from major plays will peak before 2020. Barring major new discoveries on the scale of the Bakken or Eagle Ford, production will be far below the EIA's forecast by 2040. Tight oil production from the two top plays, the Bakken and Eagle Ford, will underperform the EIA's reference case oil recovery by 28% from 2013 to 2040, and more of this production will be front-loaded than the EIA estimates. By 2040, production rates from the Bakken and Eagle Ford will be less than a tenth of that projected by the EIA. Tight oil production forecast by the EIA from plays other than the Bakken and Eagle Ford is in most cases highly optimistic and unlikely to be realized at the medium- and long-term rates projected.

Shale gas production from the top seven plays will also likely peak before 2020. Barring major new discoveries on the scale of the Marcellus, production will be far below the EIA's forecast by 2040. Shale gas production from the top seven plays will underperform the EIA's reference case forecast by 39% from 2014 to 2040, and more of this production will be front-loaded than the EIA estimates. By 2040, production rates from these plays will be about one-third that of the EIA forecast. Production from shale gas plays other than the top seven will need to be four times that estimated by the EIA in order to meet its reference case forecast.

Over the short term, U.S. production of both shale gas and tight oil is projected to be robust—but a thorough review of production data from the major plays indicates that this will not be sustainable in the long term. These findings have clear implications for medium and long term supply, and hence current domestic and foreign policy discussions, which generally assume decades of U.S. oil and gas abundance.

Even as we've become less hooked on crude, we've become more addicted to drilling.

— Randy Udall (1951-2013)

Contents

PART 1: EXECUTIVE SUMMARY

By Asher Miller, Executive Director, Post Carbon Institute

PART 1: EXECUTIVE SUMMARY - CONTENTS

PART 1: EXECUTIVE SUMMARY - FIGURES

1.1 INTRODUCTION

In recent years Americans have been hearing that the United States is poised to regain its role as the world's premier oil and natural gas producer, thanks to the widespread use of horizontal drilling and hydraulic fracturing ("fracking"). This "shale revolution," we're told, will fundamentally change the U.S. energy picture for decades to come—leading to energy independence, a rebirth of U.S. manufacturing, and a surplus supply of both oil and natural gas that can be exported to allies around the world. This promise of oil and natural gas abundance is influencing climate policy, foreign policy, and investments in alternative energy sources.

The primary source for these rosy expectations of future production is the U.S. Department of Energy (DOE). Each year the DOE's Energy Information Administration (EIA) releases its *Annual Energy Outlook* (AEO)[1], which provides a range of forecasts for energy production, consumption, and prices.

The 2014 AEO reference case projects U.S. crude oil production to rise to 9.6 million barrels of oil per day (MMbbl/d) in 2019 and slowly decline to 7.5 MMbbl/d by 2040, while natural gas production is projected to grow for at least the next 25 years and hit 37.5 trillion cubic feet per year in 2040. Tight oil (shale oil) and shale gas serve as the foundation for these optimistic forecasts.

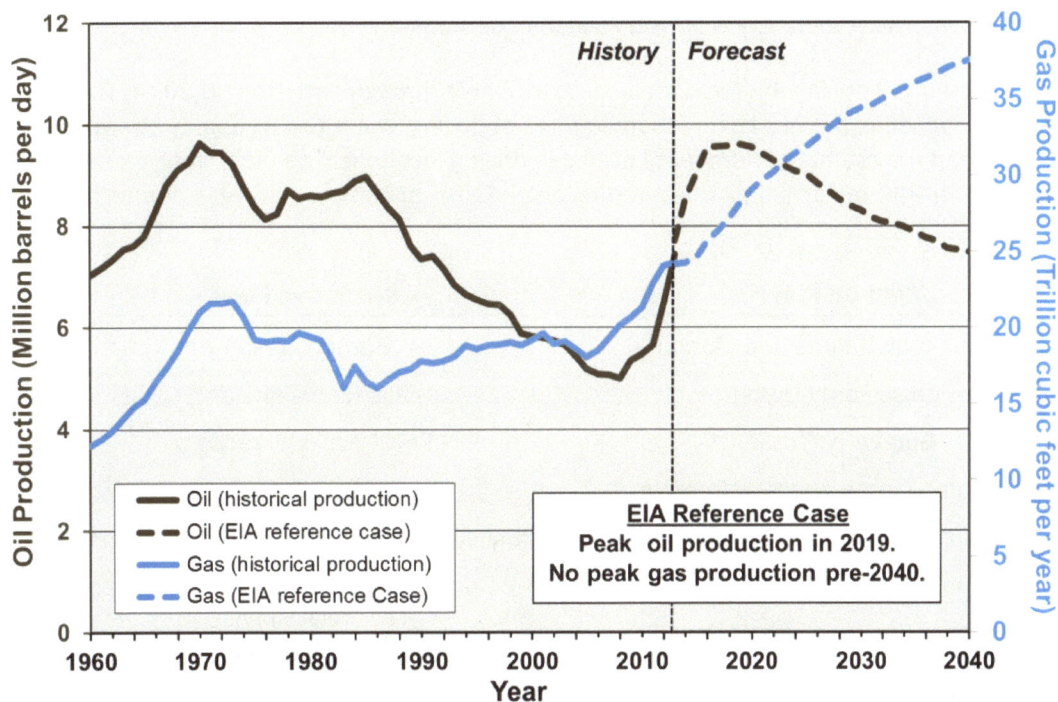

Figure 1-1. History and EIA reference case forecast of U.S. oil and natural gas production, 1960 to 2040.[2]

---

[1] EIA, *Annual Energy Outlook 2014*, http://www.eia.gov/forecasts/aeo/.
[2] EIA, *Annual Energy Outlook 2014*, http://www.eia.gov/forecasts/aeo/.

This report provides an extensive analysis of actual production data from the top seven tight oil and seven shale gas plays in the U.S. (These plays account for 89% of current tight oil production and 88% of current shale gas production, and serve as the primary sources of future production in the EIA's forecasts—82% of forecast tight oil and 88% of forecast shale gas production through 2040.) It concludes that the current boom in domestic oil and gas production is unsustainable at the rates projected by the EIA, and that the EIA's tight oil and shale gas forecasts to 2040 are extremely optimistic. What this means is that the country's current energy policy—which is largely based on the expectation of domestic oil and natural gas abundance far into the future—is badly misguided and is setting the country up for a painful, costly, and unexpected shock when the boom ends.

1.2 ABOUT THE REPORT

Drilling Deeper: A Reality Check on U.S. Government Forecasts for a Lasting Tight Oil & Shale Gas Boom was authored by J. David Hughes on behalf of Post Carbon Institute. The report investigates whether the EIA's expectation of long-term domestic oil and natural gas abundance is founded. It aims to gauge the likely future of U.S. tight oil and shale gas production based on an in-depth assessment of actual well production data from the major shale plays. The primary source of data for this analysis is Drillinginfo, a commercial database of well production data widely used by industry and government, including the EIA.[3] Drillinginfo also provides a variety of analytical tools which proved essential for the analysis.

This analysis is based on all drilling and production data available through early- to mid-2014. The report determined future production profiles given assumed rates of drilling, average well quality by area, well- and field-decline rates, and the estimated number of available drilling locations. The plays analyzed (which collectively account for 89% of current tight oil production and 88% of current shale gas production) are as follows:

Tight Oil Plays[4]	Shale Gas Plays
Bakken (North Dakota and Montana)	Barnett (Texas)
Eagle Ford (Texas)	Haynesville (Louisiana and Texas)
Spraberry (Texas)	Fayetteville (Arkansas)
Wolfcamp (Texas and New Mexico)	Woodford (Oklahoma)
Bone Spring (Texas and New Mexico)	Marcellus (Pennsylvania and West Virginia)
Austin Chalk (Gulf Coast Region)	Bakken (North Dakota and Montana; associated gas)
Niobrara (Colorado and Wyoming)	Eagle Ford (Texas; associated gas)

---

[3] See http://info.drillinginfo.com.
[4] The Monterey tight oil play in California was assessed in a previous report by this same author: J. David Hughes, *Drilling California: A Reality Check on the Monterey Shale*, Post Carbon Institute, 2013, http://www.postcarbon.org/publications/drilling-california.

The EIA's Poor Track Record

Policymakers, media, investors, and the general public typically receive the Department of Energy's EIA forecasts with little to no circumspection, despite their poor track record. In 2011, the EIA was forced to cut its estimates of technically recoverable shale gas in the Marcellus play by 80%[1] and in Poland by 99%[2] after the United States Geological Survey came out with much lower numbers. At the time of the Marcellus downgrade, an EIA spokesperson said, "We consider the USGS to be the experts in this matter... They're geologists, we're not. We're going to be taking this number and using it in our model."[3] In early 2014, the EIA slashed its estimate of technically recoverable tight oil from California's Monterey Formation by a whopping 96%.[4] Just three years previously, the agency had estimated it held fully two-thirds of all U.S. tight oil. The author of the original EIA estimate, INTEK Inc., admitted that it had been derived from oil company presentations rather than hard data.[5] The EIA's downgrade occurred after this report's author, J. David Hughes, published an analysis six months earlier that showed—using actual production data from the Monterey Formation—that the EIA's estimates were wildly optimistic.[6]

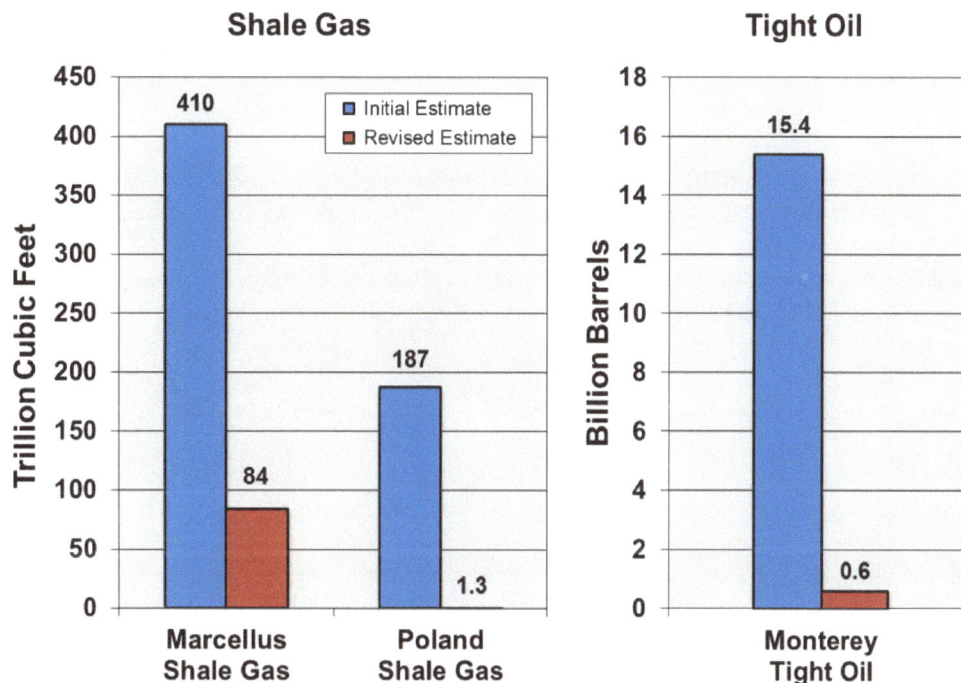

Initial EIA estimates of shale resources vs. revised estimates.

---

[1] Efstathiou, J. and Klimasinska, K., 23 August 2011, *Bloomberg*, "U.S. to Slash Marcellus Shale Gas Estimate 80%," http://www.bloomberg.com/news/2011-08-23/u-s-to-slash-marcellus-shale-gas-estimate-80-.html.

[2] Blake, M., September/October 2014, *Mother Jones*, "How Hillary Clinton's State Department Sold Fracking to the World," http://www.motherjones.com/environment/2014/09/hillary-clinton-fracking-shale-state-department-chevron.

[3] Efstathiou, J. and Klimasinska, K., "U.S. to Slash Marcellus Shale Gas Estimate 80%."

[4] Sahagun, L, 20 May 2014, *Los Angeles Times*, "U.S. officials cut estimate of recoverable Monterey Shale oil by 96%," http://www.latimes.com/business/la-fi-oil-20140521-story.html.

[5] Kern Golden Empire, 3 December 2013, "Report: Monterey Shale production 'wildly optimistic'," http://www.kerngoldenempire.com/story/report-monterey-shale-production-wildly-optimistic/d/story/VdOYdQZ-4UKgp7qNwqq8Xg.

[6] Hughes, J.D., 2013, *Drilling California: A Reality Check on the Monterey Shale*, Post Carbon Institute, http://www.postcarbon.org/publications/drilling-california.

1.3 KEY FINDINGS

The seven tight oil plays and seven shale gas plays analyzed in this report account for 82% of projected tight oil production and 88% of projected shale gas production through 2040 in the EIA's *Annual Energy Outlook 2014* reference case forecast. A detailed analysis of well production data from these plays resulted in these key findings:

1) Tight oil production from major plays will peak before 2020. Barring major new discoveries on the scale of the Bakken or Eagle Ford, production will be far below EIA's forecast by 2040.

    a) Tight oil production from the two top plays, the Bakken and Eagle Ford, will underperform EIA's reference case oil recovery by 28% from 2013 to 2040, and more of this production will be front-loaded than the EIA estimates.

    b) By 2040, production rates from the Bakken and Eagle Ford will be less than a tenth of that projected by EIA.

    c) Tight oil production forecast by the EIA from plays other than the Bakken and Eagle Ford is in most cases highly optimistic and unlikely to be realized at the rates projected.

2) Shale gas production from the top seven plays will likely peak before 2020. Barring major new discoveries on the scale of the Marcellus, production will be far below EIA's forecast by 2040.

    a) Shale gas production from the top seven plays will underperform EIA's reference case forecast by 39% from 2014 to 2040 period, and more of this production will be front-loaded than EIA estimates.

    b) By 2040, production rates from these plays will be about one-third that of the EIA forecast.

    c) Production from shale gas plays other than the top seven will need to be four times that estimated by EIA in order to meet its reference case forecast.

3) Over the short term, U.S. production of both shale gas and tight oil is projected to be robust—but a thorough review of the production data indicate that this will be unsustainable in the longer term. These findings have clear implications for current domestic and foreign policy discussions, which generally assume decades of U.S. oil and gas abundance.

Other factors that could limit production are public pushback as a result of health and environmental concerns, and capital constraints that could result from lower oil or gas prices or higher interest rates. As such factors have not been included in this analysis, the findings of this report represent a "best case" scenario for market, capital, and political conditions.

1.3.1 Tight Oil

The analysis shows that U.S. tight oil production cannot be maintained at the levels assumed by the EIA beyond 2020. The top two plays—Bakken and Eagle Ford—which account for more than 60% of current production, are likely to peak by 2017 and the remaining plays will make up considerably less of future production than has been forecast by the EIA. Rather than a peak in 2021 followed by a gradual decline to slightly below today's levels by 2040, total U.S. tight oil production is likely to peak before 2020 and decline to a small fraction of today's production levels by 2040.

1.3.1.1 General Findings

- The 3-year average well decline rates in the seven plays analyzed for this report (which collectively provide 89% of current U.S. tight oil production) range from 60% to 91%.

- The high decline rates of tight oil wells in these plays means that 43% to 64% of their estimated ultimate recovery (EUR) is recovered in the first three years.

- Field declines from the Bakken and Eagle Ford are 45% and 38% per year, respectively (this compares to 5% per year for large conventional fields). This is the amount of production that must be replaced each year with more drilling in order to maintain production at current levels (field decline is made up of all wells in a play—old and new—and hence is lower than first-year well declines).

- Based on production history, drilling locations, and declining well quality, this report found that 98% of the EIA's projected production from these seven plays has a "high" or "very high" optimism bias.

Play	Average 3-Year Well Decline Rate	Optimism Bias Rating of EIA's Forecast
Bakken	85%	High
Eagle Ford	79%	High
Spraberry	60%	Very High
Wolfcamp	81%	High
Bone Spring	91%	Low
Austin Chalk	85%	Very High
Niobrara	90%	High

- The EIA assumes that the equivalent of 100% of proved reserves and between 65% and 85% of its "unproved technically recoverable tight oil resources" will be recovered by 2040 for the plays analyzed. Considering that unproved, technically recoverable resources have no price constraints and only loose geological constraints, this is highly speculative.

- The EIA assumes that the U.S. will exit 2040 with tight oil production at levels only marginally less than today, at 3.2 MMbbl/d. A thorough analysis of the well production data suggests this is highly optimistic.

1.3.1.2 Forecasts for Bakken & Eagle Ford Tight Oil Plays

- The EIA's forecast of the timing of peak production in the Bakken and Eagle Ford is similar to this report, as is the rate of peak production.

- The EIA forecasts a much higher tail after peak production, with recovery of 19.2 billion barrels between 2012 and 2040, as opposed to 13.9 billion barrels forecast in this report.

- The EIA forecasts collective production from the Bakken and Eagle Ford to be a little over 1 million barrels per day in 2040. In contrast, the "Most Likely" drilling rate scenario presented in this report forecasts that production will fall to about 73,000 barrels per day by 2040.

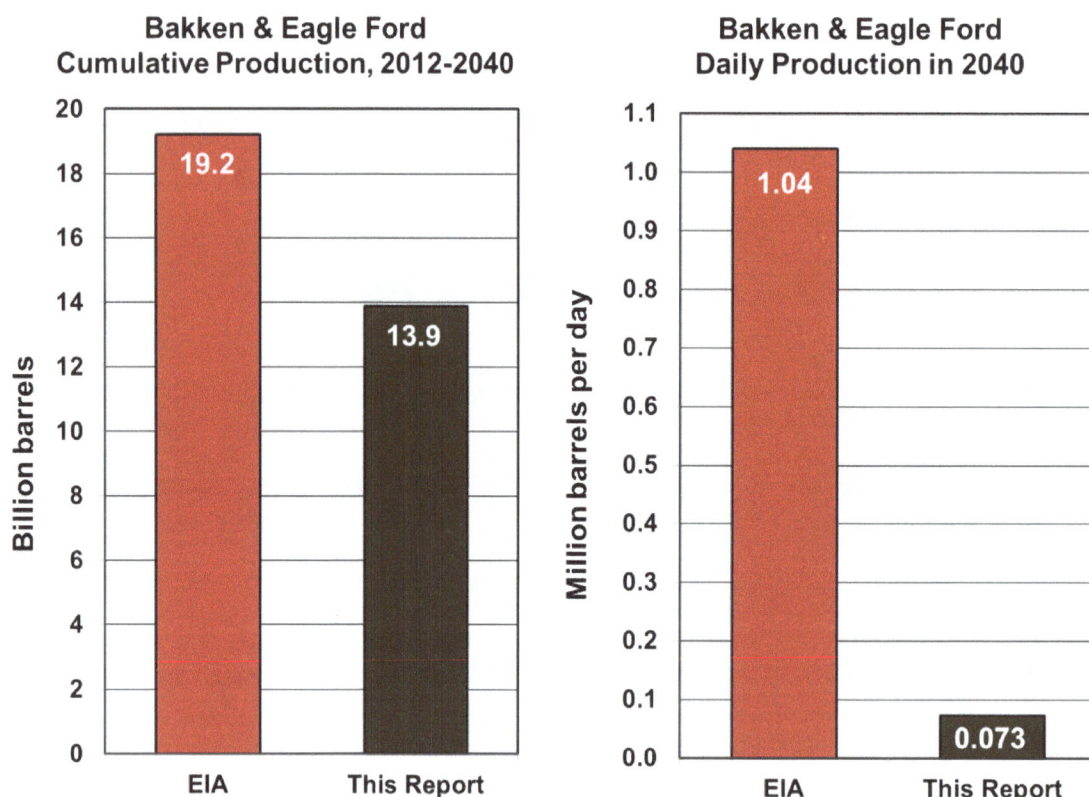

Figure 1-2. Bakken and Eagle Ford plays projected cumulative oil production from 2012 to 2040 and daily oil production in 2040, EIA projection[5] versus this report's projection.

---

[5] EIA, *Annual Energy Outlook 2014*, http://www.eia.gov/forecasts/aeo.

1.3.1.3 Forecasts for Other Tight Oil Plays

- To meet the EIA's forecasts, all other plays together would need to produce over twice as much through 2040 as what is projected for the Bakken and Eagle Ford.

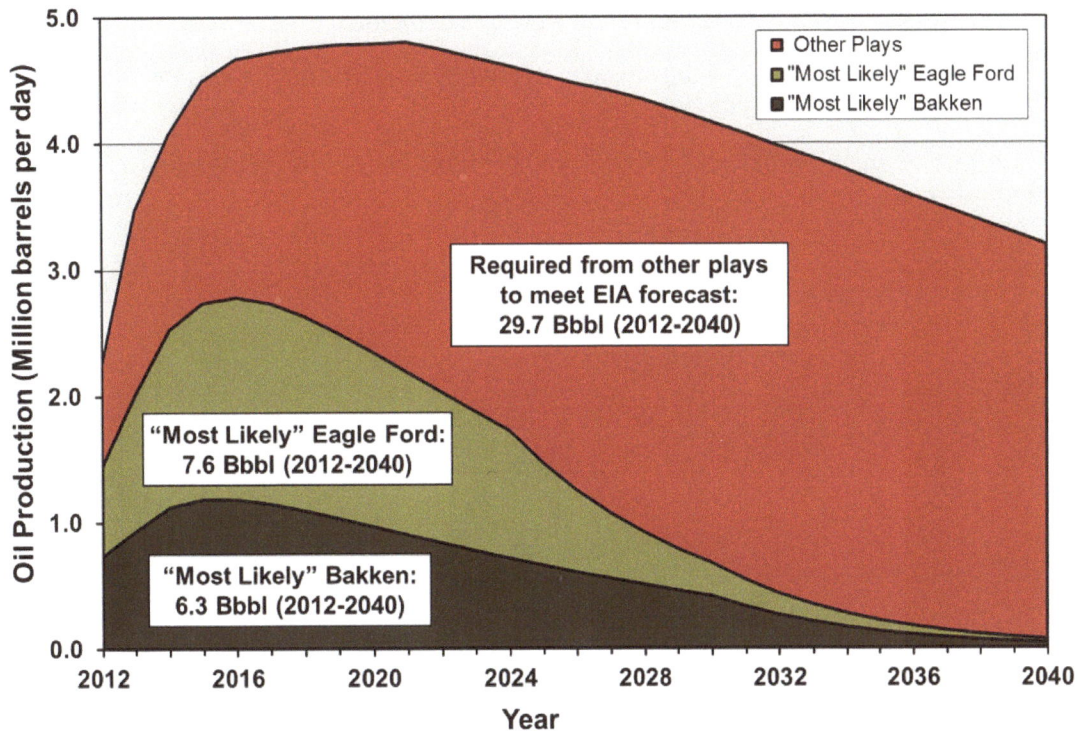

Figure 1-3. "Most Likely" scenario projections of oil production for the Bakken and Eagle Ford plays[6] with the remaining amount of production that would be required from other plays to meet the EIA's total reference case forecast.[7]

The EIA forecasts 43.6 billion barrels of U.S. tight oil will be recovered from 2012 to 2040. After subtracting the 13.9 billion barrels projected by this report for the Bakken and Eagle Ford, 29.7 billion barrels would remain to be produced from all other tight oil plays—5.3 billion barrels more than the EIA's already optimistic forecast for these plays.

---

[6] Data from Drillinginfo retrieved July 2014.
[7] EIA, *Annual Energy Outlook 2014*, Unpublished tables from AEO 2014 provided by the EIA.

- The major remaining tight oil plays are the three Permian Basin plays—Spraberry, Wolfcamp, and Avalon/Bone Spring—plus the Austin Chalk and the Niobrara. EIA forecasts expect these plays to produce four to five times their historical production in the next 26 years, but this is highly questionable, considering that:

  - These plays are already 40-60 years old, with tens of thousands of wells already drilled.

  - The Permian Basin plays' average initial well productivities are half or less the average of core counties in the Bakken or Eagle Ford.

  - The Bakken and Eagle Ford's average estimated ultimate recovery (EUR) per well is two to more than six times higher than that of these other plays.

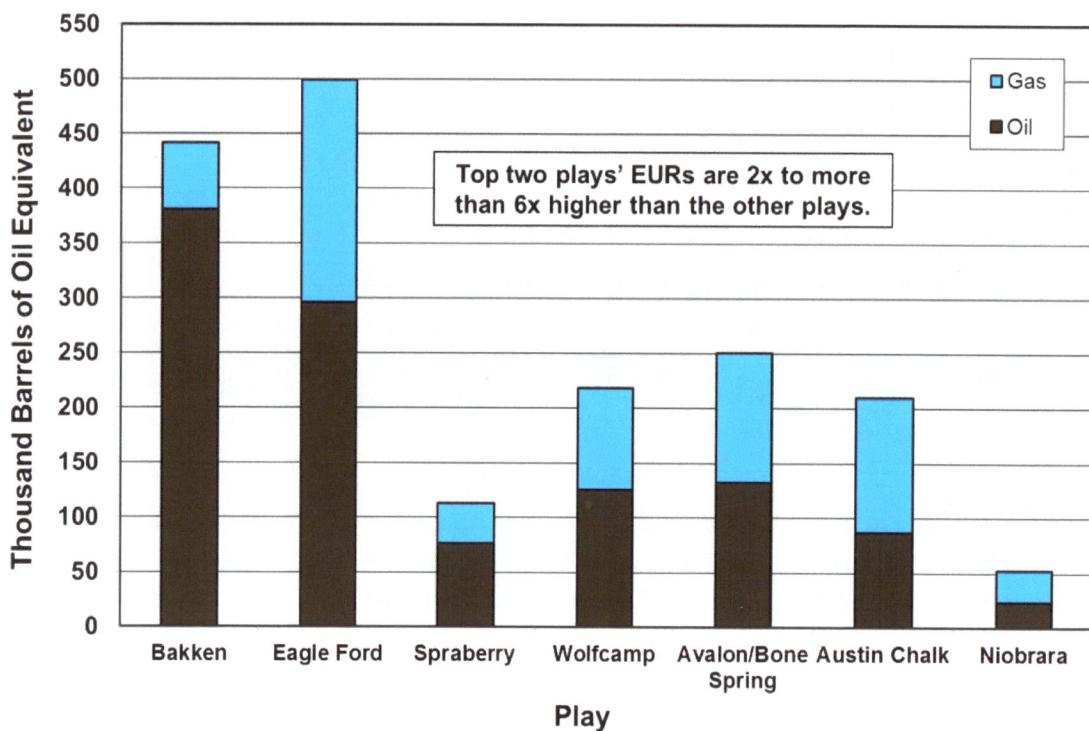

Figure 1-4. Estimated ultimate recovery (EUR) of oil and gas per well of reviewed plays, on a "barrels of oil equivalent" basis.[8]

The Bakken's and Eagle Ford's EURs per well are two to more than six times the EURs per well of the other five plays. If only horizontal wells are considered, the Bakken and Eagle Ford EURs per well are 39% to 141% higher than those of the other five plays (see discussion in Section 2).

---

[8] Based on data from Drillinginfo retrieved May-July 2014.

1.3.2 Shale Gas

The EIA now projects domestic gas production to reach nearly 38 trillion cubic feet per year by 2040, which is 55% above 2013 levels. The bulk of this production growth would come from shale gas.

This analysis shows that simply maintaining U.S. shale gas production in the medium term—let alone increasing production at rates forecast by the EIA through 2040—will be problematic. Four of the top seven shale gas plays are already in decline. Of the major plays, only the Marcellus, Eagle Ford, and Bakken (the latter two are tight oil plays producing associated gas) are growing; and yet, the EIA reference case gas forecast calls for plays currently in decline to grow to new production highs, at moderate future prices. Although significantly higher gas prices needed to justify higher drilling rates could temporarily reverse decline in some of these plays, the EIA forecast is unlikely to be realized.

1.3.2.1 General Findings

- The 3-year average well decline rates in the seven plays analyzed for this report (which collectively provide 88% of U.S. shale gas production) ranges between 74% and 82%.

- The average field decline rates for these plays ranges between 23% and 49%, meaning that between one-quarter and one-half of all production in each play must be replaced each year in order to simply maintain current production.

- Although the EIA forecast for the Marcellus play is rated as "reasonable" and its forecast for the Bakken play is rated "conservative," the deficit left by being "very highly optimistic" on some of the other plays makes finding and developing the gas required to meet the overall forecast unlikely.

Play	Average 3-Year Well Decline Rate	Average First-Year Field Decline Rate	Optimism Bias Rating of EIA's Forecast
Barnett	75%	23%	Very High
Haynesville	88%	49%	Very High
Fayetteville	79%	34%	Very High
Woodford	74%	34%	High
Marcellus	74-82%	32%	Reasonable
Eagle Ford	80%	47%	Very high
Bakken	81%	41%	Conservative

- Because productivity of shale wells declines rapidly, many new wells must be drilled just to maintain existing production levels. Of the top shale gas plays, only the Marcellus, Eagle Ford, and Bakken are currently seeing enough drilling to maintain and grow production.

- Major shale gas plays are variable in well quality. The Marcellus and Haynesville are much more productive on average than the other plays analyzed in this report. Even within plays, well quality varies considerably.

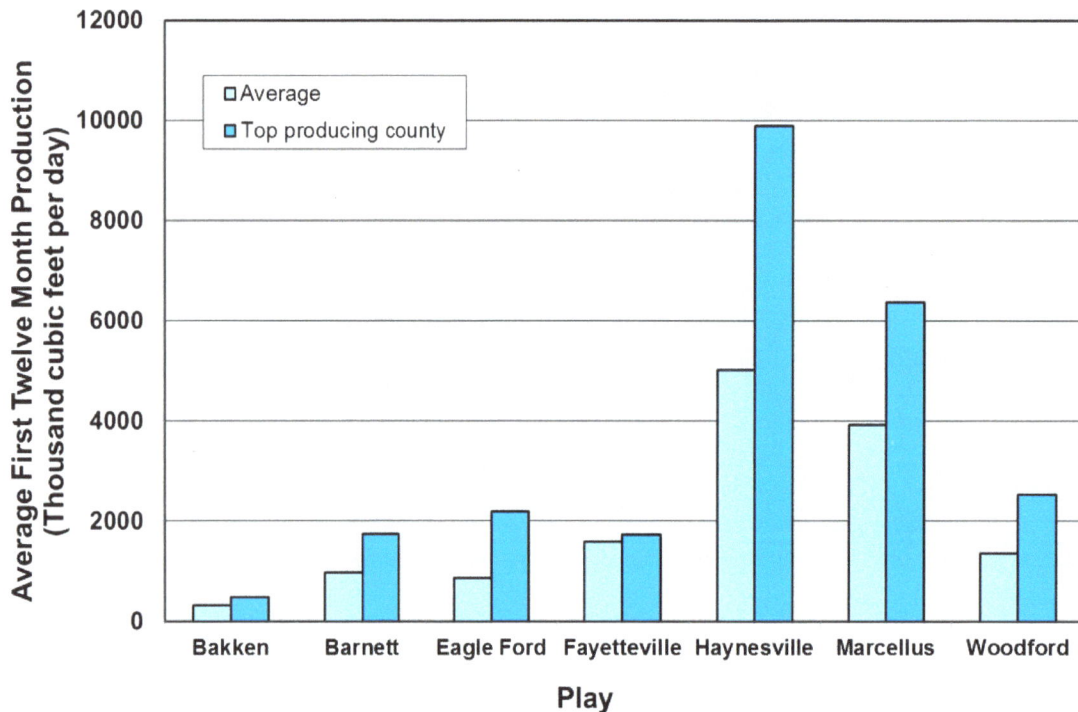

Figure 1-5. Average first-year gas production per well in 2013 from horizontal wells both play-wide and in the top-producing county for the plays analyzed in this report.[9]

---

[9] Data from Drillinginfo retrieved August to September 2014.

- Despite years of concerted efforts and claims that technological innovation can overcome steep well decline rates and the move from "sweet spots" to lower quality parts of plays, average well productivity has gone flat in all major shale gas plays except the Marcellus.

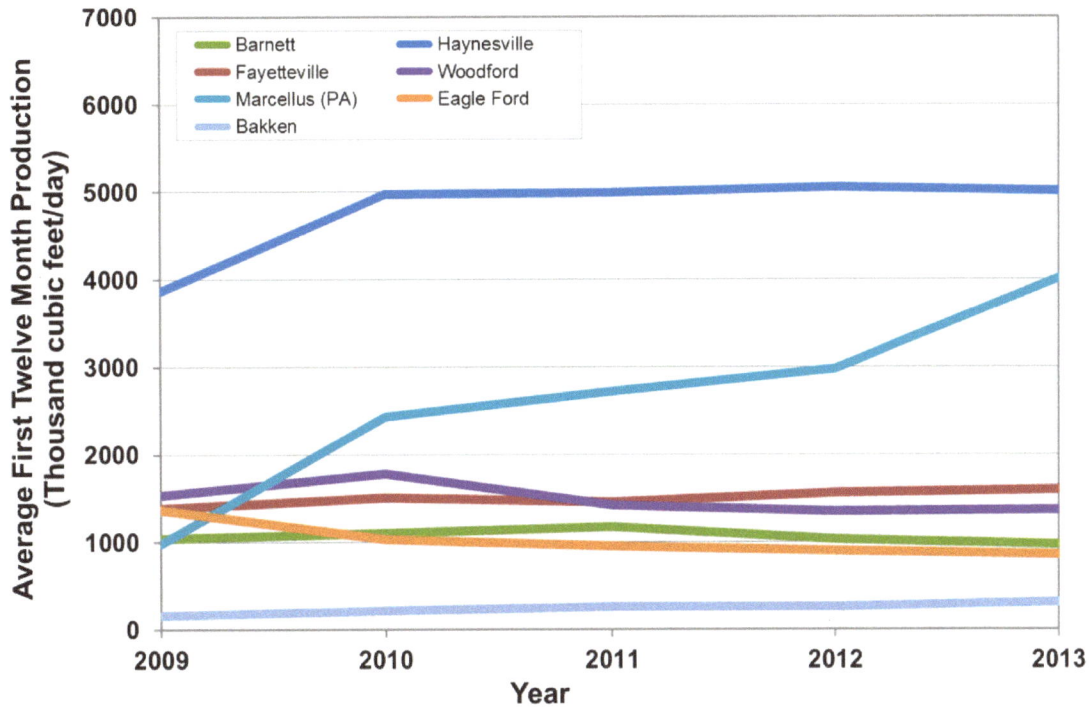

Figure 1-6. Average production over first twelve months per well for major U.S. shale gas plays.[10]

- Approximately 130,000 additional shale gas wells will need to be drilled by 2040 to meet the projections of this report, on top of the 50,000 wells drilled in these plays through 2013. Assuming an average well cost of $7 million, this would require $910 billion of additional capital input by 2040, not including leasing, operating, and other ancillary costs.

---

[10] Data from Drillinginfo retrieved August 2014.

1.3.2.2 Forecasts for Shale Gas Plays

- The EIA assumes that 74% to 110% of its "unproved technically recoverable resources" plus "proved reserves" will be recovered by 2040 for the seven major plays analyzed. Considering that unproved, technically recoverable resources have no price constraints and only loose geological constraints, this is highly speculative.

- This analysis found that the EIA reference case forecast for the top seven shale gas plays overestimates cumulative production through 2040 in this report's "Most Likely" scenario by 64%.

- The EIA further estimates that in 2040, shale gas production from the seven plays analyzed will be 182% higher (nearly 3 times) than estimated in this report—and that by 2040, another 49.6 Tcf will have been recovered from other plays not analyzed in this report.

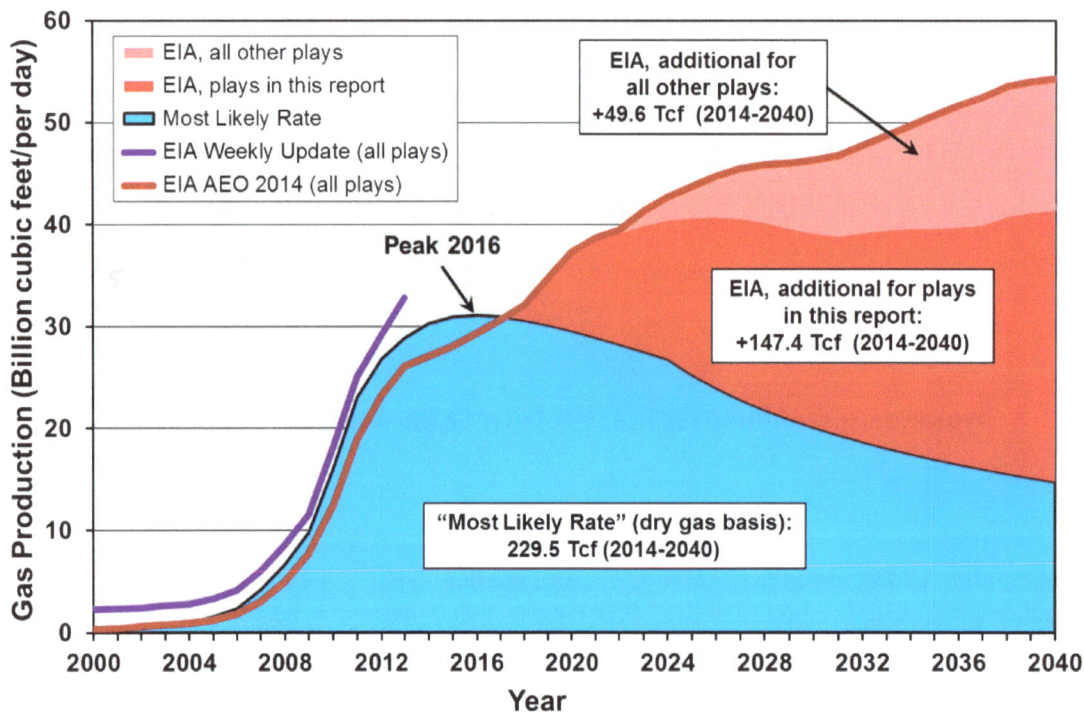

Figure 1-7. Totaled "Most Likely Rate" scenarios for the seven plays analyzed in this report, compared to the EIA's reference case forecast for these plays and for all plays.[11,12]

The "Most Likely Rate" scenario projections here are made on a "dry gas" basis. Also shown are the EIA's gas production statistics from its *Natural Gas Weekly Update*,[13] which contradict the early years of its AEO 2014 forecast.

---

[11] EIA, *Annual Energy Outlook 2014*, unpublished tables from AEO 2014 provided by the EIA.

[12] EIA, *Annual Energy Outlook 2014*, reference case forecast, Table 14, oil and gas supply, http://www.eia.gov/forecasts/aeo/excel/aeotab_14.xlsx.

[13] EIA, *Natural Gas Weekly Update*, retrieved October 2014, http://www.eia.gov/naturalgas/weekly.

- In this report's "Most Likely" scenario, cumulative dry shale gas production over the 2014-2040 period is 229.5 trillion cubic feet (Tcf)—46% lower than the EIA Reference Case (377 Tcf).

- In this report's "Most Likely" scenario, shale gas production from the seven plays analyzed peaks in the 2016-2017 timeframe and declines by more than half, to 14.8 billion cubic feet per day (Bcf/d) by 2040. In contrast, the EIA expects production from these plays to keep growing through 2040, with shale gas production in that year at 41.8 Bcf/d—nearly three times higher than this report finds justifiable.

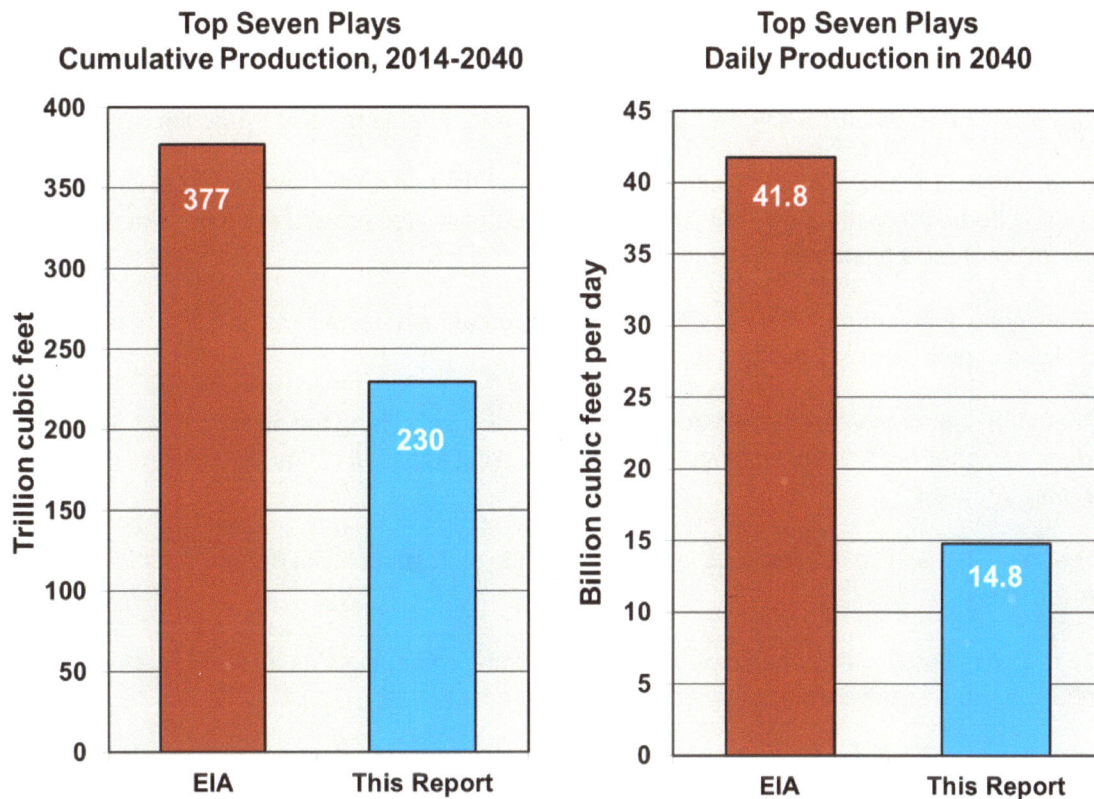

Figure 1-8. Projected cumulative gas production to 2040 and daily gas production in 2040, EIA projection[14] versus this report's projection.

The values given here are for the seven plays analyzed in this report. These plays constitute 88% of cumulative U.S. shale gas production from 2014 to 2040 in the EIA's reference case forecast.

---

[14] EIA, *Annual Energy Outlook 2014*, http://www.eia.gov/forecasts/aeo.

1.4 Implications

This report shows that the EIA's optimistic forecasts for future U.S. tight oil and shale gas production are based on a set of *false premises*, namely that:

- High-quality shale plays are ubiquitous, and there will be always be new discoveries and production from emerging plays to fill the gap left by declining production from major existing plays.

- Technological advances can overcome steep decline rates and declining well quality as drilling moves from sweet spots to poorer quality rock, in order to maintain high production rates.

- Large estimated resources underground imply high and durable rates of extraction over decades.

Actual production data from the past decade of shale gas and tight oil drilling clearly do not support these assumptions. Unfortunately, the EIA's rosy forecasts have led policymakers and the American public to believe a number of *false promises*:

- That cheap and abundant natural gas supplies can create a domestic manufacturing resurgence and millions of new jobs over the long term.[15]

- That abundant domestic oil and natural gas resources justify lifting the oil export ban (imposed 40 years ago after the Arab oil embargo)[16] and fast-tracking approval of liquefied natural gas (LNG) export terminals.[17]

- That the U.S. can use its newfound energy strength to shift geopolitical trends in our long-term favor.[18]

- That we can easily limit carbon dioxide emissions from power plants as a result of natural gas replacing coal as the primary source of electricity production.[19]

The promises associated with the expectation of robust and relatively cheap shale gas and high-cost but rising tight oil production have also led to a tempering of investments in renewable energy and nuclear power.[20] If, as this report shows, these premises and promises are indeed false, the implications are profound. It calls into question plans for LNG and crude oil exports and the benefits of the shale boom in light of the amount of drilling and capital investment that would be required, along with the environmental and health impacts associated with it. Conventional wisdom holds that the shale boom will last for decades, leaving the U.S. woefully unprepared for a painful, costly, and unexpected shock when the shale boom winds down sooner than expected. Rather than planning for a future where domestic oil and natural gas production is maintained at current or higher levels, we would be wise to harness this temporary fossil fuel bounty to quickly develop a truly sustainable energy policy—one that is based on conservation, efficiency, and a rapid transition to distributed renewable energy production.

---

[15] Nelson Schwartz, "Boom in Energy Spurs Industry in the Rust Belt," *New York Times*, September 8, 2014, http://nyti.ms/1qHoxXz.

[16] Jay Fitzgerald, "Pressure builds to allow US exports of crude," *Boston Globe*, September 21, 2014, http://bit.ly/1uDI0sP.

[17] Amy Harder, "House Passes Bill Speeding Up Liquefied Natural-Gas Exports," *Wall Street Journal*, June 25, 2014, http://on.wsj.com/1lsgKqN.

[18] Robert Blackwill and Meghan O'Sullivan, "America's Energy Edge: The Geopolitical Consequences of the Shale Revolution," *Foreign Affairs,* March/April 2014, http://www.foreignaffairs.com/articles/140750/robert-d-blackwill-and-meghan-l-osullivan/americas-energy-edge.

[19] Isaac Arnsdorf, "Fracking Sucks Money From Wind While China Eclipses U.S.," Bloomberg, May 29, 2014, http://bloom.bg/1iu9Y3m.

[20] Ibid.

PART 2: TIGHT OIL

DRILLING DEEPER   17   PART 2: TIGHT OIL

PART 2: TIGHT OIL - CONTENTS

PART 2: TIGHT OIL - FIGURES

PART 2: TIGHT OIL - TABLES

2.1 INTRODUCTION

2.1.1 Overview

The widespread adoption of hydraulic fracturing ("fracking") and horizontal drilling in the United States to extract oil and natural gas from previously inaccessible shale formations has been termed the "shale revolution." In just the last few years, U.S. oil production—universally held to be in terminal decline a mere decade ago—has grown rapidly and significantly thanks to oil produced from shales ("tight oil"). The U.S. Energy Information Administration (EIA) now projects domestic oil production to reach the previous 1970 peak of 9.6 million barrels per day (MMbbl/d) by 2019 and decline gradually to 7.5 MMbbl/d by 2040.[1]

The environmental, health, and quality of life impacts of shale development have stoked controversy across the country. In contrast, the expectation of long-term domestic oil abundance—driven by optimistic forecasts from industry and government—has been widely reported and little questioned, despite the myriad economic and policy consequences.

This report investigates whether the EIA's expectation of long-term domestic oil abundance is founded. It aims to gauge the likely future production of U.S. tight oil, based on an in-depth assessment of actual well production data from the major shale plays. It determines future production profiles given assumed rates of drilling, average well quality by area, well- and field-decline rates, and the estimated number of available drilling locations. This analysis is based on all drilling and production data available through early- to mid-2014.

The analysis shows that U.S. tight oil production cannot be maintained at the levels assumed by the EIA beyond 2020. The top two plays, which account for more than 60% of production, are likely to peak by 2017 and the remaining plays will make up considerably less of future production than has been forecast by the EIA. Rather than a peak in 2021 followed by a gradual decline to slightly below today's levels by 2040, U.S. tight oil is likely to peak before 2020 and decline to a small fraction of today's production levels by 2040. The analysis also underscores the amount of drilling, the amount of capital investment, and the associated scale of environmental and community impacts that will be required to meet these projections. These findings call into question plans for crude oil exports and highlight the real risks to long-term U.S. energy security.

---

[1] Per the EIA's "reference case" in *Annual Energy Outlook 2014*, http://www.eia.gov/forecasts/aeo.

2.1.2   Methodology

This report analyzes the top two U.S. tight oil plays—the Bakken and the Eagle Ford—in depth, followed by an assessment of five additional tight oil plays that make up most of the balance of the EIA's tight oil forecasts in its 2014 Annual Energy Outlook (AEO 2014).

The Bakken and Eagle Ford are investigated in depth as they account for nearly two-thirds of U.S. tight oil production and now have an extensive drilling history with which to assess key parameters; the report develops projections of their likely production levels given various scenarios of drilling and investment. The other tight oil plays are assessed based on their drilling and production history in comparison to the EIA forecasts of future production; they differ from the Bakken and Eagle Ford in that most of them have a long history of conventional oil and gas production stretching back decades. In total, all these plays account for 82% of the 2014-2040 tight oil production in the EIA's reference case forecast, and hence provide a solid basis for assessing its credibility. The remaining 18% comes from a number of smaller plays whose ultimate contribution remains highly speculative.

The primary source of data for this analysis is Drillinginfo, a commercial database of well production data widely used by industry and government, including the EIA.[2] Drillinginfo also provides a variety of analytical tools which proved essential for the analysis.

A detailed analysis of well production data for the major tight oil plays reveals several fundamental characteristics that will determine future production levels:

1. **Rate of well production decline:** Tight oil plays have high well production decline rates, typically in the range of 80-85% in the first three years.

2. **Rate of field production decline:** Tight oil plays have high field production declines, typically in the range of 40-45% per year, which must be replaced with more drilling to maintain production levels. This compares to field declines in the range of 5-6% per year in major conventional oil fields.[3]

3. **Average well quality:** All tight oil plays invariably have "core" areas or "sweet spots", where individual well production is highest and hence the economics are best. Sweet spots are targeted and drilled off early in a play's lifecycle, leaving lesser quality rock to be drilled as the play matures (requiring higher oil prices to be economic); thus the number of wells required to offset field decline inevitably increases with time. Although technological innovations including longer horizontal laterals, more fracturing stages, more effective additives and higher-volume frack treatments have increased well productivity in the early stages of the development of all plays, they have provided diminishing returns over time, and cannot compensate for poor quality reservoir rock.

4. **Number of potential wells:** Plays are limited in area and therefore have a finite number of locations to be drilled. Once the locations run out, production goes into terminal decline.

5. **Rate of drilling:** The rate of production is directly correlated with the rate of drilling, which is determined by the level of capital investment.

---

[2] See http://info.drillinginfo.com.
[3] IEA, *World Energy Outlook 2008*, http://www.worldenergyoutlook.org/media/weowebsite/2008-1994/weo2008.pdf.

The basic methodology used is as follows:

- Historical production, number of currently producing- and total-wells drilled, the split between horizontal- and vertical/directional-wells, and the overall play area were determined for all plays. Average well production decline for both horizontal and vertical/directional wells, and the average estimated ultimate recovery (EUR), were also assessed for all plays. For the Bakken and Eagle Ford, these parameters were assessed at the county- as well as at the play-level (the top counties in terms of the number of producing wells were analyzed individually, whereas counties with few wells were aggregated).

- Field decline rates and the number of available drilling locations were determined at the county- and play-level for the Bakken and Eagle Ford.

- First-year average production was established from type decline curves (i.e., average well decline profiles) constructed for all wells drilled in the year in question; 2013 was the year used as representative of future average first-year production levels per well. Average first-year production is used to determine the number of wells needed to offset field decline each year, and to determine the production trajectory over time given various drilling rates. In determining future production rates, the current trends in well productivity over time were considered; for example if recent well quality trends were increasing, it was assumed for plays in early stages of development that well quality would increase somewhat in the future before declining as drilling moves into lower quality outlying portions of plays.

- Projections of future production profiles were made for the Bakken and Eagle Ford based on various drilling rate scenarios. These projections assume a gradation over time from the well quality observed in the current top counties of a play to the well quality observed in the outlying counties as available drilling locations are used up. The different drilling rate scenarios were prepared so that the effect of a high drilling rate, presumably due to favorable economic conditions, compared to a low or a "most likely" drilling rate, could be assessed, both in terms of production over time and cumulative oil recovery from the play by 2040.

- Production history for all plays and production projections (in the case of the Bakken and Eagle Ford) were then compared to the EIA forecasts to assess the likelihood that these forecasts could be met.

- All plays were then compared to each other in terms of well quality and other parameters and an overall assessment of the likely long-term sustainability of tight oil production was determined.

Although public pushback against fracking due to health and environmental concerns has limited access to drilling locations in states like New York and Maryland and several municipalities, as well as triggered lawsuits, this report assumes there will be no restrictions to access due to environmental concerns. It also assumes there will be no restrictions on access to the capital required to meet the various drilling rate scenarios. In these respects, it presents a "best case," as any restrictions on access to drilling locations or to the capital needed to drill wells would reduce forecast production levels.

2.2 THE CONTEXT OF U.S. OIL PRODUCTION

2.2.1 U.S. Oil Production Forecasts

The EIA's *Annual Energy Outlook 2014* provides various scenarios of future U.S. oil production, as well as price projections and stated assumptions in terms of available technically recoverable reserves and resources, play areas, well productivity, and so forth.

Figure 2-1 illustrates the range of the EIA's oil production forecasts through 2040 compared to historical production. Most scenarios project the U.S. to meet or exceed its all-time peak production, which occurred in 1970. These scenarios assume cumulative production of between 77 and 123 billion barrels of oil between 2013 and 2040, which is 2.7-4.2 times the *proved reserves* (i.e., economically recoverable with current technology) that were thought to exist as of 2012.[4] Adding in *unproved resources*, which are uncertain estimates without price constraints, between a third and a half of remaining potentially recoverable oil in the U.S. will be consumed over the next 26 years according to the EIA projection. This amounts to the equivalent of 54-84% of all the oil produced over the 54 years between 1960 and 2013.

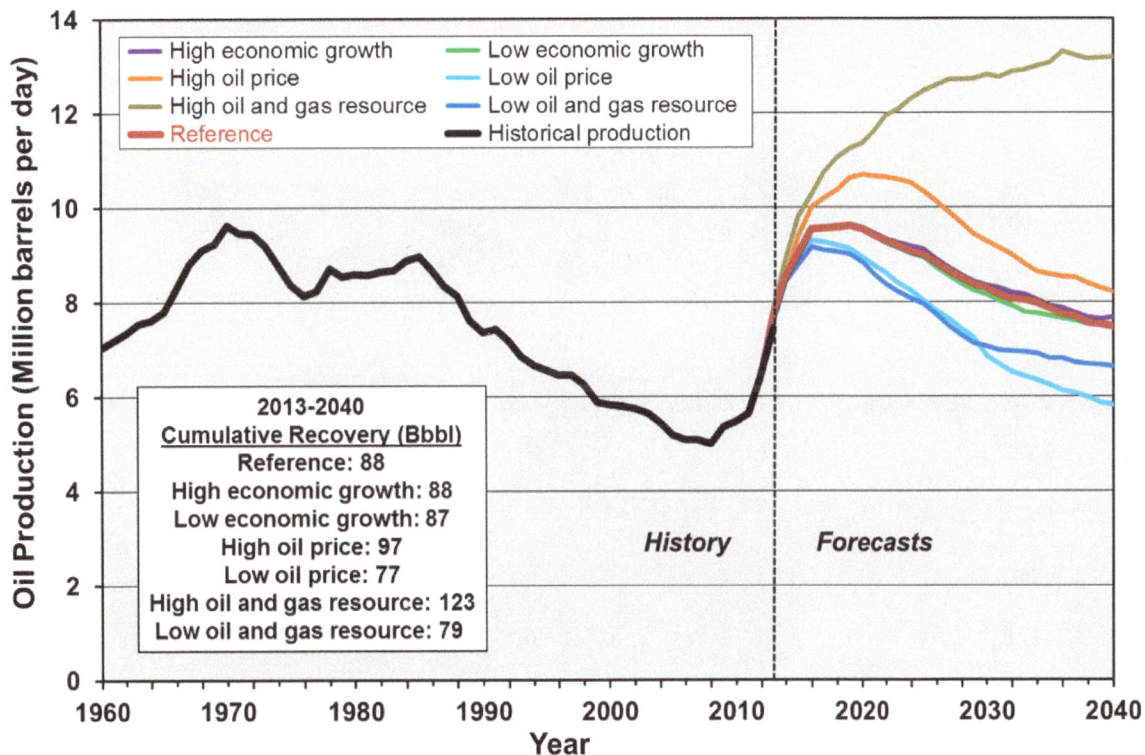

Figure 2-1. Scenarios of U.S. oil production through 2040 from the EIA's *Annual Energy Outlook 2014*,[5] compared to historical production from 1960.

Oil production includes both crude oil and lease condensates.

---

[4] EIA, *Assumptions to the Annual Energy Outlook 2014*, http://www.eia.gov/forecasts/aeo/assumptions/pdf/oilgas.pdf.
[5] EIA, *Annual Energy Outlook 2014*, http://www.eia.gov/forecasts/aeo.

The source of this optimism in future oil production is the application of high-volume, multi-stage, hydraulic fracturing technology in horizontal wells, which has unlocked previously inaccessible oil trapped in highly impermeable shales and tight source rocks. Figure 2-2 illustrates the EIA's reference case projection for oil production by source through 2040. Although conventional production is forecast to be flat or declining over the period, tight oil production increases rapidly to a peak early in the next decade, amounting to roughly half of all U.S. oil production. Oil prices in this reference case are forecast to remain below $140 per barrel over the period. Notwithstanding talk of U.S. energy independence, this scenario implies that U.S. oil production, even with tight oil, will amount to only 40% of projected 2040 demand.

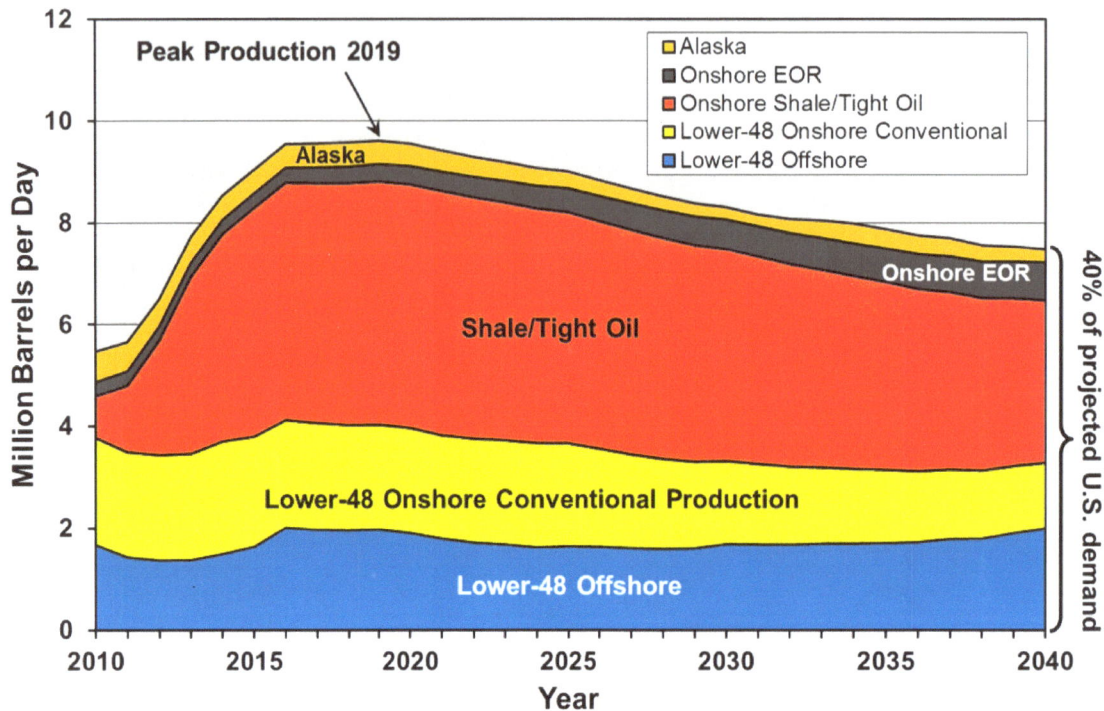

Figure 2-2. EIA reference case projection of U.S. oil production by source through 2040.[6]

[6] EIA, *Annual Energy Outlook 2014*, http://www.eia.gov/forecasts/aeo

Figure 2-3 illustrates EIA's projections for tight oil production in several cases. These assume the extraction of between 37 (low oil price case) and 47 billion barrels (high oil price case) by 2040. This amounts to all of the 7.15 billion barrels of proved tight oil reserves and between 50% and 67% of the EIA's estimated 59.2 billion barrels of unproved tight oil resources (unproved resources have no implied price required for extraction and are highly uncertain, as evidenced by the EIA's recent 96% downgrade of resources in the Monterey Shale of California[7]).

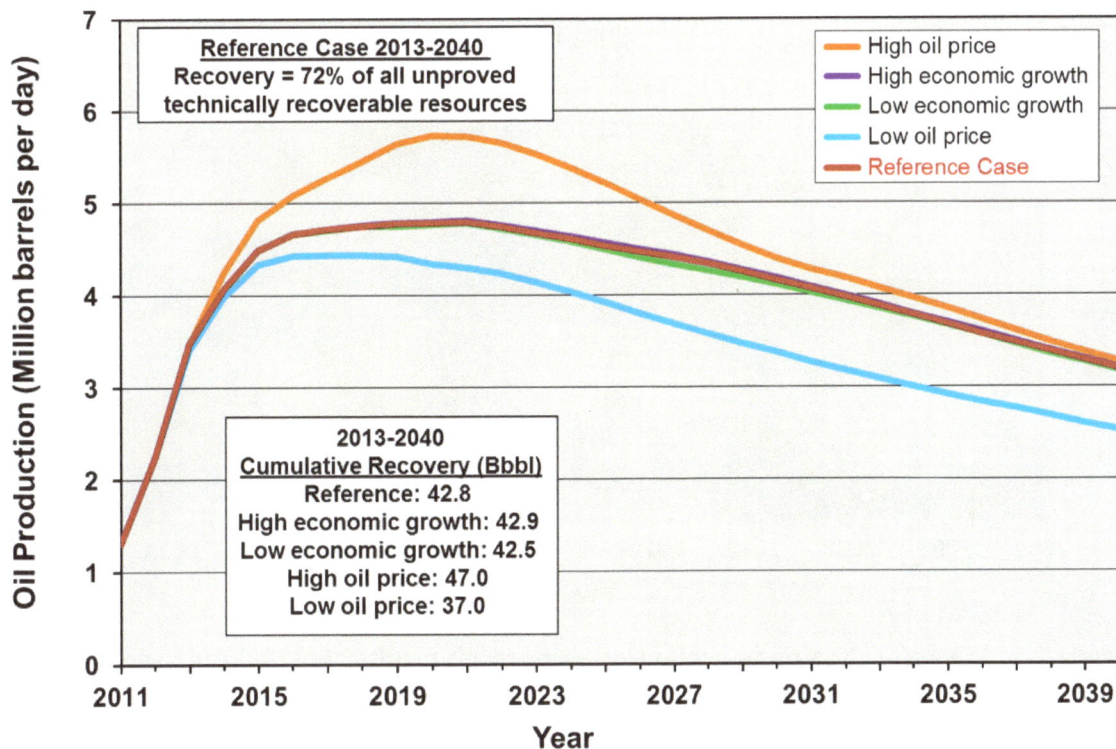

Figure 2-3. EIA scenarios of U.S. tight oil production through 2040.[8]

According to the EIA, proved reserves of tight oil are 7.15 billion barrels and unproved technically recoverable resources are estimated at 59.2 billion barrels, as of January 1, 2012.[9]

---

[7] Louis Sahagun, "U.S. officials cut estimate of recoverable Monterey Shale oil by 96%," *Los Angeles Times*, May 20, 2014, http://www.latimes.com/business/la-fi-oil-20140521-story.html.
[8] EIA, *Annual Energy Outlook 2014*, http://www.eia.gov/forecasts/aeo.
[9] EIA, *Assumptions to the Annual Energy Outlook 2014*, http://www.eia.gov/forecasts/aeo/assumptions/pdf/oilgas.pdf

Figure 2-4 illustrates how the EIA reference case projections for tight oil production are divided between the Bakken, the Eagle Ford, and all other plays.

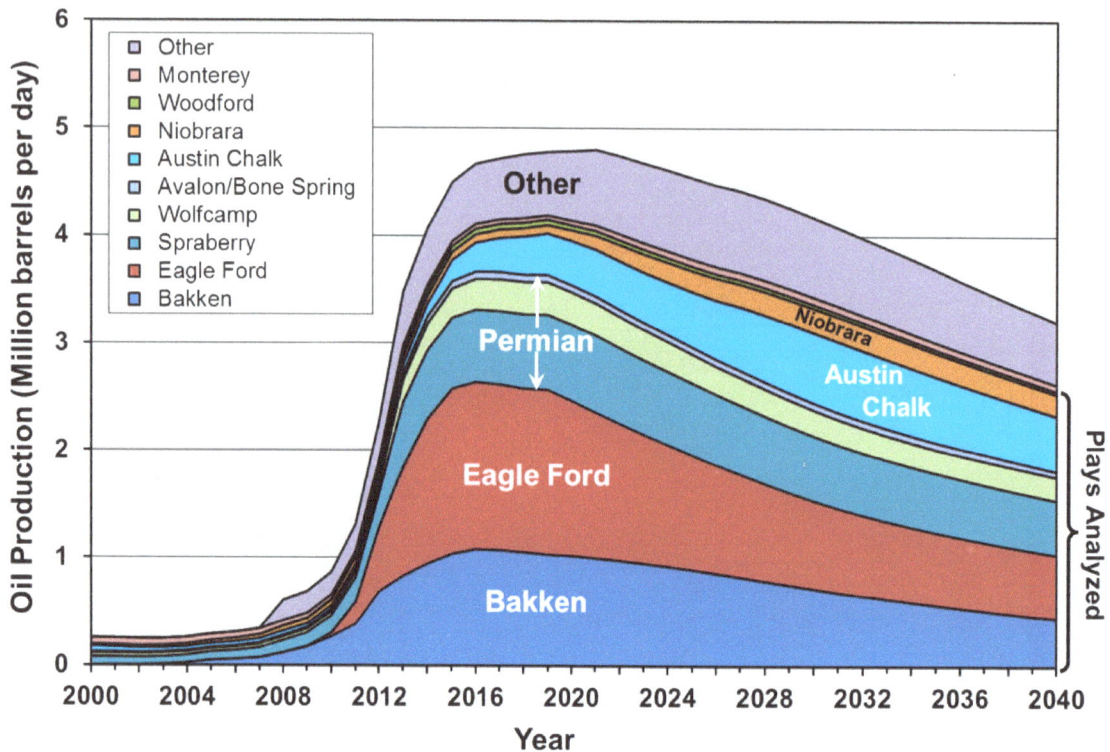

Figure 2-4. EIA reference case projection of tight oil production divided among Bakken, Eagle Ford, and all other plays, 2011-2040.[10]
This report analyzed the seven most productive plays, which account for 82% of EIA's tight oil production forecast to 2040.

The EIA reference case clearly expects the Bakken and Eagle Ford to provide a slowly declining but significant foundation of tight oil production for the next few decades. The Bakken and Eagle Ford are relatively new plays, with substantial tight oil resources that have only recently been unlocked by directional drilling and hydraulic fracturing.

Tight oil production in all these plays has risen quickly due to rapid increases in drilling rates and sustained high levels of capital input. However, high well- and field-decline rates, coupled with a finite number of drilling locations, suggest that production will drop off sharply when sweet spots are depleted; therefore, the projected long slow production decline of these plays warrants further scrutiny. Section 3 of this report explores the realistic production potential for the Bakken and Eagle Ford in depth.

---

[10] EIA, *Annual Energy Outlook 2014*, unpublished tables from AEO 2014 provided by the EIA.

The remainder of tight oil production is expected to come from seven major plays as well as numerous emerging plays, as illustrated in Figure 2-5.

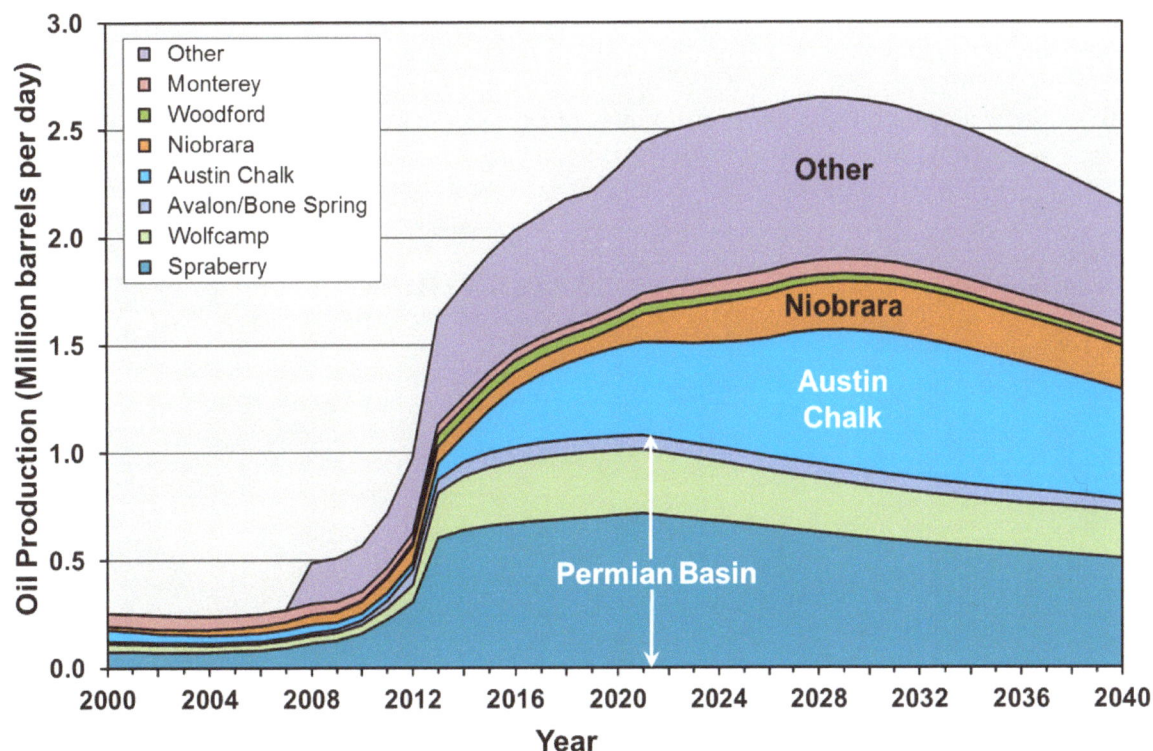

Figure 2-5. EIA reference case projections of tight oil production from plays other than the Bakken and Eagle Ford, through 2040.[11]

Of the Permian Basin plays, only the top three are labeled here; the remaining are minor plays included in "Other."

Unlike the Bakken and Eagle Ford, most of these plays have been known for a long time; their growing production reflects the successful application of new technology to extract additional resources. They are projected by the EIA to account for two-thirds of tight oil production in 2040; therefore, sustained production projected from these mature plays warrants further scrutiny. Sections 2.4 and 2.5 of this report explore the realistic production potential of these plays in depth.

---

[11] EIA, *Annual Energy Outlook 2014*, unpublished tables from AEO 2014 provided by the EIA.

2.2.2   Current U.S. Tight Oil Production

Production of tight oil began in the Bakken Field of Montana and North Dakota in the early 2000s. With the widespread application of horizontal drilling and hydraulic fracturing beginning in 2005, production grew rapidly. The Eagle Ford Field of southern Texas was unknown as recently at 2007, and now is the single largest producer of tight oil in the U.S. The distribution of tight oil and shale gas plays in the lower 48 states is illustrated in Figure 2-6.

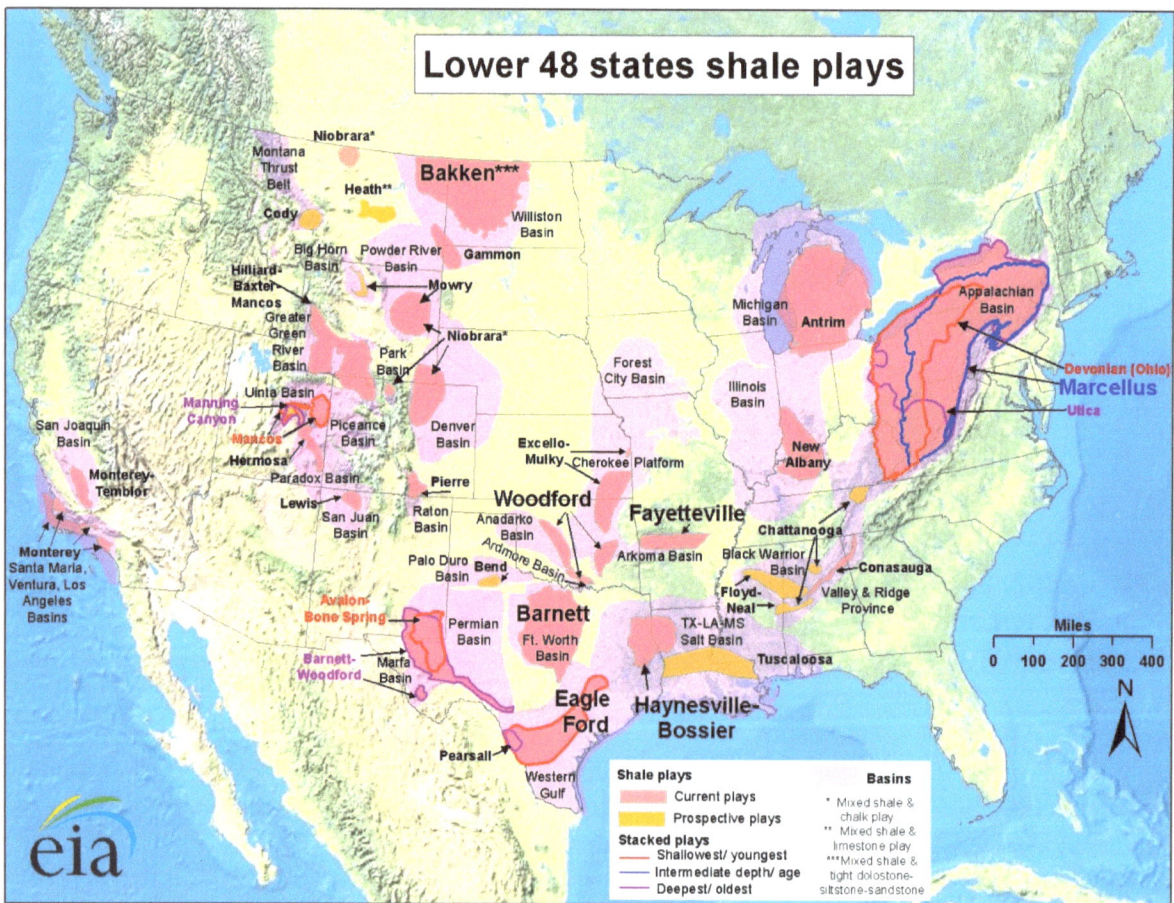

Figure 2-6. Distribution of lower 48 states shale gas and oil plays.[12]

---

[12] EIA, *Annual Energy Outlook 2014*, http://www.eia.gov/forecasts/aeo/.

Current production from U.S. tight oil plays is estimated by the EIA at 3.7 MMbbl/d. Despite the apparent widespread nature of shale plays as shown in Figure 4, 62% of this production comes from just the top two plays: the Bakken and Eagle Ford. A further 25% comes from the five plays of the Permian Basin in Texas and New Mexico. Figure 2-7 illustrates tight oil production by play from 2000 through May, 2014, according to the EIA.

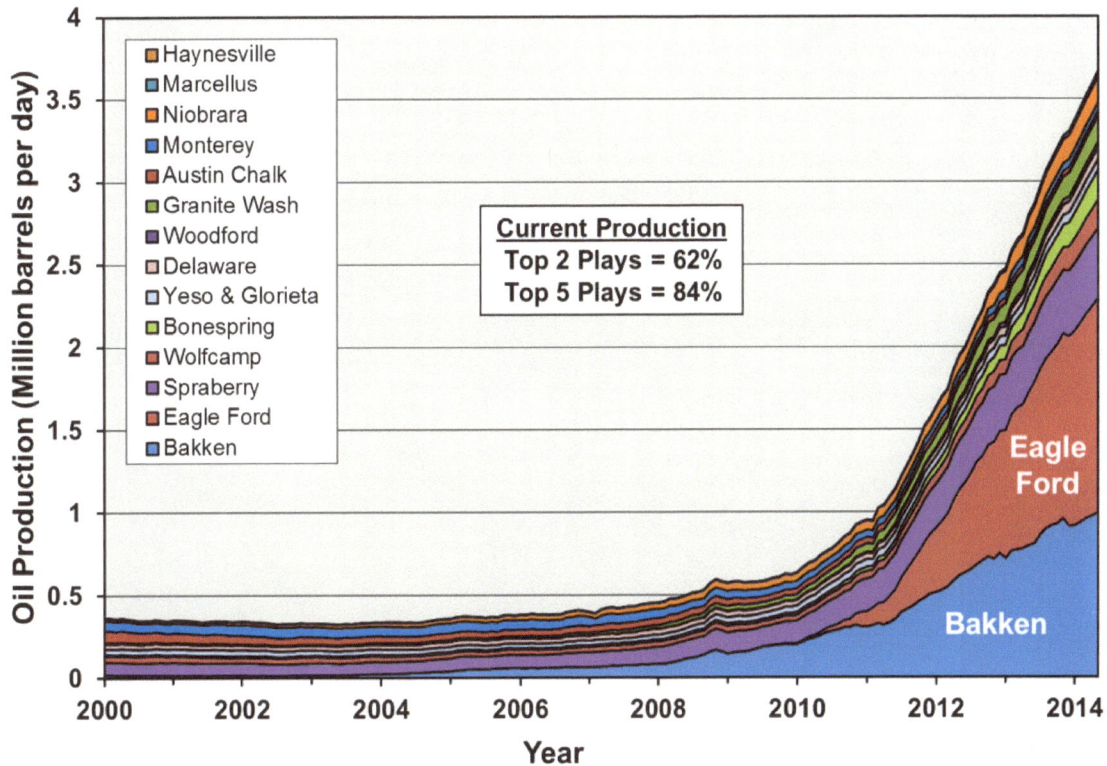

Figure 2-7. U.S. tight oil production by play, 2000 through May 2014.[13]

The Permian Basin, which is made up of several plays (the largest of which are noted), is the third largest projected source of tight oil.

---

[13] EIA estimates obtained in June 2014 from http://www.eia.gov/naturalgas/weekly, where it appears to have been mistakenly posted; no longer available at this location.

2.3   THE BAKKEN AND EAGLE FORD PLAYS

This report investigates the Bakken play and Eagle Ford play in depth because they are the foundation of the U.S. tight oil "shale revolution." They are the two most productive U.S. tight oil plays, accounting for 62% of current production, and are projected to account for over half of total tight oil production well into the next decade.

Moreover, the Bakken and Eagle Ford are new tight oil plays, having only recently been unlocked by directional drilling and fracking. In comparison, most of the other major U.S. tight oil plays are decades old with tens of thousands of conventional wells. Thus, the Bakken and Eagle Ford are the best representatives of what may be expected from future tight oil discoveries.

2.3.1   Bakken Play

The EIA forecasts recovery of 8.8 billion barrels of oil from the Bakken play by 2040. The analysis of actual production data presented below suggests that this forecast is unlikely to be realized.

The Bakken play is where tight oil production got its start—first in the Elm Coulee Field of Montana, then in the western counties of North Dakota. The Bakken Formation is underlain by the Three Forks Formation, which is also productive and is separated from the Bakken by as little as 30 feet. The analysis herein encompasses both the Bakken and Three Forks.

The U.S. Geological Survey (USGS) produced a new assessment of the Bakken and Three Forks in 2013 in which they estimated a mean technically recoverable resource of 7.4 billion barrels.[14] They broke the play into six "assessment units" (AUs) as illustrated in Figure 2-8. The EIA has apparently used this breakdown in its estimates of the play area used to calculate an unproved recoverable resource of 9.2 billion barrels (54% of which are in the Three Forks Formation) in its 2014 reference case; however, it does not provide an updated map showing the areas it has included.[15] In the EIA's analysis, the Bakken play is comprised of five contiguous units totaling 14,594 square miles plus a single underlying Three Forks unit totaling 17,652 square miles (USGS areas for these units, shown in Figure 2-8, are somewhat larger).

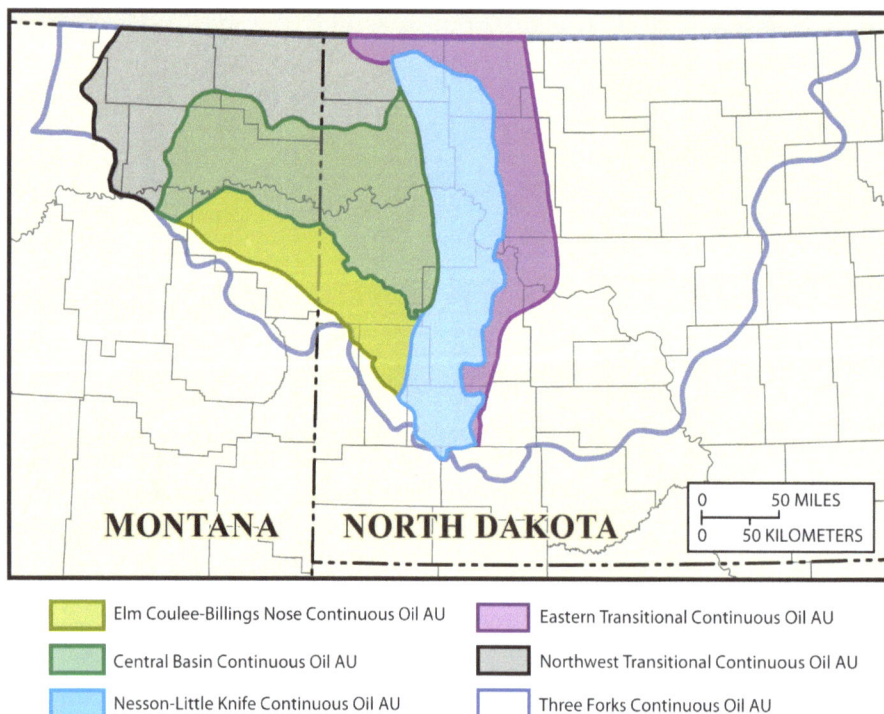

Figure 2-8. USGS demarcation of Bakken and Three Forks tight oil assessment units.[16]
The USGS demarcates five contiguous Bakken units and one underlying, much larger, Three Forks unit.

[14] USGS, *Assessment of Undiscovered Oil Resources in the Bakken and Three Forks Formations, Williston Basin Province, Montana, North Dakota, and South Dakota*, 2013, http://pubs.usgs.gov/fs/2013/3013/fs2013-3013.pdf.
[15] EIA, *Assumptions to the Annual Energy Outlook 2014*, http://www.eia.gov/forecasts/aeo/assumptions/pdf/oilgas.pdf. At publication, the most recent shapefile for the EIA play area was dated May 2011, available at http://www.eia.gov/pub/oil_gas/natural_gas/analysis_publications/maps/maps.htm.
[16] Map by Post Carbon Institute, using data from USGS, *Assessment of Undiscovered Oil Resources in the Bakken and Three Forks Formations, Williston Basin Province, Montana, North Dakota, and South Dakota*, 2013, http://pubs.usgs.gov/fs/2013/3013/fs2013-3013.pdf.

Figure 2-9 illustrates the distribution of wells in the Bakken as of early 2014. Over 9,200 wells have been drilled to date, of which 8,534 were producing oil at the time of writing. Although the play covers parts of 15 counties, most drilling is concentrated in McKenzie, Mountrail, Dunn, Williams, and Divide counties in North Dakota and Richland County in Montana. The functional prospective limits of the play are well defined by wells with little or no productivity, and encompass approximately 12,700 square miles; this is a markedly smaller area than the play area demarcated by the EIA.

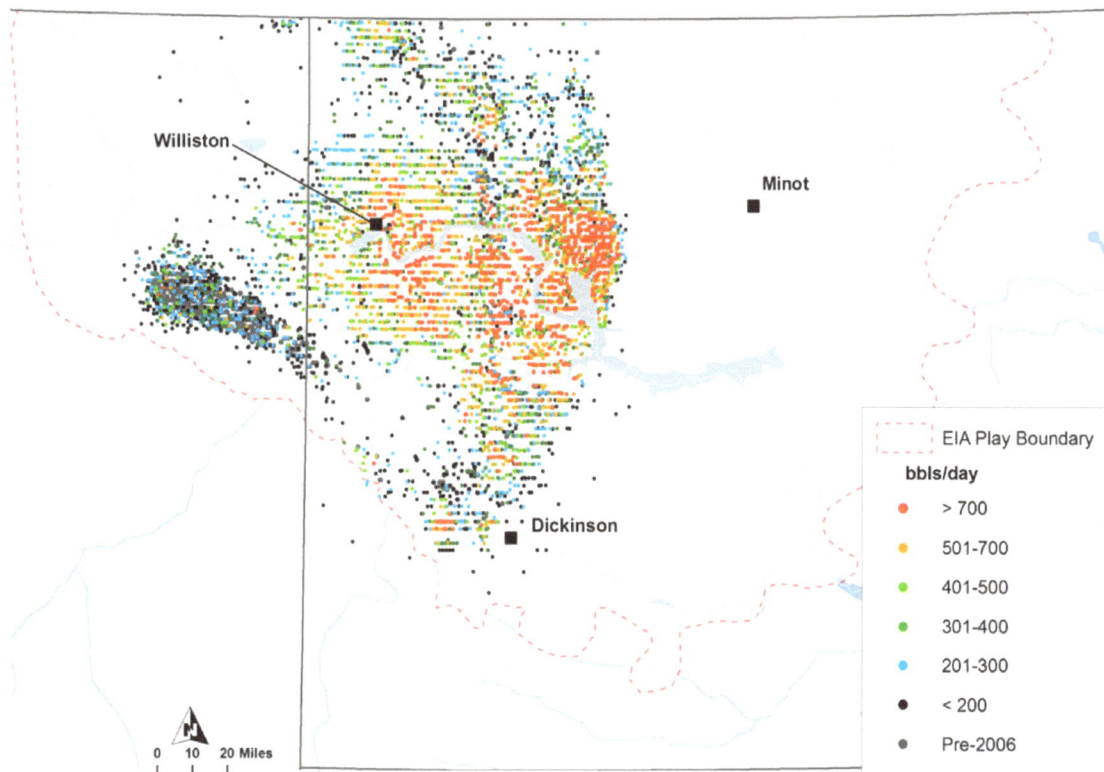

Figure 2-9. Distribution of wells in the Bakken play as of mid-2014 illustrating highest one-month oil production (initial productivity, IP),[17] with EIA play boundary.[18]

The size of the Bakken play as defined by the extent of where productive drilling has actually occurred is approximately 12,700 square miles, in contrast to the much larger area designated as the play by the EIA (2011). Well IPs are categorized approximately by percentile; see Appendix.

The case for such a smaller Bakken play area than what the EIA and USGS claim is further underlain by observing where operators actually have acreage and where drilling is occurring. For example, the leaseholdings of Continental Resources, one of the largest operators in the Bakken, are notably concentrated in the productive area of the play.[19]

---

[17] Data from Drillinginfo retrieved September 2014.

[18] At publication, the most recent shapefile for the EIA's play area for the Bakken was dated May 2011, available at http://www.eia.gov/pub/oil_gas/natural_gas/analysis_publications/maps/maps.htm#geodata.

[19] Continental Resources, September 2014 investor presentation, http://phx.corporate-ir.net/External.File?item=UGFyZW50SUQ9NTU0MDg2fENoaWxkSUQ9MjUwMTQyfFR5cGU9MQ==&t=1.

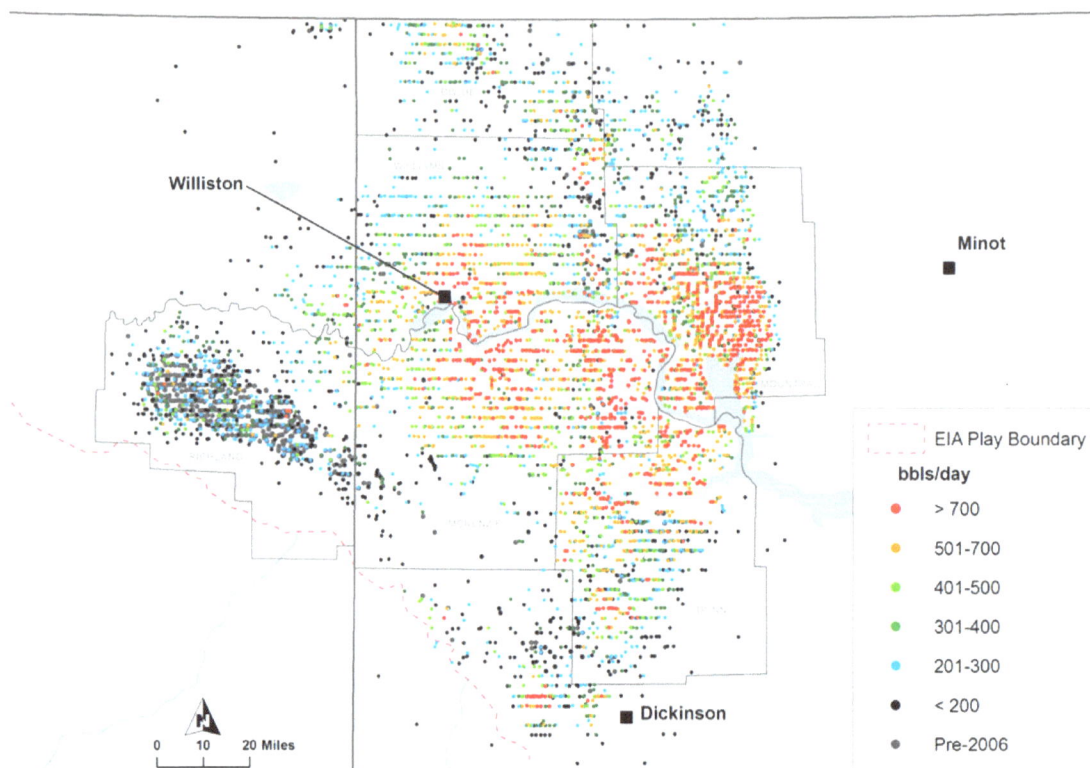

Figure 2-10. Detail of Bakken play showing distribution of wells as of early 2014, and illustrating highest one-month oil production (initial productivity, IP),[20] with EIA play boundary.[21]

The top six producing counties are indicated. Well IPs are categorized approximately by percentile; see Appendix.

[20] Data from Drillinginfo retrieved August 2014.
[21] At publication, the most recent shapefile for the EIA's play area for the Bakken was dated May 2011, available at http://www.eia.gov/pub/oil_gas/natural_gas/analysis_publications/maps/maps.htm#geodata.

Production in the Bakken was nearly one million barrels of oil per day and 1.1 billion cubic feet of gas per day at the time of writing, as illustrated in Figure 2-11.[22] Gas production is expressed in Figure 2-11 as barrels of oil equivalent (6,000 cubic feet of gas equals approximately one barrel of oil on an energy equivalent basis). Ninety-eight percent of this production is from horizontally drilled, hydraulically fractured ("fracked") wells. The rate of drilling has grown from about 500 wells per year in 2009 to about 2,000 wells per year in mid-2012, where it has remained.

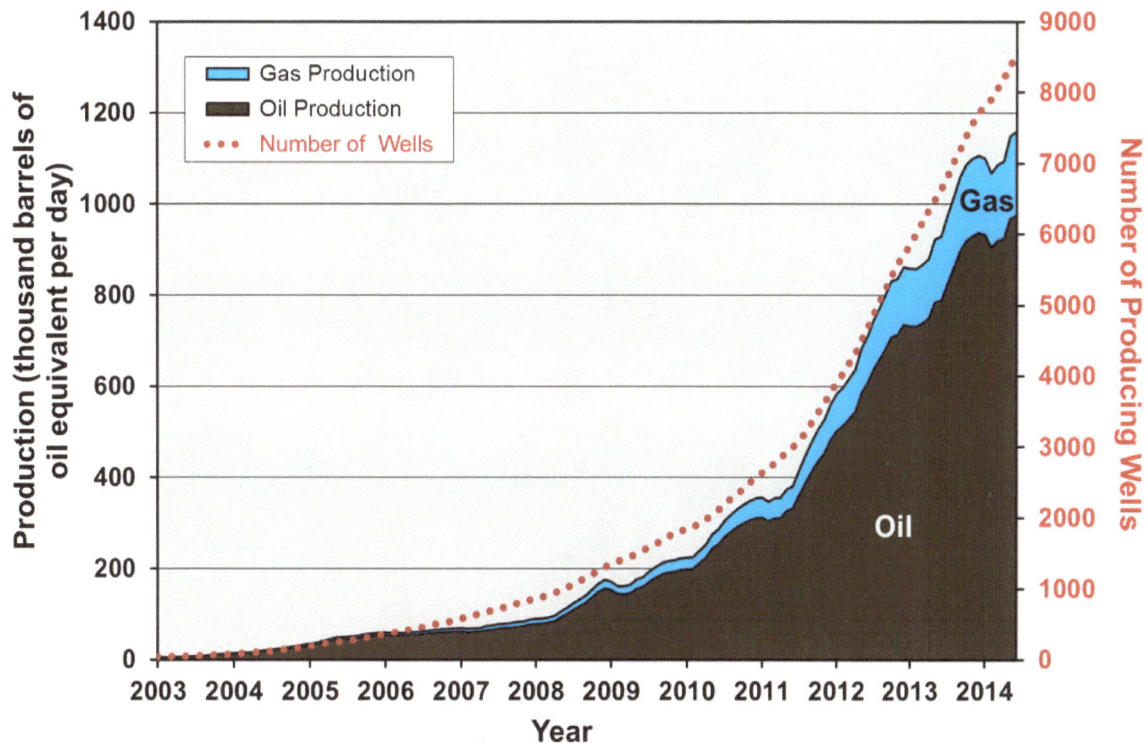

Figure 2-11. Bakken play tight oil and gas production and number of producing wells, 2003 to 2014.[23]

[22] Although the EIA's widely cited Drilling Productivity Report (DPR) states as of September 2014 that Bakken has produced over 1 million barrels per day of oil since February 2014, it must be noted that the DPR's figures for the Bakken seem to overstate production and recent months are based on estimates. See http://www.pphb.com/images/pdfs/musings2014/Musings040114.pdf.
[23] Data from Drillinginfo retrieved September 2014. Three-month trailing moving average.

The amount of oil added to total play production by each new well has been declining since early 2012 as illustrated in Figure 2-12. This is due to the fact that the higher production grows, the more intrinsic decline must be offset by new wells.

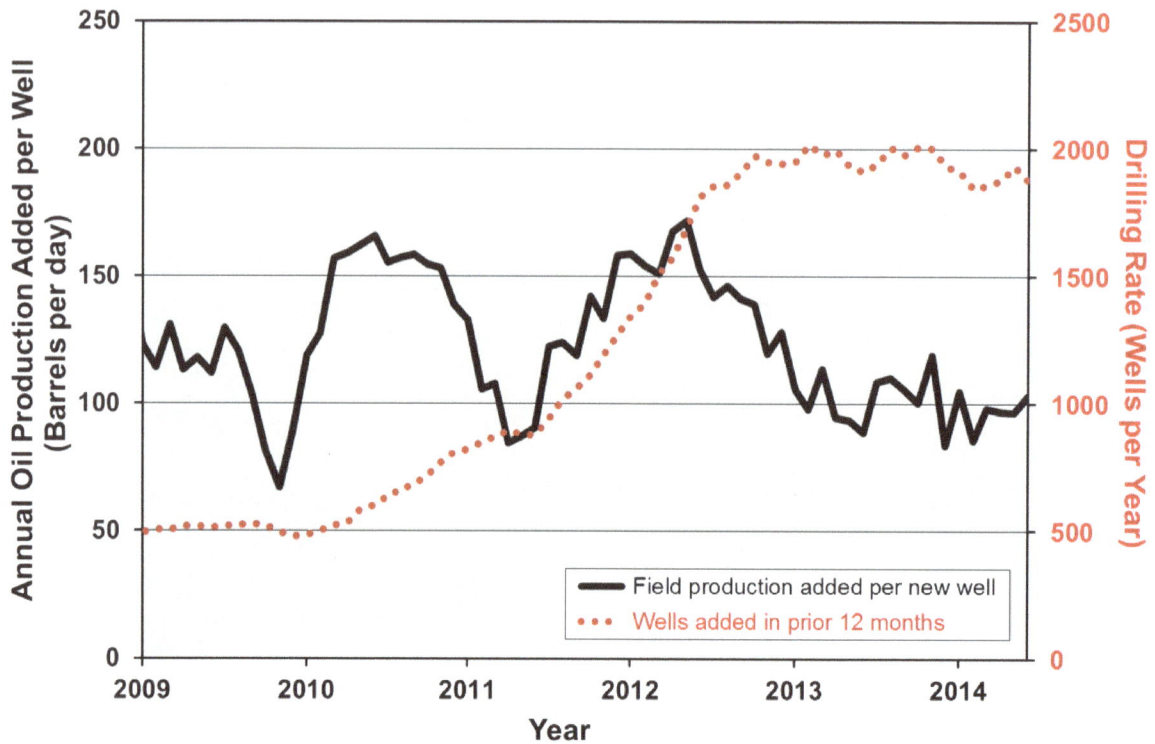

Figure 2-12. Annual oil production added per new well and annual drilling rate in the Bakken play, 2009 through 2014.[24]

---

[24] Data from Drillinginfo retrieved September 2014. Three-month trailing moving average.

2.3.1.1 Well Decline

The first key fundamental in determining the life cycle of Bakken production is the *well decline rate*. Bakken wells exhibit high decline rates in common with all shale plays. Figure 2-13 illustrates the average decline profile of Bakken horizontal wells. Decline rates are steepest in the first year and are progressively less in the second and subsequent years. The average decline rate over the first three years of well life is 85%.

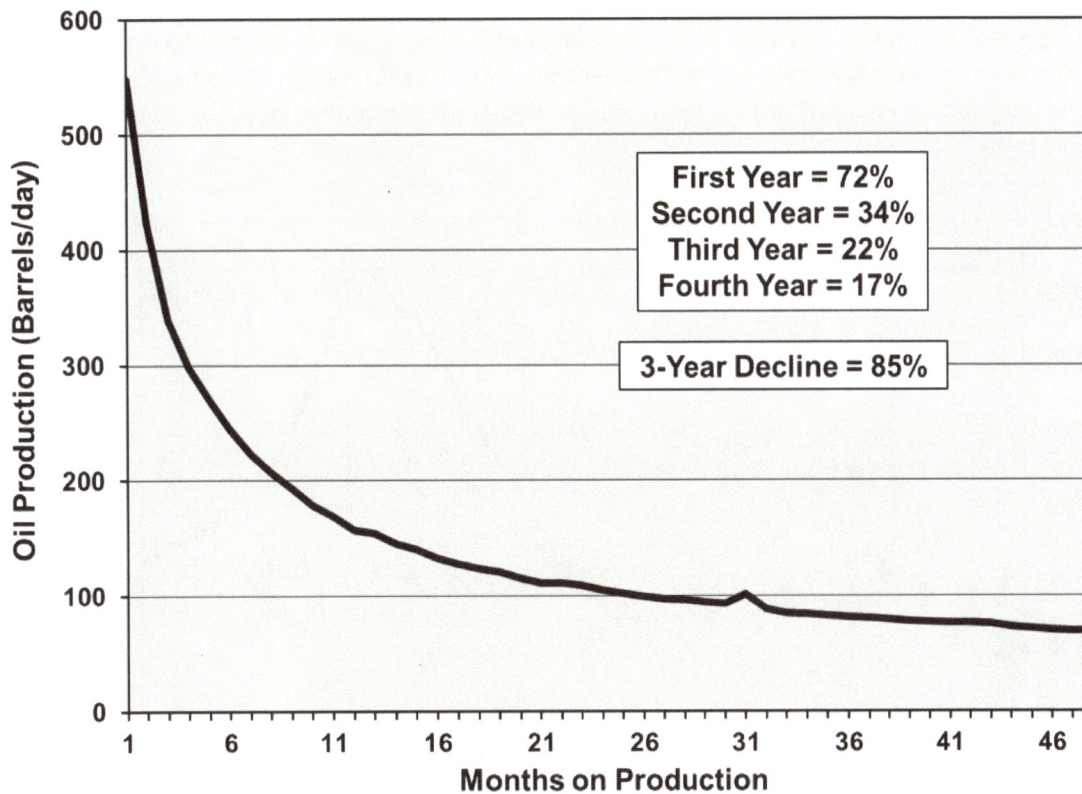

Figure 2-13. Average decline profile for horizontal tight oil wells in the Bakken play.[25]
Decline profile is based on all horizontal tight oil wells drilled since 2009.

---

[25] Data from Drillinginfo retrieved April 2014.

2.3.1.2   Field Decline

The second key fundamental is the overall *field decline rate*, which is the total amount of production in a given play that would be lost in a year without more drilling. Figure 2-14 illustrates production from the 5,300 wells drilled in the Bakken prior to 2013. The field decline rate of the first year without new drilling is 45%. This is lower than the well decline rate as the field decline is made up of new wells, declining at high rates, and older wells, declining at lesser rates. The field decline has been relatively constant at 45% for the past three years in the Bakken. Assuming new wells will produce in their first year at the first-year rates observed for wells drilled in 2013, 1,470 new wells would need to be drilled each year to offset field decline at current production levels. At an average cost of $8 million per well,[26] this would represent a capital input of about $11.8 billion per year, exclusive of leasing, operating, and other infrastructure costs, just to keep production flat at 2013 levels.

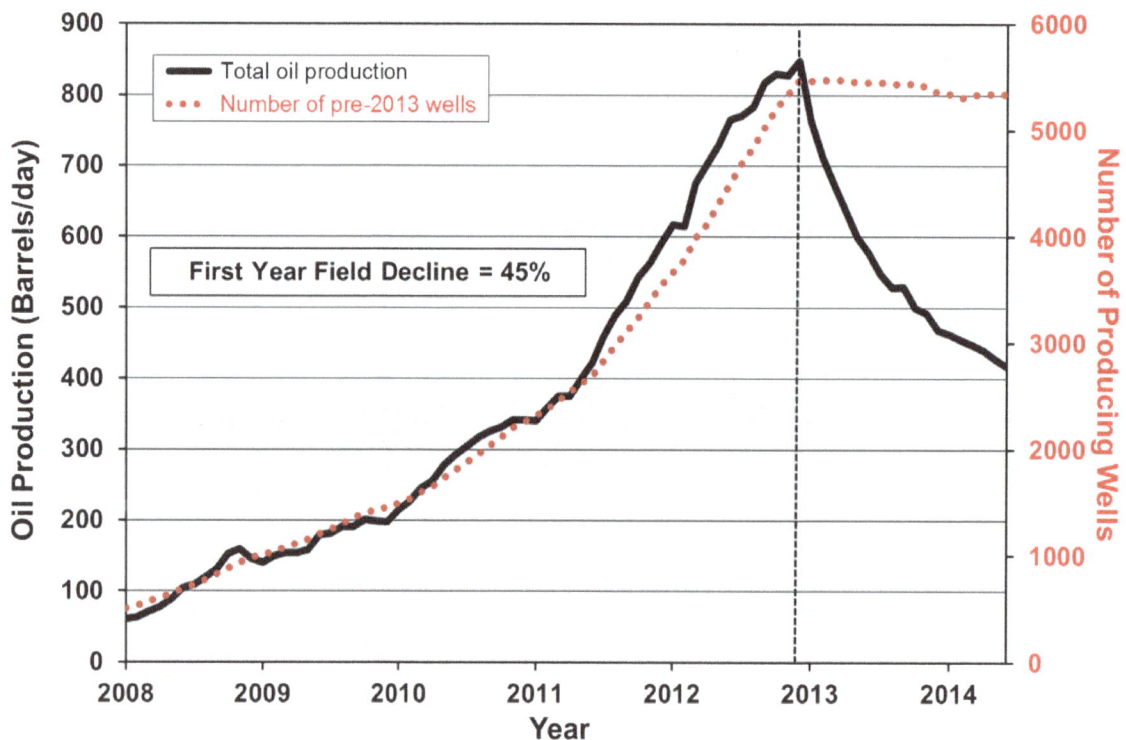

Figure 2-14. Production rate and number of horizontal tight oil wells in the Bakken play prior to 2013.[27]

In order to offset the 45% field decline rate, 1,470 new wells per year producing at 2013 levels would be required.

---

[26] Ingrid Pan, "Most operators are seeing declining well costs in the Bakken," *Market Realist*, December 12, 2013, http://marketrealist.com/2013/12/operators-seeing-declining-well-costs-bakken.
[27] Data from Drillinginfo retrieved September 2014.

2.3.1.3 Well Quality

The third key fundamental is the trend of *average well quality* over time. Petroleum engineers tell us that technology is constantly improving, with longer horizontal laterals, more frack stages per well, more sophisticated mixtures of proppants and other additives in the frack fluid injected into the wells, and higher-volume frack treatments. This has certainly been true over the past few years, which, along with multi-well pad drilling, has reduced well costs. In the Bakken, however, technological improvements appear to be approaching the limits of diminishing returns: improvements in average well quality are flat to slightly increasing at best. The average first-year production rate of Bakken wells is only 7% above its last-highest point, in 2011, as illustrated in Figure 2-15. Moreover, it is likely that this slight rise in average well quality is in part a result of concentrating drilling in the sweet spots, as discussed in the following section, rather than significant technology improvements.

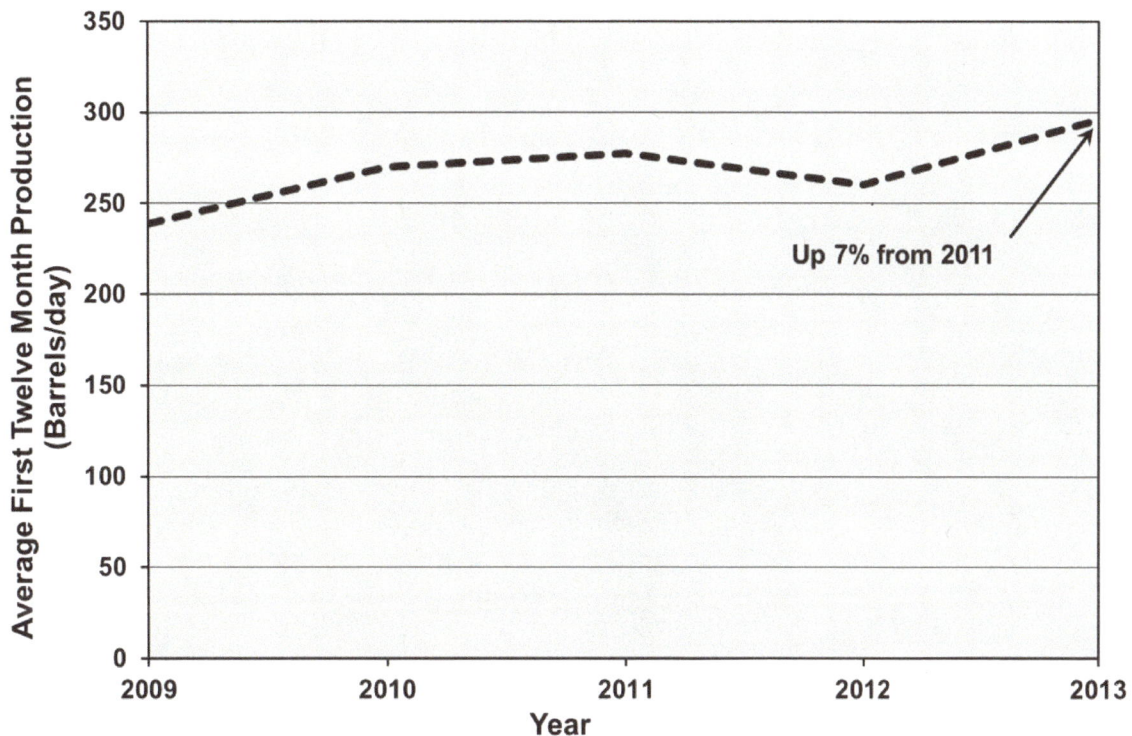

Figure 2-15. Average first-year production rates for Bakken tight oil wells, 2009 to 2013.[28]

The slight improvement over 2011 is likely as much a result of focusing drilling in sweet spots as significant technology improvements.

---

[28] Data from Drillinginfo retrieved September 2014.

Another measure of well quality is cumulative production and well life. Figure 2-16 illustrates the cumulative production of all wells that were producing in the Bakken as of March 2014. Eighty-two percent of these wells are less than 5 years old, and knowing that production will be down more than 90% after 5 years, their economic lifespan is uncertain. Although it can be seen that there are a few very good wells that recovered more than 600,000 barrels of oil in the first few years, and undoubtedly were great economic successes, the average well has produced just 127,765 barrels over a lifespan averaging 35 months. Only 1% of these wells are more than 10 years old. The lifespan of wells is another key parameter as many operators assume a minimum life of 30 years and longer—this is conjectural at this point given the lack of long-term well-performance data.

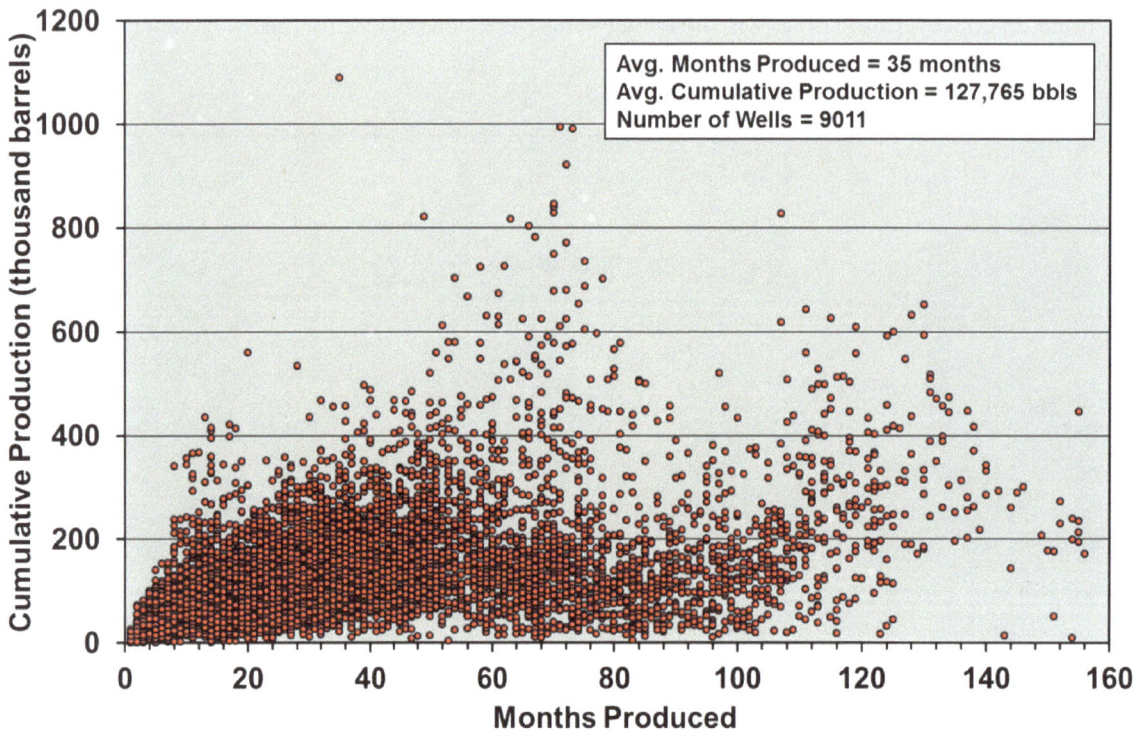

Figure 2-16. Cumulative oil production and length of time produced for Bakken wells that were producing as of March 2014.[29]

Very few wells are greater than ten years old, with a mean age of 35 months and a mean cumulative recovery of 127,765 barrels.

---

[29] Data from Drillinginfo retrieved September 2014.

Cumulative production of course depends on how long a well has been producing, so looking at young wells in not necessarily a good indication of how much oil these wells will produce over their lifespan (although production is heavily weighted to the early years of well life). A measure of well quality independent of age is initial productivity (IP), which is often focused on by operators. Figure 2-17 illustrates the average daily output over the first six months of production (six-month IP) for all wells in the Bakken play. Again, as with cumulative production, there are a few exceptional wells—4% of wells produced more than 600 barrels per day over the first six months—but the average for all wells drilled between 2008 and 2014 is just 262 barrels per day. The trend line on Figure 2-17 shows the average over time, which is declining as of the first half of 2014 as drilling moves into lower-quality areas. Figure 2-9 and Figure 2-10 illustrate the distribution of IPs in map form.

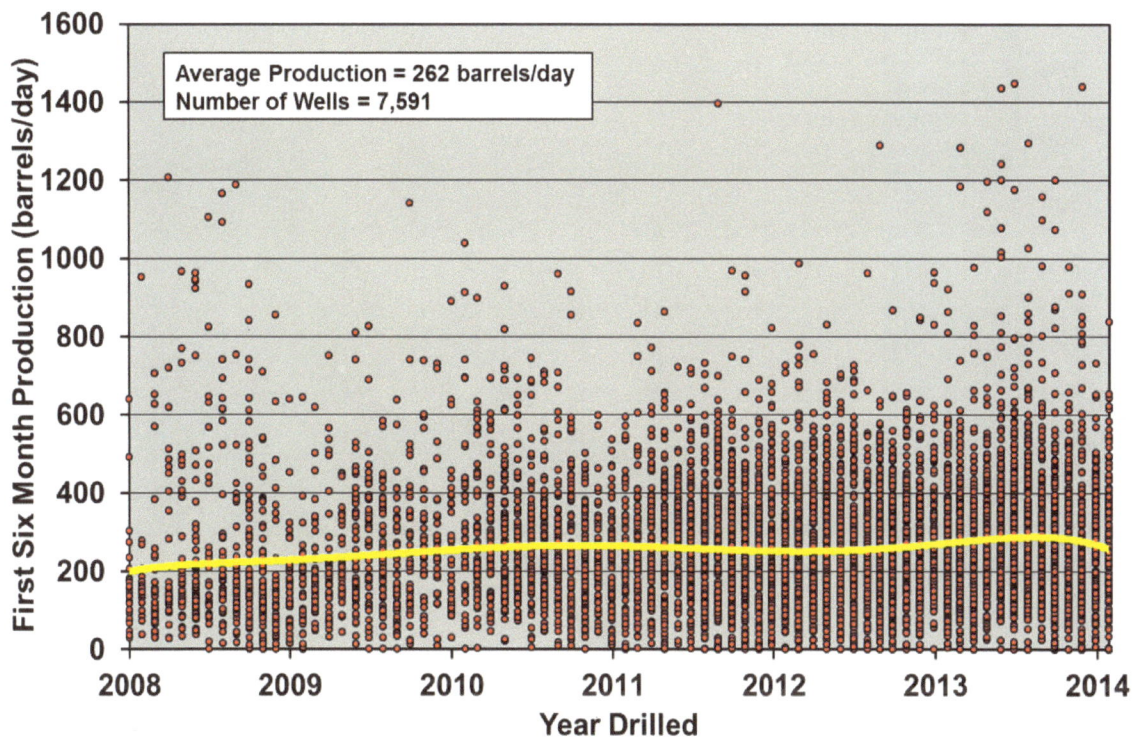

Figure 2-17. Average oil production over the first six months for all wells drilled in the Bakken play, 2008-2014.[30]

Although there are a few exceptional wells, the average well produced 262 barrels per day over this period.

---

[30] Data from Drillinginfo retrieved September 2014.

Different counties in the Bakken display markedly different well production rate characteristics, which are critical in determining the most likely production profile in the future. Figure 2-18, which illustrates production over time by county, shows that the top two counties produce 55% of the total, the top four produce 87%, and the remaining eleven produce just 13%. Clearly, years of widespread drilling (see Figure 2-19 for number of wells drilled per county) have not resulted in significant production increases outside the top four counties.

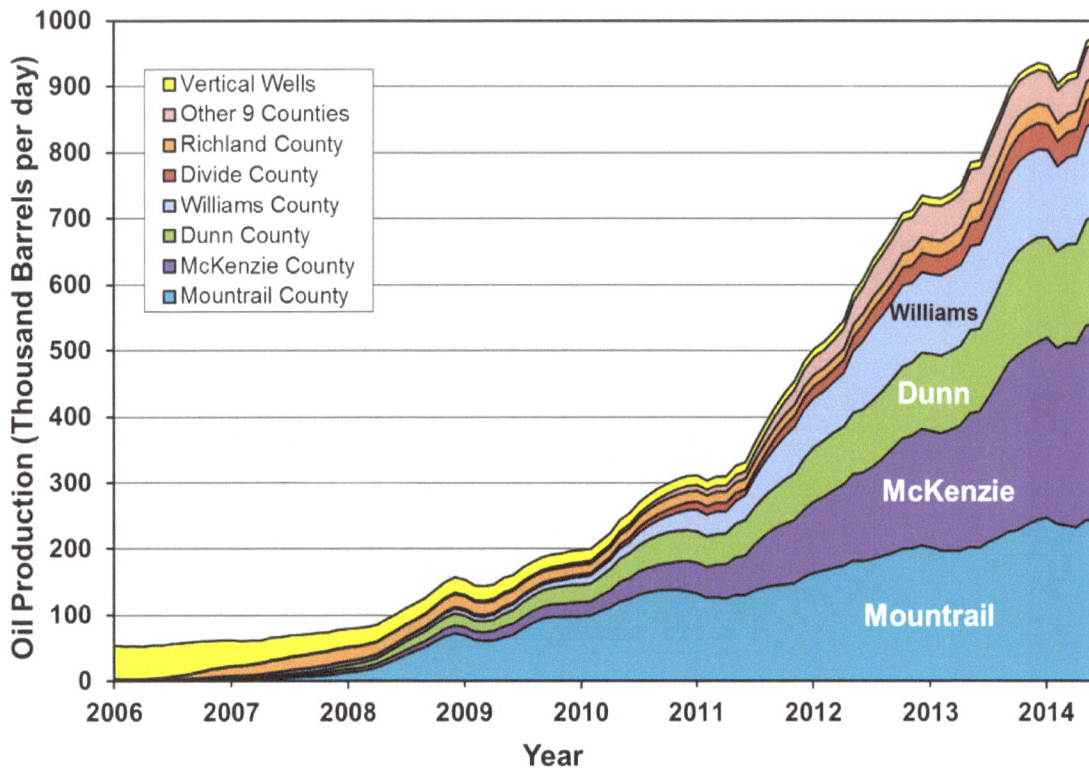

Figure 2-18. Oil production by county in the Bakken play, 2006 through 2014.[31]
The top four counties produced 87% of production in 2014.

---

[31] Data from Drillinginfo retrieved September 2014. Three-month trailing moving average.

The same trend holds in terms of cumulative production since the field commenced. As illustrated in Figure 2-19, the top two counties have produced half of the oil and the top four more than three-quarters. This trend will likely become even more pronounced given that the production rate share from these counties is increasing as noted above.

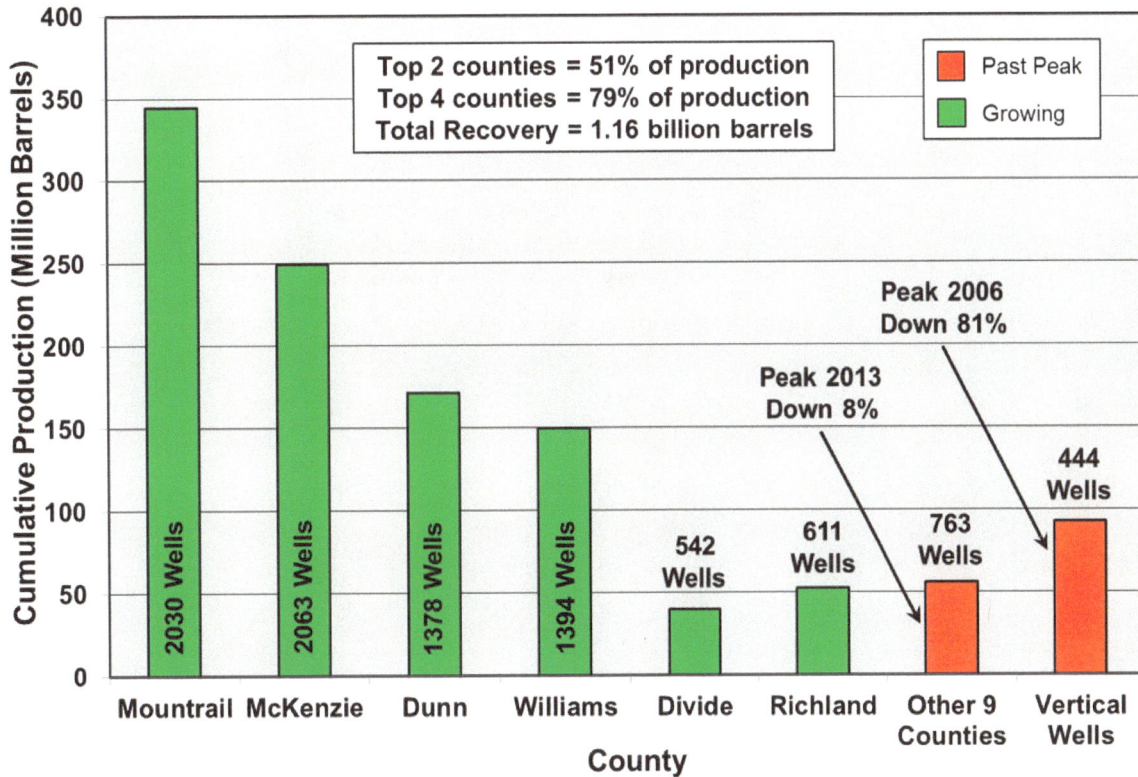

Figure 2-19. Cumulative oil production by county in the Bakken play through 2014.[32]

The top four counties have produced 79% of the 1.16 billion barrels produced to date. Note that production from vertical wells in all counties is grouped at right; the cumulative tallies by county are for horizontal wells only.

---

[32] Data from Drillinginfo retrieved September 2014.

The Bakken also produces significant amounts of natural gas (see the Bakken section in *Part3: Shale Gas* of this report for a full discussion). As with oil, cumulative production of natural gas is concentrated in the top four counties as illustrated in Figure 2-20. Although natural gas does add value for operators and amounts to 18% of the energy produced from the play, the high discount of natural gas price compared to the price of oil and the lack of gathering infrastructure (particularly in remote regions) have resulted in the flaring of some 30% of production. This has attracted considerable attention, including the enactment of new regulations.[33] The Bakken currently produces about 1.1 billion cubic feet per day and has produced more than one trillion cubic feet since 2006.

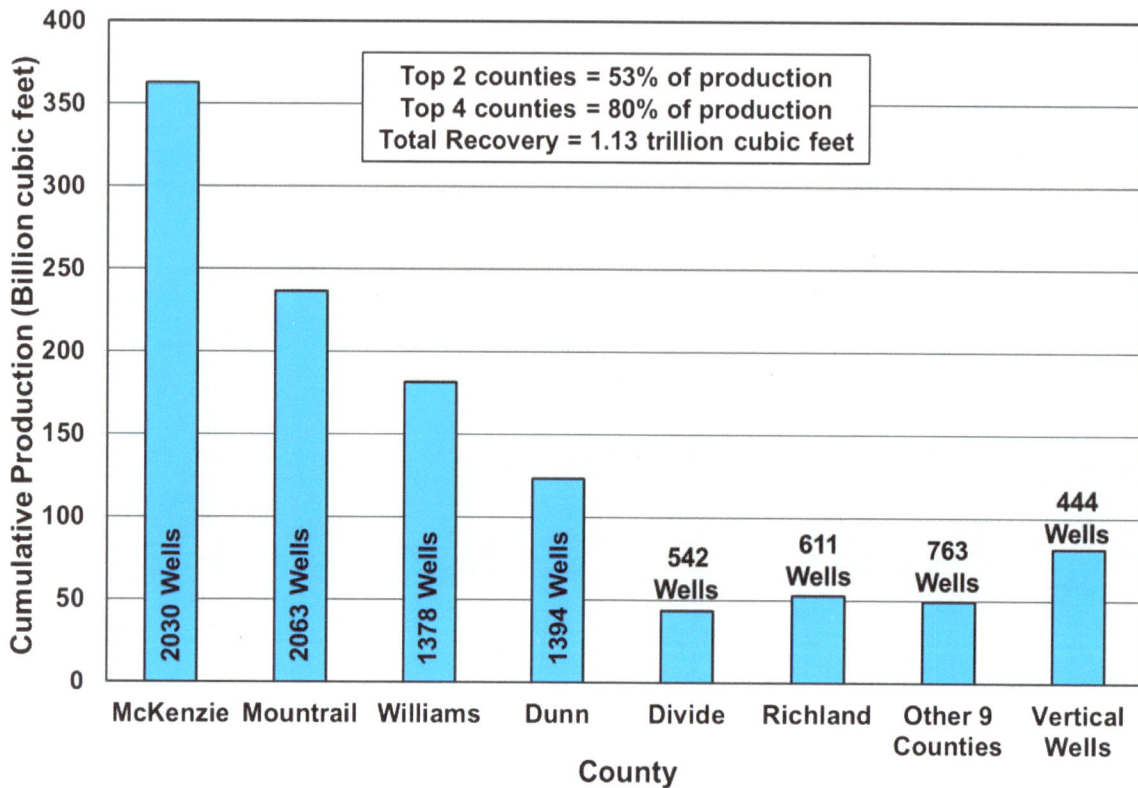

Figure 2-20. Cumulative gas production by county in the Bakken play through 2014.[34]
The top four counties have produced 80% of the 1.13 trillion cubic feet produced to June 2014.

[33] Anna Driver and Emest Scheyder, "North Dakota flaring crackdown may slow oil field growth," Reuters, June 5, 2014, http://www.reuters.com/article/2014/06/05/bakken-flaring-idUSL1N0OK2AI20140605.
[34] Data from Drillinginfo retrieved September 2014.

Operators are highly sensitive to the economic performance of the wells they drill, which typically cost in the order of $8 million or more each, not including leasing costs and other expenses. The areas of highest quality—the "core" or "sweet spots"—have now been well defined. Figure 2-21 illustrates average well decline profiles by county; these can be seen as a measure of well quality. The well decline profiles from the top three counties are all above the Bakken average, hence these counties are attracting the bulk of the drilling and investment.

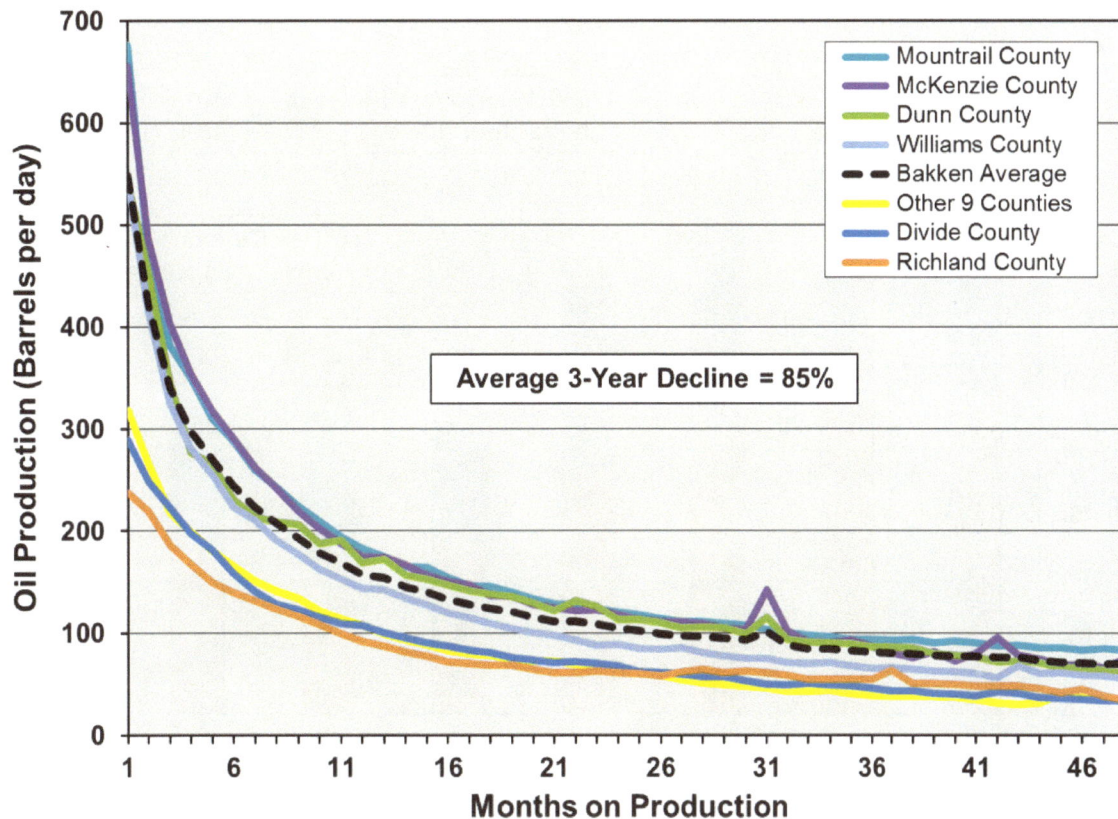

Figure 2-21. Average tight oil well decline profiles by county for the Bakken play.[35]

The top four counties which have produced most of the oil and gas in the Bakken are clearly superior. If natural gas is included, on a "barrels of oil equivalent basis," average initial production in counties like Mountrail and McKenzie is over 800 barrels per day. Well decline profiles are based on horizontal wells drilled since 2009.

Another measure of well quality is "estimated ultimate recovery" (EUR), the amount of oil a well will recover over its lifetime. To be clear, no one knows what the lifespan of a Bakken well is, given that few of them are more than seven years old. Operators fit hyperbolic and/or exponential curves to data such as presented in Figure 2-21, assuming well life spans of 30-50 years (as is typical for conventional oil wells), but so far this is speculation given the nature of the extremely low permeability reservoirs and the completion technologies used in the Bakken. Nonetheless, for comparative well quality purposes only, one can use the data in Figure 2-21, which show that wells exhibit steep initial decline rates with progressively more gradual decline rates, and assume a constant terminal decline rate thereafter to develop a theoretical EUR.

---

[35] Data from Drillinginfo retrieved April 2014.

Figure 2-22 illustrates theoretical EURs for horizontal wells by county for the Bakken, for comparative purposes of well quality; these range from 203,000 to 442,000 barrels per well. This compares to EURs of 13,000 to 340,000 barrels per well assumed by the EIA (the EIA weighted mean EUR—based on potential number of wells—is 146,000 barrels).[36] EURs in the top four counties are 50% to more than 100% higher than in the remaining parts of the play. The steep well production declines mean that well payout (if it is achieved) comes in the first few years of production, as between 52% and 62% of an average well's lifetime production occurs in the first four years.

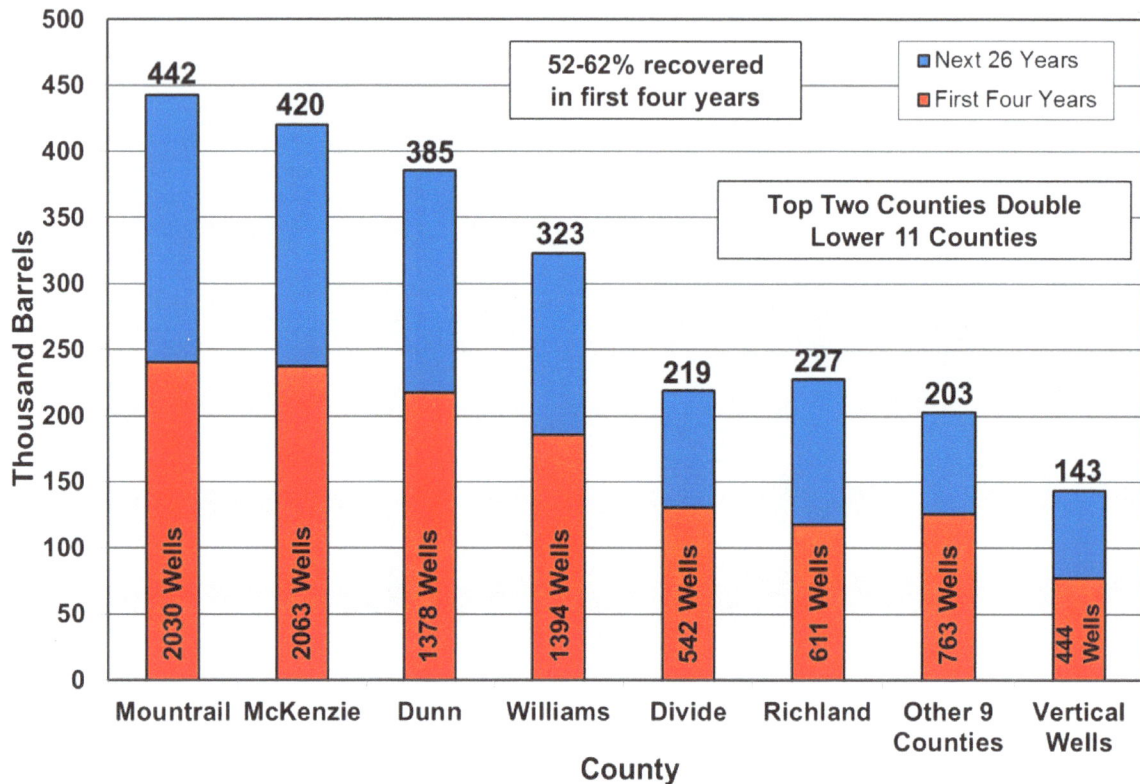

Figure 2-22. Estimated ultimate recovery of oil per well by county for the Bakken play.[37]

EURs are based on average well decline profiles (Figure 2-21) and a terminal decline rate of 13%. These are for comparative purposes only as it is highly uncertain if wells will last for 30 years, as are the decline rates at the end of well life. The EURs by county are for horizontal wells only; the EUR for vertical wells is shown at right. The steep decline rates mean that most production occurs early in well life.

---

[36] EIA, *Assumptions to the Annual Energy Outlook 2014*, http://www.eia.gov/forecasts/aeo/assumptions/pdf/oilgas.pdf.
[37] Data from Drillinginfo retrieved September 2014.

Well quality can also be expressed as the average rate of production over the first year of well life. If we know both the field's decline rate and the average well's first-year production rate, we can calculate the number of wells that need to be drilled each year in order to offset field decline and maintain production. Given that drilling is currently focused on the highest-quality counties, the average first-year production rate per well will necessarily fall as drilling moves into lower-quality counties over time (i.e., as the best locations are drilled off). As average well quality falls, the number of wells that must be drilled to offset field decline must rise, until the drilling rate can no longer offset decline and the field peaks.

Figure 2-23 illustrates the average first-year oil production rate of wells by county. Notwithstanding modest gains in the top four counties, which are also those that are most densely drilled, future technology improvements are unlikely to postpone for long the inevitable decline in average overall well quality as drilling moves into lower quality counties.

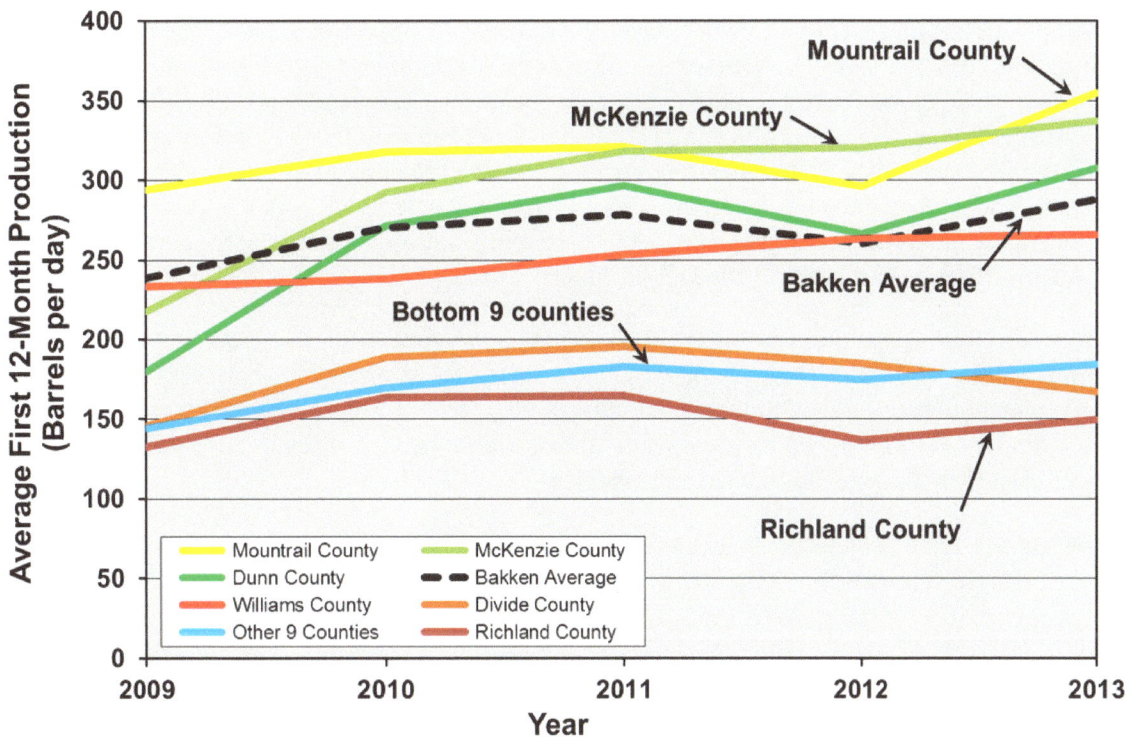

Figure 2-23. Average first-year oil production rates of wells in the Bakken play by county, 2009 to 2013.[38]

Well quality is rising most rapidly in Mountrail County, which is also the county with the current highest well density. First year production rate in the lowest-producing 11 counties, where the bulk of remaining drilling locations are, is flat. The top four counties have roughly double the well quality of the lowest 11.

[38] Data from Drillinginfo retrieved April 2014.

2.3.1.4   Number of Wells

The fourth key fundamental is the number of wells that can ultimately be drilled in the Bakken, a function of (a) the size of the area worth drilling and (b) the density of drilling that will likely occur. This issue is hotly debated in investor presentations. One of the most optimistic views comes from Continental Resources, one of the first companies to drill in the Bakken, whose CEO claims 100,000 wells may ultimately be drilled.[39] The North Dakota Industrial Commission is bullish, but less so, at 40,000 wells[40] in addition to the 9,225 already drilled. In contrast, the EIA estimates 73,697 wells, 29,186 of which are in the Bakken with the remainder in the Three Forks (obtained from the product of well density and play area in the EIA assumptions[41]).

Determining the likely density at which operators will drill wells requires consideration of both the geology of the play and the mechanics of hydraulic fracturing. Typical wells in the Bakken have horizontal laterals of 10,000 feet in length with 25 or more frack stages. The EIA suggests that the Bakken may be drilled at a density of 2 wells per square mile[42] which would space horizontal laterals 1,320 feet from each other. One operator, Enerplus, suggests (based on a drilling pilot in one of the best areas) that 3.5 wells can be drilled per square mile, including both the Bakken and Three Forks.[43] Continental is testing downspacing of horizontal laterals to just 660 feet apart on four layers of the Bakken and Three Forks, which, if successful, could be up to a staggering 16 wells per square mile.[44] There is no confirmation if this actually worked over a period of time long enough to assess the degree of interference between wells, which would only become apparent after 6-12 or more months of production history.

Wells spaced less than 2,000 feet apart in the Bakken may undergo interference, meaning that wells cannibalize each other's oil over time, as noted by Thuot, based on an empirical analysis of Bakken data.[45] This means that although oil can be produced more quickly by spacing wells closer together than 2,000 feet, the ultimate amount of oil produced per well will be reduced, and the total amount of oil recovered per unit area will not be substantially increased. Thuot concludes:

1. *Well interference in the Bakken appears to occur for hydraulically fractured horizontal wellbores spaced closer than roughly 2,000 feet.*

2. *The magnitude of well interference on production appears to increase over time.*

3. *The full impact of well interference in the analysis above is likely somewhat masked since operators become more proficient in drilling and completion techniques over time. As we saw, secondary wells over-perform when spaced wider than 2,000 feet.*

---

[39] Christopher Helman, "Harold Hamm: The Billionaire Oilman Fueling America's Recovery," *Forbes*, April 16, 2014, http://www.forbes.com/sites/christopherhelman/2014/04/16/harold-hamm-billionaire-fueling-americas-recovery.

[40] North Dakota Industrial Commission, *Development of the Bakken Resource*, June 11, 2014, https://www.dmr.nd.gov/oilgas/presentations/ActivityUpdate2014-06-11NCSLBismarck.pdf.

[41] EIA, *Assumptions to the Annual Energy Outlook 2014*, http://www.eia.gov/forecasts/aeo/assumptions/pdf/oilgas.pdf.

[42] EIA, *Assumptions to the Annual Energy Outlook 2014*, http://www.eia.gov/forecasts/aeo/assumptions/pdf/oilgas.pdf.

[43] Enerplus, "A Deeper Look into the Williston Basin" investor presentation, June 18, 2014, http://www.enerplus.com/files/pdf/investor-relations/Williston%20Basin%20Deck_June%2018_FINAL%202.pdf.

[44] Continental Resources, July 2014 Investor presentation, retrieved August 2014 from http://investors.clr.com/phoenix.zhtml?c=197380&p=irol-presentations.

[45] Kevin Thuot, "There Will Be Blood: Well Spacing & The Bakken Shale Oil Milkshake," Drillinginfo, November 26, 2013, http://info.drillinginfo.com/well-spacing-bakken-shale-oil.

This implies that fractures propagated from a wellbore drain in the order of 1,000 feet from the well. Given that 2 wells per square mile places 10,000 foot laterals 1,320 feet apart, a 2,000-foot spacing would require considerably lower well densities.

Given that the four layers ("benches") of the Three Forks lie between 80 and 250 feet below the middle Bakken target zone, it is likely that wells drilled in the middle Bakken are also draining oil from at least some of the underlying Three Forks benches, ultimately limiting the number of wells needed to effectively recover the oil. Therefore, there are practical limits to well downspacing..

Determining the area actually conducive to drilling is comparatively straightforward. After years of exploration and thousands of wells drilled, operators have delineated the limits of the play and focused their efforts on those areas with proven potential; thus by identifying the farthest-lying wells with little to no production as the likely edge of the play, and estimating the size of the area within that edge which is clearly attracting industry interest, the functional area of the Bakken play can be calculated. By this method, the area likely to be conducive to drilling is approximately 12,700 square miles (see Figure 2-9).

Based on the above parameters, and given the fact that much of the area covered by the Bakken is of much lower quality than the top four counties, an estimate of two wells per square mile may be reasonable for the whole area, with an estimate of three wells per square mile being on the optimistic upside. This translates to approximately 25,400 wells if drilled at a density of two wells per square mile, and 38,100 wells locations if drilled at a density of three wells per square mile. Allowing three wells per square mile on average over the whole region would provide for greater density in the best quality parts of the play and lower density in the outlying lower quality areas.

Of course, these estimates assume that the entire area designated as the Bakken play is available for drilling—failing to account for parks, towns, rivers, reservoirs, and other areas not conducive to drilling. A slightly more conservative but possibly more realistic calculation would include a "risk" that 20% of the play's remaining area will be undrillable. After accounting for wells already drilled, this risk would reduce the total number of potential wells to approximately 21,400 and 31,500 for the two- and three-well per section cases, respectively. Either way, the Bakken play could experience somewhere between three and four times the number of wells drilled to date.

2.3.1.5 Rate of Drilling

The fifth key fundamental is the *rate of drilling*. As noted earlier, the Bakken play has a field decline of 45% per year, meaning that 45% of production has to be replaced with new wells each year to keep production flat. As the amount of oil produced from an average well in its first year of production is known from the data, the number of wells needed to offset field production decline each year at a given production level can be easily calculated. For the Bakken, at current production levels, some 1,470 wells must be drilled each year just to keep production flat. Since drilling rates in the Bakken are now at about 2,000 wells per year, production will keep growing as long as these rates are sustained. However, the higher production grows, the more wells are needed to offset the 45% field decline. And as drilling moves into lower quality parts of the play, even more wells will be needed, for as illustrated above (Figure 2-23), well quality in 11 of the 15 counties is at least 40% lower than in the best four.

2.3.1.6   Future Production Scenarios

Based on the five key fundamentals outlined above, several production projections for the Bakken play were developed to illustrate the effects of changing the rate of drilling and the number of drilling locations. These production projections intentionally ignore questions of economics (e.g., the amount of capital required and whether oil prices would support drilling in less productive areas) or politics (e.g., community opposition, new government regulation) in order to analyze what is technically possible.

The projections are given in three cases, differentiated by the number of drilling locations:

1. A "Low Well Density Case" of 100% of the play area being drillable, at 2 wells per square mile. (The EIA assumes that 2 wells can be drilled per square mile in the Bakken and 2.5 wells per square mile can be drilled in the underlying Three Forks.)

2. An "Optimistic Case" of 100% of the play area being drillable at 3 wells per square mile.

3. A "Realistic Case" of 80% of the remaining play area being drillable (i.e., the remaining play area is "risked" at 80% to account for undrillable areas like parks, towns, rivers, etc.), at 3 wells per section.

Each case includes three scenarios, differentiated by the rate of drilling:

1. MOST LIKELY RATE scenario: Drilling continues at the current rate of 2,000 wells per year and then declines to 1,000 wells per year as drilling moves into the lower quality counties.

2. EXPANDED RATE scenario: Drilling increases to 2,500 wells per year and then declines to 1,500 wells per year as drilling moves into the lower quality counties.

3. FASTEST RATE scenario: Drilling is increased 50% over the current rate to 3,000 wells per year, and held constant until locations run out.

The critical parameters used for determining production rates in these scenarios are given in Table 2-1.

Parameters	Counties							Total
	Divide	Dunn	McKenzie	Mountrail	Richland	Williams	Other 9	
Production Jan 2014 (Kbbl/d)	38	165	296	245	29	143	50	966
% of Field Production	4	17	31	25	3	15	5	100
Cumulative Oil (million bbls)	40	172	249	344	52	150	56	1063
Cumulative Gas (Bcf)	44	123	363	237	53	182	49	1050
Number of Wells	542	1378	2063	2030	611	1394	763	8781
Number of Horizontal Producing Wells	524	1282	1875	1896	565	1318	693	8153
Average EUR per well (Kbbls)	219	385	420	443	227	323	203	378
Field Decline (%)	51	38	49	40	30	50	54	45
3-Year Well Decline (%)	85	84	88	86	73	88	88	85
Average First Year Production in 2013 (bbl/d)	169	308	344	376	148	271	180	296
New Wells Needed to Offset Field Decline	115	202	418	258	60	266	150	1468
Area in square miles	1259	2010	2742	1824	2084	2071	18000	29990
% Prospective	60	60	75	65	55	90	25	39
Net square miles	755	1206	2057	1186	1146	1864	4500	12714
Well Density per square mile	0.72	1.14	1.00	1.71	0.53	0.75	0.17	0.75
Additional locations to 2/sq. Mile	969	1034	2050	341	1681	2334	8237	16646
Additional locations to 3/sq. Mile	1724	2240	4107	1527	2828	4198	12737	29360
Population	2071	3536	6360	7673	9667	22398	N/A	N/A
Total Wells 2/sq. Mile	1511	2412	4113	2371	2292	3728	9000	25427
Total Wells 3/sq. Mile	2266	3618	6170	3557	3439	5592	13500	38141
Total Wells 2/sq. Mile Risked at 80%	1317	2205	3703	2303	1956	3261	7353	22098
Total Wells 3/sq. Mile Risked at 80%	1921	3170	5348	3251	2873	4752	10953	32269

Table 2-1. Parameters for projecting Bakken tight oil production, by county
Area in square miles under "Other" is estimated.

Low Well Density Case

In the "Low Well Density Case" (Figure 2-24), assuming 100% of the area is drillable, approximately 17,700 wells remain to be drilled on top of the more than 8,500 wells currently producing, for a total of 25,500 wells (including wells no longer producing).

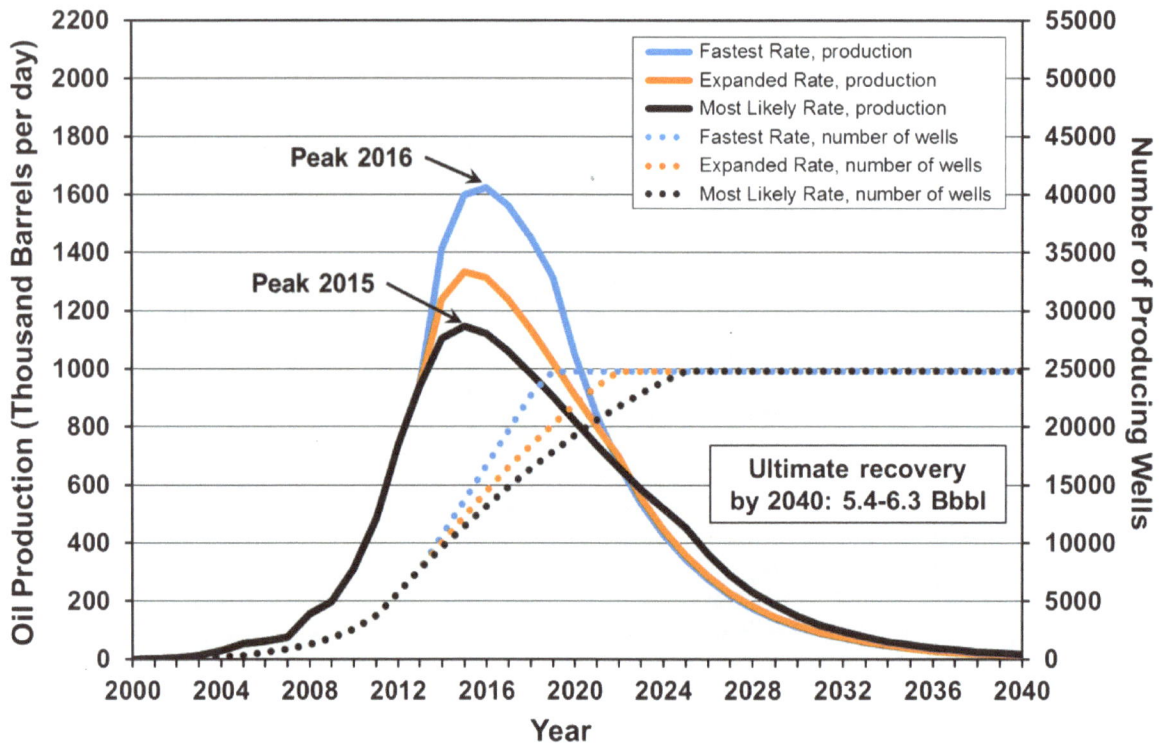

Figure 2-24. Three drilling rate scenarios of Bakken tight oil production, in the "Low Well Density Case" (100% of play area is drillable at two wells per square mile).[46]

"Most Likely Rate" scenario: drilling continues at 2,000 wells/year, declining to 1,000 wells/year;
"Expanded Rate" scenario: drilling increases to 2,500 wells/year, declining to 1,500 wells/year;
"Fastest Rate" scenario: drilling increases to 3,000 wells/year, holding constant.

The drilling rate scenarios in this case have the following results:

1. MOST LIKELY RATE scenario: Peak production occurs in 2015 at 1.15 MMbbl/d. Drilling continues until 2025, and total oil recovery by 2040 is 5.4 billion barrels.

2. EXPANDED RATE scenario: Peak production occurs in 2015 at 1.33 MMbbl/d. Drilling continues until 2022, and total oil recovery by 2040 is 5.7 billion barrels. Production would be lower after 2023 than in the "Most Likely Rate" case as faster drilling would recover the oil sooner.

3. FASTEST RATE scenario: Peak production occurs in 2016 at 1.63 MMbbl/d. Drilling continues until 2019, and total oil recovery by 2040 is 6.3 billion barrels. As in the "Expanded Rate" scenario, production would be lower after 2023 than in the "Most Likely Rate" case.

---

[46] Data from Drillinginfo retrieved September 2014.

Optimistic Case

If technological advances allow for a denser drilling footprint of three wells per section, ultimate recovery increases somewhat—but the timing of production peaks remain virtually the same (pushed back by only a year). This case would see the drilling of 29,400 wells on top of the more than 8,500 currently producing wells for a total of 38,100 wells (including wells no longer producing). Figure 2-25 illustrates this "Optimistic Case."

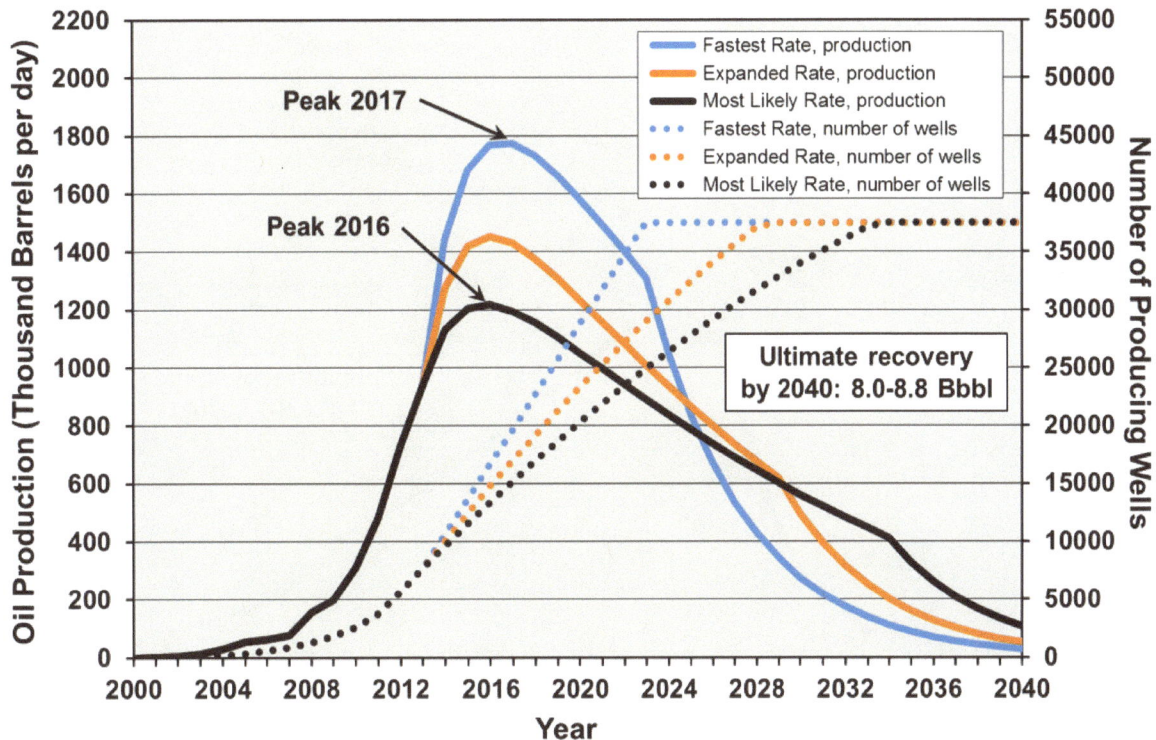

Figure 2-25. Three drilling rate scenarios of Bakken tight oil production, in the "Optimistic Case" (100% of play area is drillable at three wells per square mile).[47]

"Most Likely Rate" scenario: drilling continues at 2,000 wells/year, declining to 1,000 wells/year; "Expanded Rate" scenario: drilling increases to 2,500 wells/year, declining to 1,500 wells/year; "Fastest Rate" scenario: drilling increases to 3,000 wells/year, holding constant.

The drilling rate scenarios in this case have the following results:

1. MOST LIKELY RATE scenario: Peak production occurs in 2016 at 1.22 MMbbl/d. Drilling continues until 2034, and total oil recovery by 2040 is 8.0 billion barrels.

2. EXPANDED RATE scenario: Peak production occurs in 2016 at 1.45 MMbbl/d. Drilling continues until 2029, and total oil recovery by 2040 is 8.3 billion barrels.

3. FASTEST RATE scenario: Peak production occurs in 2017 at 1.77 MMbbl/d. Drilling continues until 2023, and total oil recovery by 2040 is 8.8 billion barrels. In this scenario, production would be considerably lower after 2026 than in the "Most Likely Rate" scenario.

---

[47] Data from Drillinginfo retrieved September 2014.

Realistic Case

A more realistic case (Figure 2-26) is that 80% of the remaining play area will be drillable at three wells per square mile (i.e., the case includes a "risk" that 20% of the play remaining area will be undrillable). This allows for surface features that preclude drilling, such as towns, rivers, reservoirs, parks and other surface features which may limit access for drilling. This scenario would see the drilling of 23,500 wells on top of the more than 8,500 currently producing wells for a total of 32,300 wells (including wells no longer producing).

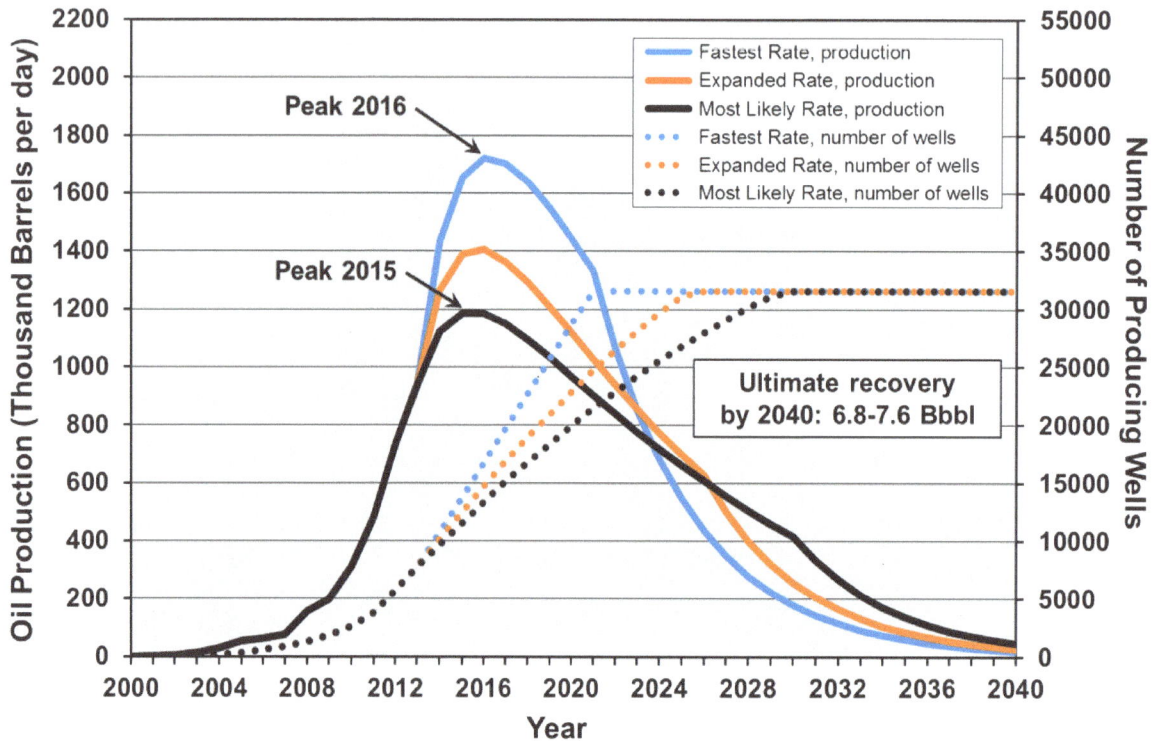

Figure 2-26. Three drilling rate scenarios of Bakken tight oil production, in the "Realistic Case" (80% of the remaining play area is drillable at three wells per square mile).[48]
"Most Likely Rate" scenario: drilling continues at 2,000 wells/year, declining to 1,000 wells/year;
"Expanded Rate" scenario: drilling increases to 2,500 wells/year, declining to 1,500 wells/year;
"Fastest Rate" scenario: drilling increases to 3,000 wells/year, holding constant.

The drilling rate scenarios in this case have the following results:

1. MOST LIKELY RATE scenario: Peak production occurs in 2015 at 1.19 MMbbl/d. Drilling continues until 2030, and total oil recovery by 2040 is 6.8 billion barrels.

2. EXPANDED RATE scenario: Peak production occurs in 2016 at 1.41 MMbbl/d. Drilling continues until 2026, and total oil recovery by 2040 is 7.1 billion barrels.

3. FASTEST RATE scenario: Peak production occurs in 2016 at 1.72 MMbbl/d. Drilling continues until 2021, and total oil recovery by 2040 is 7.6 billion barrels. In this scenario, production would be considerably lower after 2024 than in the "Most Likely Rate" scenario.

[48] Data from Drillinginfo retrieved September 2014.

2.3.1.7 Comparison to EIA Forecast

Figure 2-27 compares the EIA's reference case projection for Bakken tight oil production to the "Most Likely Rate" scenario of the "Realistic" case presented above.

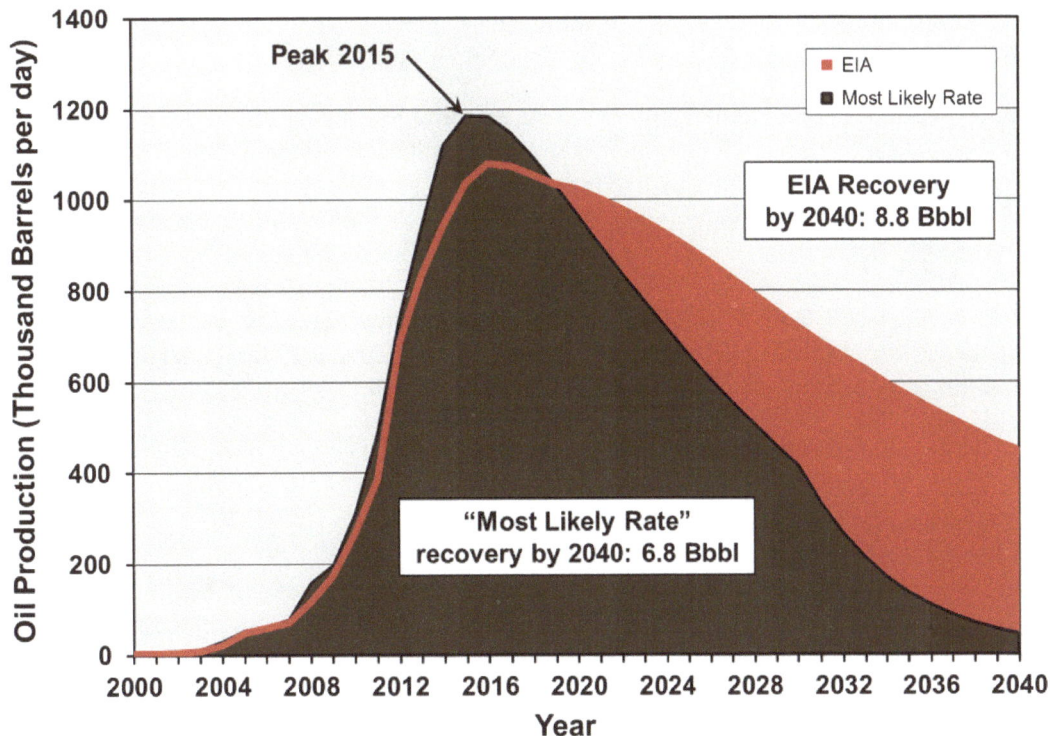

Figure 2-27. "Most Likely Rate" scenario ("Realistic" case) of Bakken tight oil production compared to the EIA reference case, 2000 to 2040.[49]

In this "Most Likely Rate" scenario, drilling continues at 2,000 wells/year, declining to 1,000 wells/year.

This comparison reveals:

- The EIA's forecast of the timing of peak production (2016) in the Bakken is similar to the projection of this report.

- The EIA's forecast of the production rate at peak (1.08 million bpd) is lower than the projection of this report (1.19 million bpd), but only slightly.

- The EIA projects a higher tail of production after peak, with estimated ultimate recovery (EUR) of 8.8 billion barrels by 2040 (7.9 billion for 2014-2040) as opposed this report's projection of 6.8 billion barrels by 2040 (5.7 billion for 2014-2040).

In short, the EIA is forecasting 2.2 billion additional barrels of future Bakken production than this report finds substantiated.

---

[49] EIA, *Annual Energy Outlook 2014.*

2.3.1.8   Bakken Play Analysis Summary

Several conclusions can be made from the foregoing analysis of the Bakken play:

1. High well- and field-decline rates mean a continued high rate of drilling is required to maintain, let alone increase, production. The observed 45% per year field decline rate requires the drilling of 1,470 wells per year just to maintain current production levels.

2. The production profile is most dependent on drilling rate and to a lesser extent the number of drilling locations (i.e., greatly increasing the number of drilling locations would not change the production profile nearly as much as changing the drilling rate). Drilling rate is determined by capital input, which currently is about $16 billion per year to drill 2,000 wells, not including leasing and other ancillary costs.

3. Peak production is highly likely to occur in the 2015 to 2017 timeframe and will occur at between 1.15 and 1.77 MMbbl/d. The most likely peak is between 1.15 and 1.22 MMbbl/d in the 2015 to 2016 timeframe.

4. Increased drilling rates will raise the level of peak production and move it forward a few months but do not appreciably increase cumulative oil recovery through 2040. Increased drilling rates effectively recover the oil sooner, making the supply situation worse later.

5. The projected recovery of 6.8 billion barrels by 2040 in the "Most Likely Rate" scenario (2,000 wells/year declining to 1,000 wells/year) of the "Realistic" case (80% of play drillable, at 3 wells per square mile), agrees fairly well with the mean estimate of latest USGS assessment of the Bakken (including the Three Forks) of 7.4 billion barrels.[50]

6. These projections are optimistic in that they assume the capital will be available for the drilling "treadmill" that must be maintained (roughly $188 billion is needed to drill more than 23,500 wells, exclusive of leasing and ancillary costs). This is not a sure thing as drilling in the poorer-quality parts of the play will require much higher oil prices to be economic. Failure to maintain drilling rates will result in a steeper drop-off in production.

7. Nearly four times the current number of wells will be required to recover 6.8 billion barrels by 2040 in the "Realistic" case.

8. Projections that the Bakken will continue to grow and then maintain a plateau followed by a gentle decline for the foreseeable future[51] are unlikely to be realized.

[50] USGS, *Assessment of Undiscovered Oil Resources in the Bakken and Three Forks Formations, Williston Basin Province, Montana, North Dakota, and South Dakota*, 2013, http://pubs.usgs.gov/fs/2013/3013/fs2013-3013.pdf.
[51] North Dakota Industrial Commission, *Development of the Bakken Resource*, June 11, 2014, https://www.dmr.nd.gov/oilgas/presentations/ActivityUpdate2014-06-11NCSLBismarck.pdf.

2.3.2 Eagle Ford Play

The EIA forecasts recovery of 10.8 billion barrels of oil from the Eagle Ford play by 2040. The analysis of actual production data presented below suggests that this forecast is unlikely to be realized.

The Eagle Ford play of southern Texas is now the largest tight oil play in the U.S; it was unknown prior to 2007. In the EIA's analysis, the Eagle Ford play totals 11,165 square miles.[52] This report considers a surface area for the Eagle Ford defined by where productive drilling has actually occurred; after years of exploration, Eagle Ford producers have presumably focused their efforts on those areas with proven potential. By identifying the farthest-lying wells with little to no production as the likely edge of the play, and estimating the size of the area within that edge that is clearly attracting industry interest, the functional prospective area of the Eagle Ford play is calculated at approximately 7,200 square miles. Forecasts of production outside this area cannot substantiated by currently available drilling information. Figure 2-28 illustrates the distribution of tight oil wells as of mid- 2014 as well as the significantly larger EIA play boundary. More than 10,500 wells have been drilled to date, of which 10,088 were producing oil at the time of writing.

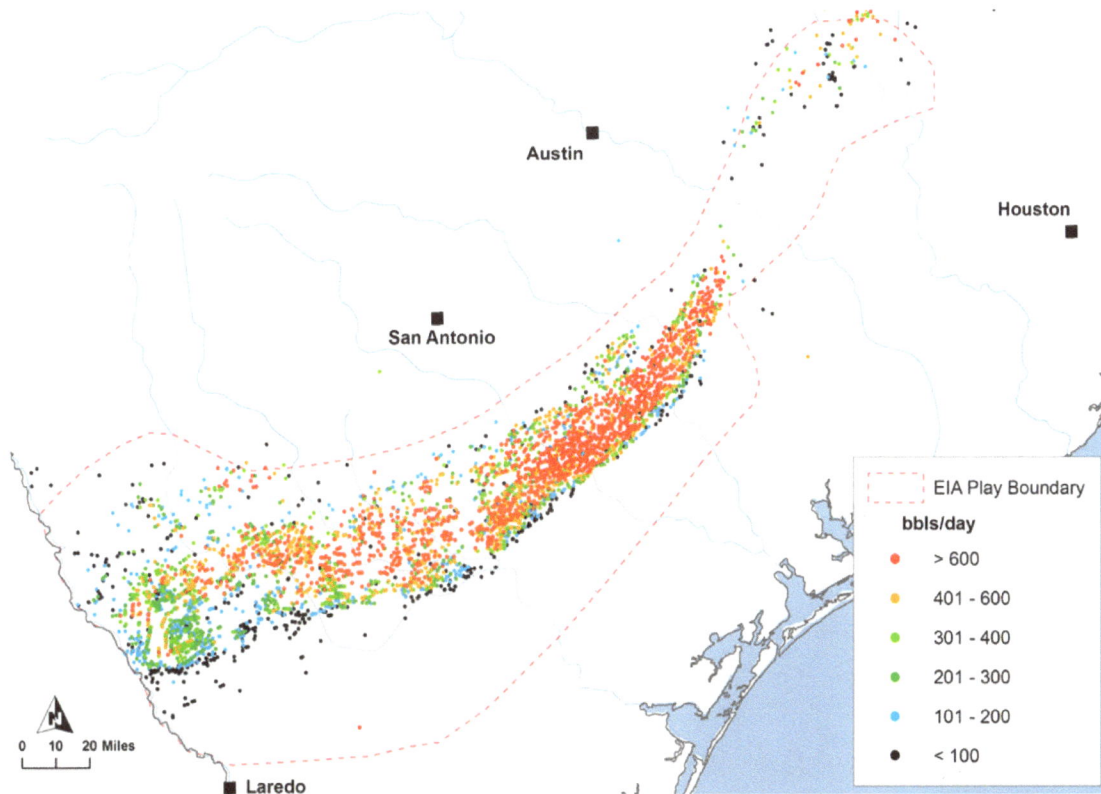

Figure 2-28. Distribution of wells in the Eagle Ford as of mid-2014 illustrating highest one-month oil production (initial productivity, IP),[53] with EIA play boundary.[54]

The size of the Eagle Ford play as defined by the extent of where productive drilling has actually occurred is approximately 7,200 square miles, in contrast to the much larger area designated as the play by the EIA. Well IPs are categorized approximately by percentile; see Appendix.

---

[52] EIA, *Assumptions to the Annual Energy Outlook 2014*, http://www.eia.gov/forecasts/aeo/assumptions/pdf/oilgas.pdf .
[53] Data from Drillinginfo retrieved August 2014.
[54] At publication, the most recent shapefile for the EIA's play area for the Eagle Ford was dated May 2011, available at http://www.eia.gov/pub/oil_gas/natural_gas/analysis_publications/maps/maps.htm#geodata.

The play covers parts of 28 counties although most drilling is concentrated in six counties which account for 81% of production.

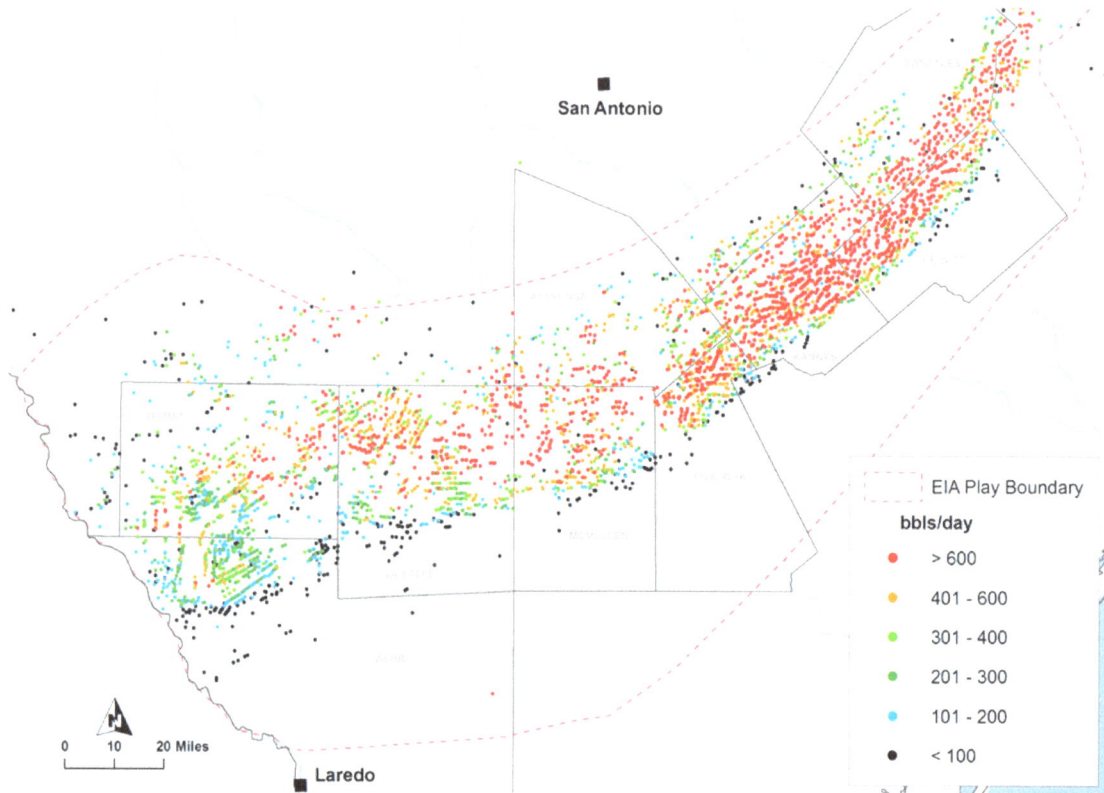

Figure 2-29. Detail of Eagle Ford play showing distribution of wells as of mid-2014 illustrating highest one-month oil production (initial productivity, IP),[55] with EIA play boundary.[56]

The top six producing counties are indicated. Well IPs are categorized approximately by percentile; see Appendix.

[55] Data from Drillinginfo retrieved August 2014.
[56] At publication, the most recent shapefile for the EIA's play area for the Eagle Ford was dated May 2011, available at http://www.eia.gov/pub/oil_gas/natural_gas/analysis_publications/maps/maps.htm#geodata.

The Eagle Ford is both a prolific oil producer and a natural gas producer. It has oil, wet gas and dry gas windows, with oil being produced up dip (i.e., in the shallower part of the formation) along the northwestern portion of the field and gas in the down dip (i.e., in the deeper part of the formation) southeastern portion. Figure 2-30 illustrates the distribution of wells classified as "oil" and "gas" in the main part of the field stretching northeast from the Mexican border.

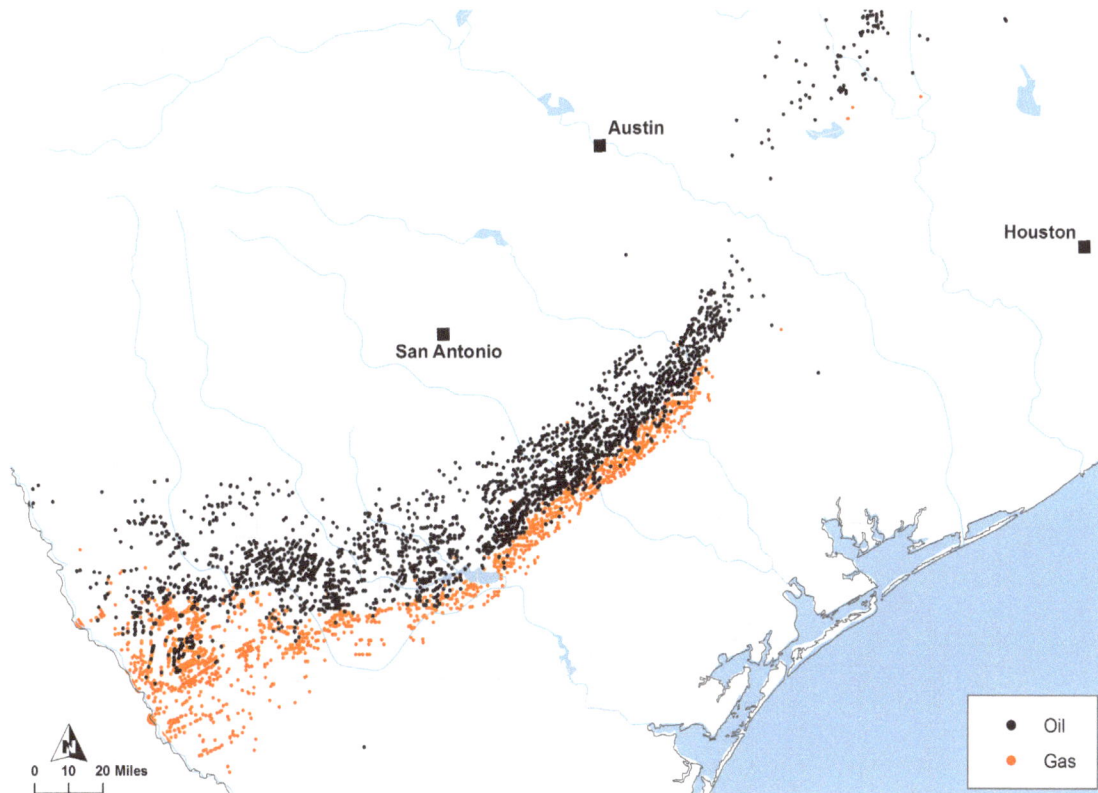

Figure 2-30. Distribution of oil and gas wells in the main portion of the Eagle Ford play as of early 2014.[57]

The Mexican border is on the left. Orange wells are classified as "gas" and black wells are classified as "oil".

---

[57] Data from Drillinginfo retrieved August 2014.

Production in the Eagle Ford was nearly 1.3 million barrels of oil and 4.9 billion cubic feet of gas per day at the time of writing, as illustrated in Figure 23. Gas production is expressed in Figure 2-31 as barrels of oil equivalent (6,000 cubic feet of gas equals approximately one barrel of oil on an energy equivalent basis). Ninety-eight percent of this production is from horizontal fracked wells. The rate of drilling has grown from about 500 wells per year in early 2011 to about 3,500 wells per year in 2014.

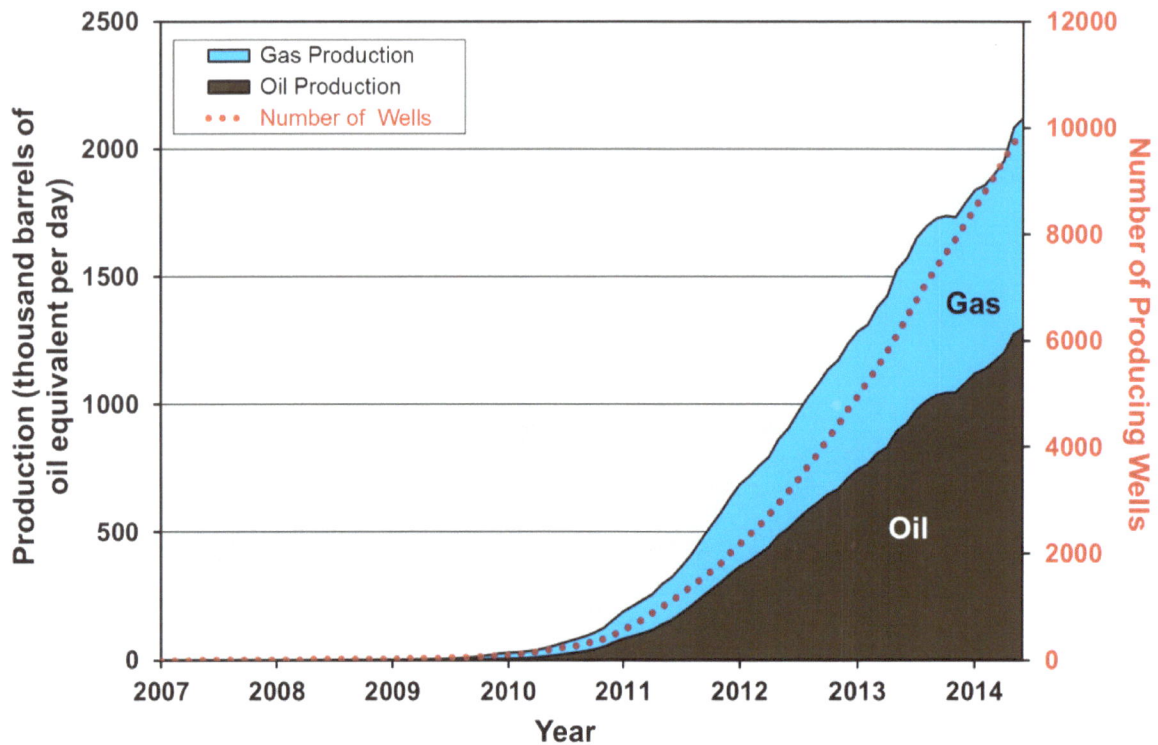

Figure 2-31. Eagle Ford play tight oil and gas production and number of producing wells, 2007 to 2014.[58]

Gas production is expressed as "barrels of oil equivalent" (6,000 cubic feet of gas is approximately equivalent to one barrel of oil on an energy basis).

---

[58] Data from Drillinginfo retrieved September 2014. Three-month trailing moving average.

The amount of oil added to total play production by each new well has been declining since mid-2011 as illustrated in Figure 2-32.

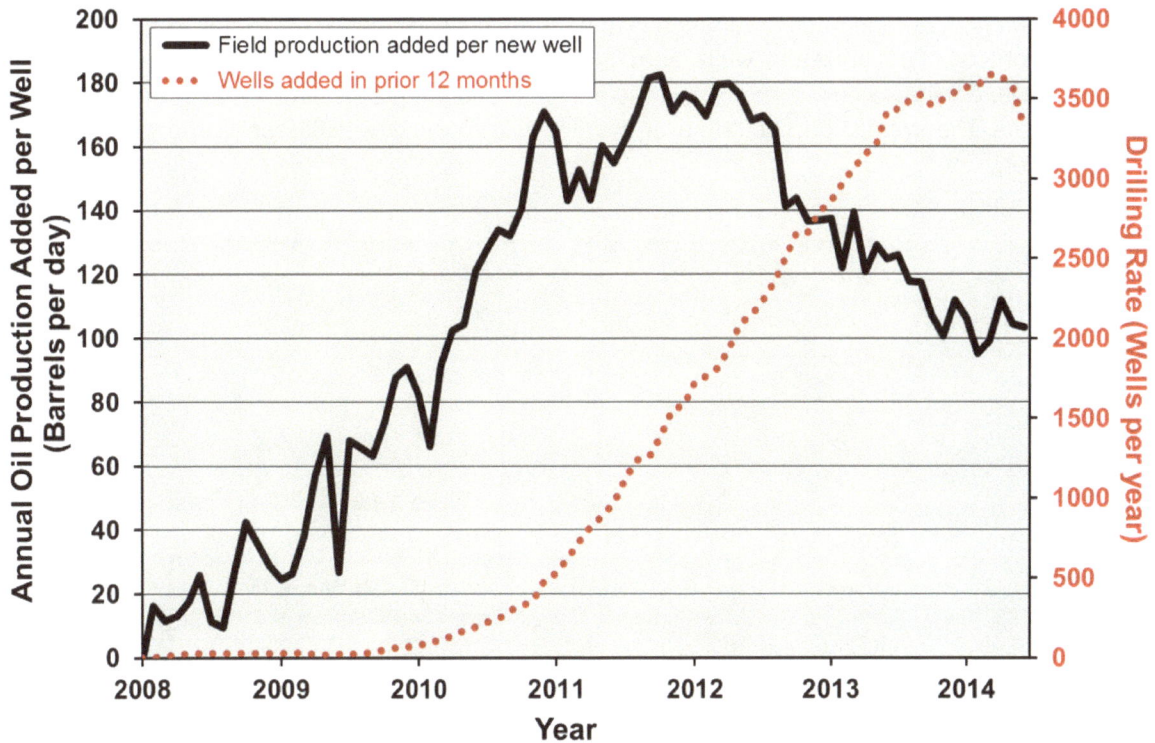

Figure 2-32. Annual oil production added per new well and annual drilling rate in the Eagle Ford play, 2008 through 2014, 2008 to 2014.[59]

---

[59] Data from Drillinginfo retrieved September 2014. Three-month trailing moving average.

2.3.2.1   Well Decline

The first key fundamental in determining the life cycle of Eagle Ford production is the *well decline rate*. Eagle Ford wells exhibit high decline rates in common with all shale plays. Figure 2-33 illustrates the average decline profile of Eagle Ford horizontal wells, both for oil alone and for oil and gas on a "barrels of oil equivalent" basis. Decline rates are steepest in the first year and are progressively less in the second and subsequent years. The average decline rate over the first three years of well life for oil and gas is 79% and 80%, respectively.

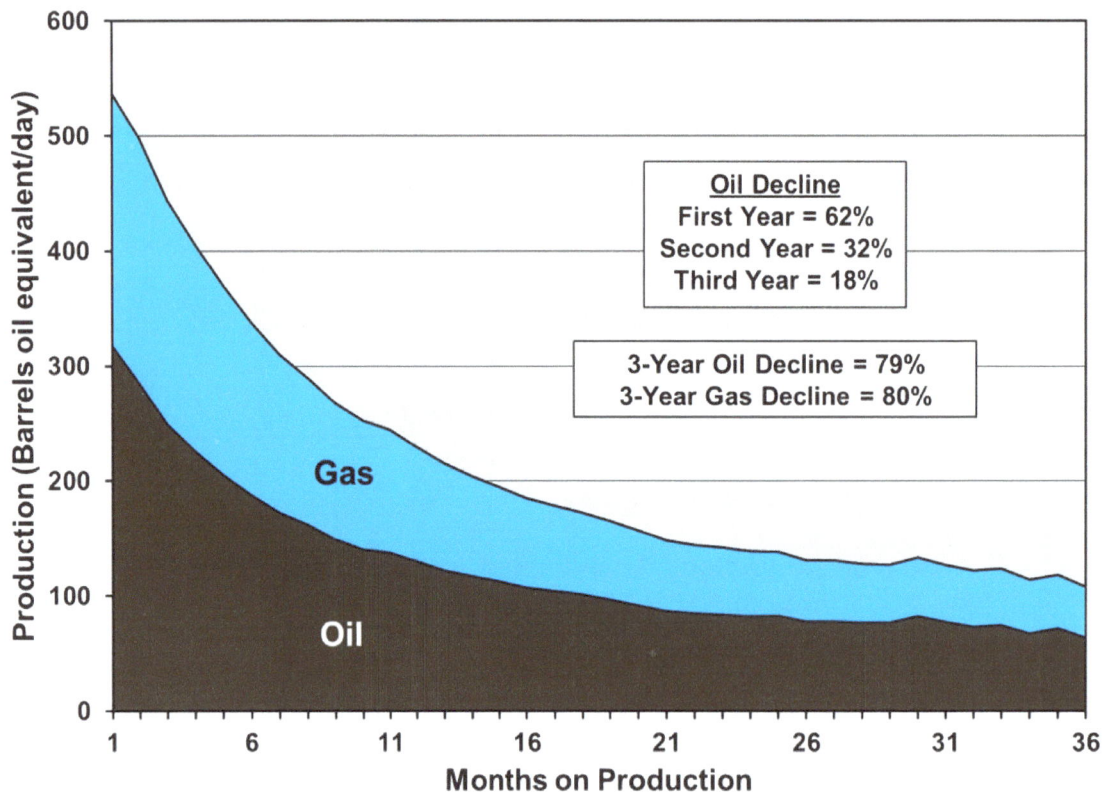

Figure 2-33. Average decline profile for horizontal tight oil and shale gas wells in the Eagle Ford play.[60]

Gas has been converted to barrels of oil on an energy equivalent basis. Decline profile is based on all horizontal wells drilled since 2009.

---

[60] Data from Drillinginfo retrieved May 2014.

2.3.2.2   Field Decline

A second key fundamental is the overall *field decline rate*, which is the amount of production that would be lost in a year without more drilling. Figure 2-34 illustrates oil production from the 5,800 horizontal wells spudded (i.e., drilling was started) prior to 2013, and the 4,964 wells actually producing prior to 2013 (wells are being drilled at such a high rate that many wells drilled prior to 2013 were not connected and producing until well into 2013). The first-year decline for producing wells is 38%. This is lower than the well decline rate as the field decline is made up of new wells, declining at high rates, and older wells, declining at lesser rates. As will be shown later, a field decline of 38% requires 2,285 wells to offset at current production levels, representing capital input of $18.3 billion assuming an average well cost of $8 million.

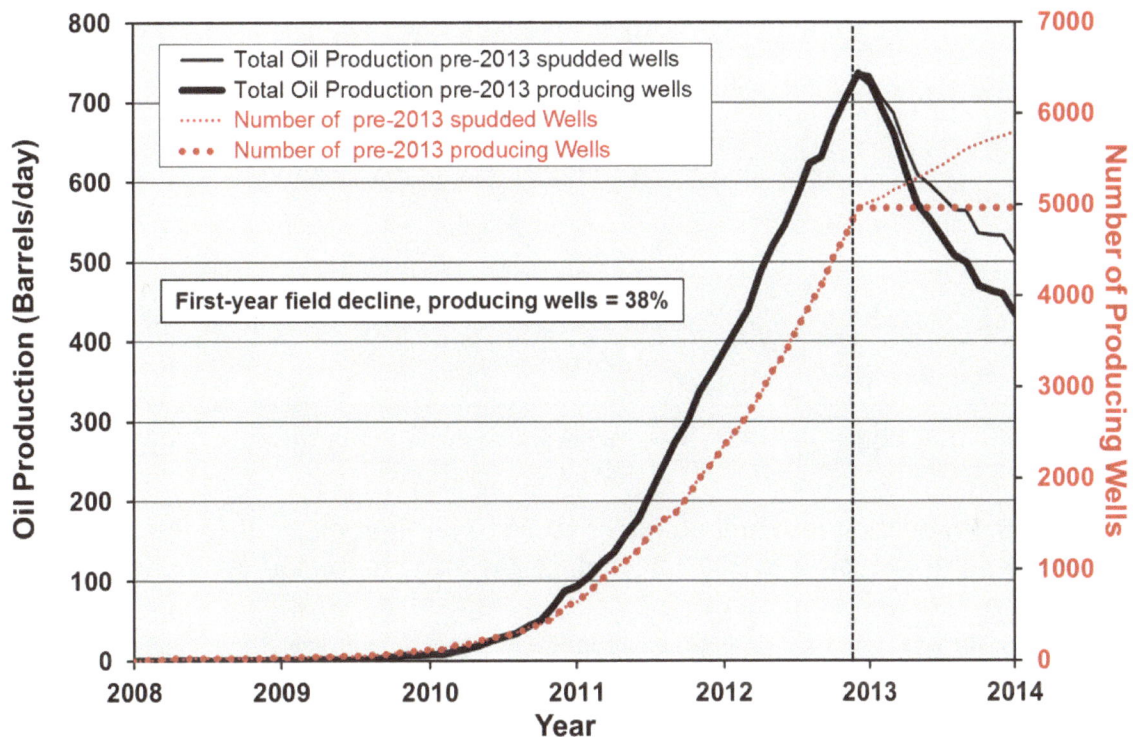

Figure 2-34. Production rate and number of horizontal tight oil wells in the Eagle Ford spudded or producing prior to 2013.[61]

Many of the spudded wells were not connected and producing until well into 2013. In order to offset the 38% field decline rate, 2,285 new wells per year producing at 2013 levels would be required.

---

[61] Data from Drillinginfo retrieved May 2014.

Figure 2-35 illustrates the same analysis on a "barrels of oil equivalent" basis to account for the large amounts of gas also produced. Field decline for wells producing prior to 2013 is 42% in the first year on a barrels oil equivalent basis, and for gas on a standalone basis is 47%.

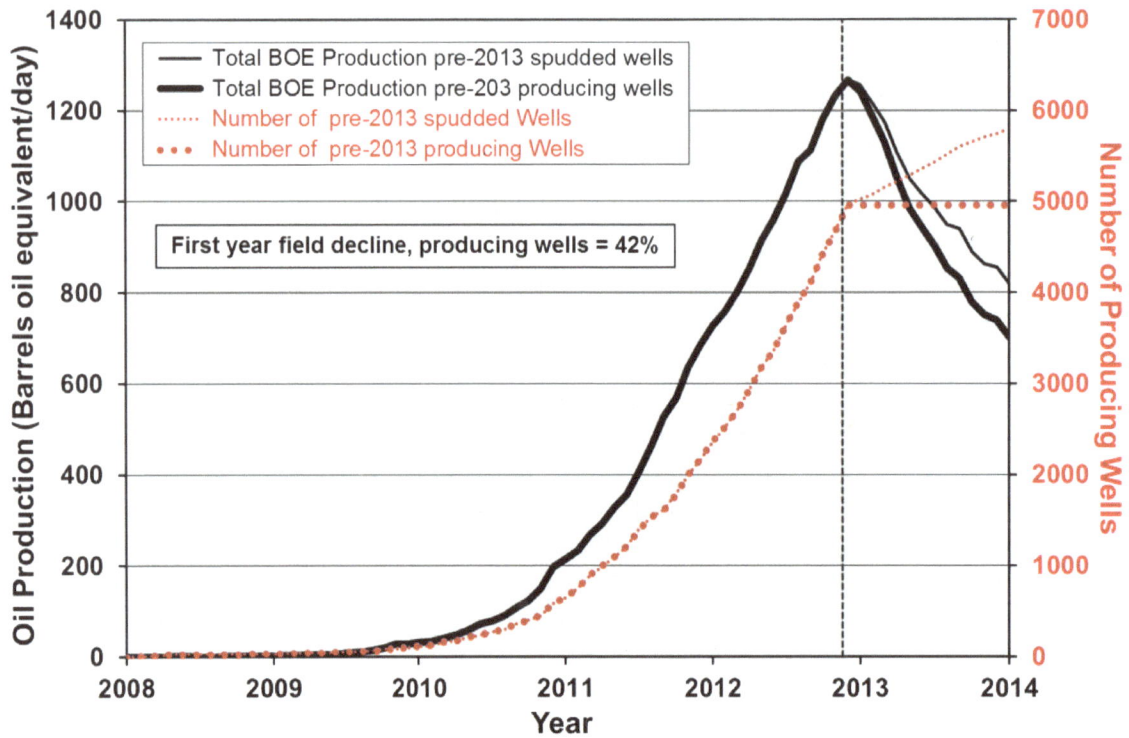

Figure 2-35. Production rate and number of horizontal tight oil wells in the Eagle Ford spudded or producing prior to 2013, including gas on a "barrels of oil equivalent" basis.[62]

Field decline is 42% per year for oil and gas on a "barrels of oil equivalent" basis, and for gas on a standalone basis is 47%.

[62] Data from Drillinginfo retrieved May 2014.

2.3.2.3   Well Quality

The third key fundamental is the trend of *average well quality* over time. As noted earlier, petroleum engineers tell us that technology is constantly improving, with longer horizontal laterals, more frack stages per well, more sophisticated mixtures of proppants and other additives in the frack fluid injected into the wells, and higher-volume frack treatments. This has certainly been true over the past few years, which along with multi-well pad drilling has reduced well costs. It is, however, approaching the limits of diminishing returns and improvements in average well quality are flat to very slightly increasing at best.

Figure 2-36 illustrates production rate trends in oil, gas and "barrels of oil equivalent" from 2009 to 2013 based on the average first year production of wells. On a barrels of oil equivalent basis (BOE) there has been no improvement since 2012, whereas there has been a 4% improvement in oil productivity and a decrease in gas productivity. These trends reflect the shift in operator emphasis to liquids production with the low price of gas, focusing drilling in the oil window of the Eagle Ford, as well as concentrating on the sweet spots defined in the initial wave of drilling. The lack of improvement on a BOE basis suggests better technology is having a very limited, if any, effect; there appears to still be room for significant numbers of new wells in sweet spots, so operators have not yet been forced to move into lower quality parts of the play.

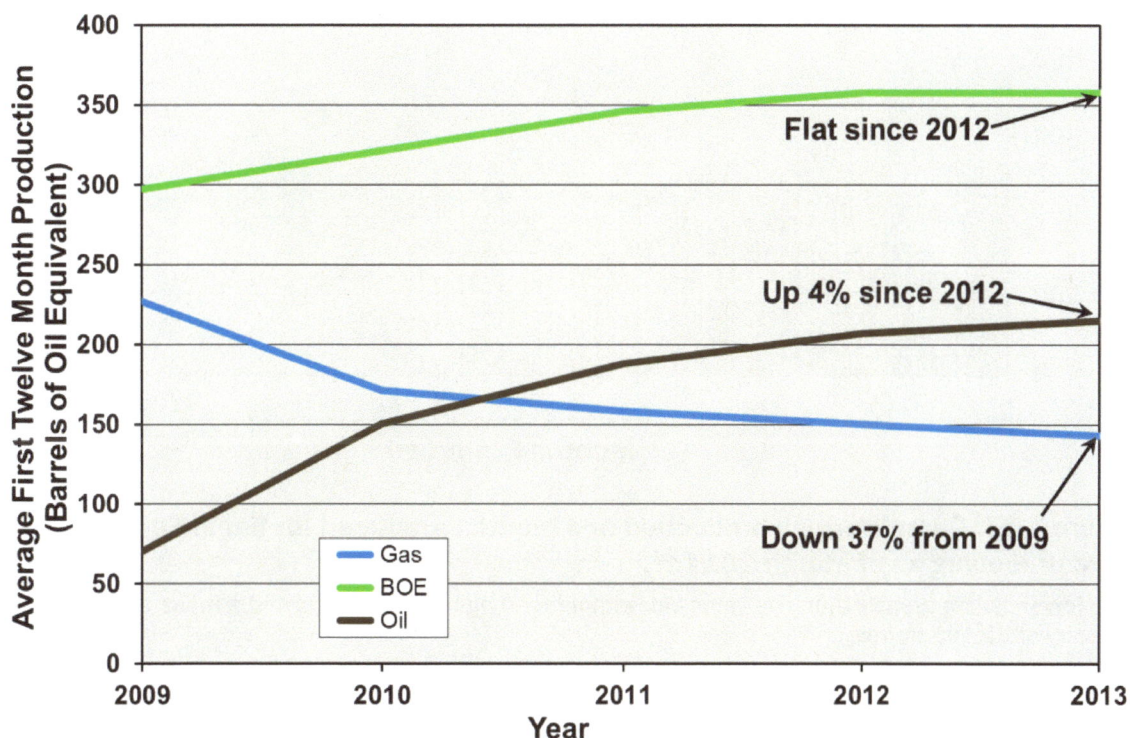

Figure 2-36. Average first year production rates for Eagle Ford wells from 2009 to 2013.[63]

Total production on a "barrels of oil equivalent" basis is unchanged since 2012, whereas oil has risen slightly and gas has fallen. This reflects the focus on liquids production over gas and the concentration of drilling in the oil window of the field, as well as the focus on proven sweet spots, along with likely limited gains from technological improvements in the most recent year.

---

[63] Data from Drillinginfo retrieved May 2014.

Another measure of well quality is cumulative production and well life. Figure 2-37 illustrates the cumulative production of all oil wells that were producing in the Eagle Ford as of March 2014. Eighty-nine percent of these wells are less than 3 years old, and knowing that production will be down nearly 80% after 3 years, their economic lifespan is uncertain. Although it can be seen that there are a few very good wells that recovered more than 400,000 barrels of oil in the first few years, and undoubtedly were great economic successes, the average well has produced just 72,145 barrels over a lifespan averaging 20 months. Less than 1% of these wells are more than 5 years old. The lifespan of wells is another key parameter as many operators assume a minimum life of 30 years and longer—this is conjectural at this point given the lack of long-term well-performance data.

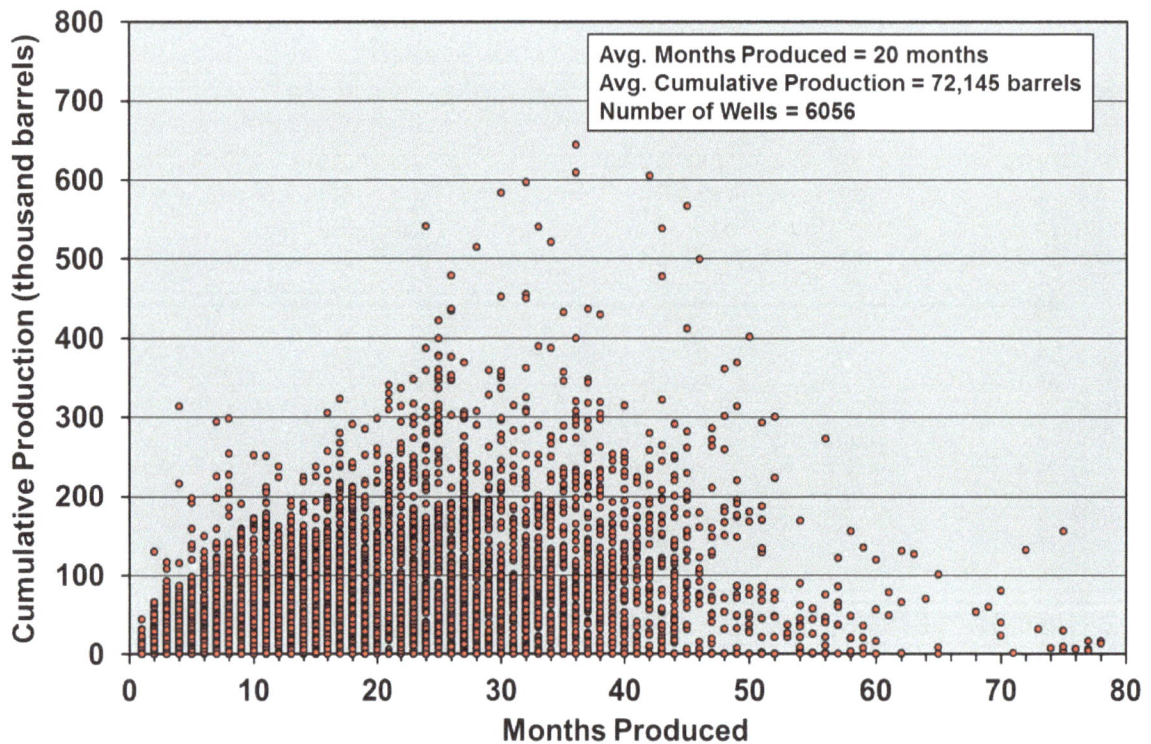

Figure 2-37. Cumulative oil production and months produced for Eagle Ford wells that were producing as of March 2014.[64]

Very few wells are greater than five years old, with a mean age of 20 months and a mean cumulative recovery of 72,145 barrels.

---

[64] Data from Drillinginfo retrieved September 2014. Note that only leases with one well and individual wells are included in this figure (Texas has a practice of lumping production from multi-well leases with production from individual wells).

Cumulative production of course depends on how long a well has been producing, so looking at young wells in not necessarily a good indication of how much oil these wells will produce over their lifespan (although production is heavily weighted to the early years of well life). A measure of well quality independent of age is initial productivity (IP) which is often focused on by operators. Figure 2-38 illustrates the average daily output over the first six months of production (six-month IP) for all oil wells in the Eagle Ford play. Again, as with cumulative production, there are a few exceptional wells—4% of wells produced more than 600 barrels per day over the first six months—but the average for all wells drilled between 2008 and 2014 is just 262 barrels per day. The trend line on Figure 2-38 shows the average over time, which has been increasing slightly over the period, owing to both better technology and the focus of drilling on sweet spots. Figure 2-28 and Figure 2-29 illustrate the distribution of IPs in map form.

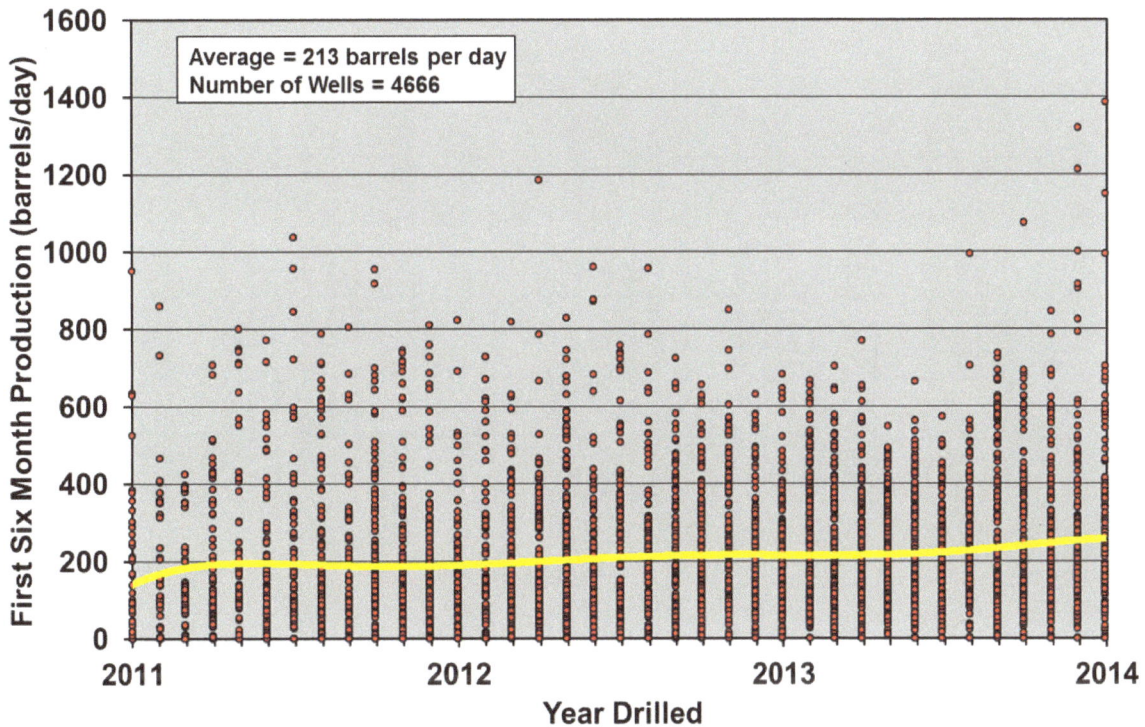

Figure 2-38. Average oil production over the first six months for all wells drilled in the Eagle Ford play.[65]

Although there are a few exceptional wells, the average well produced 213 barrels per day over this period. The trend line indicates variation in mean productivity over time.

---

[65] Data from Drillinginfo retrieved September 2014. Note that only leases with one well and individual wells are included in this figure (Texas has a practice of lumping production from multi-well leases with production from individual wells).

Drilling has focused on liquids-rich parts of the play given the low price of gas in recent years, however the Eagle Ford still produces large amounts of gas which adds to the economic viability of wells. Figure 2-39 illustrates the average production of wells over the first six months on a "barrels of oil equivalent" basis (converting natural gas to its oil equivalent on an energy basis—6000 cubic feet of natural gas equals one barrel of oil). The trend line in this case, combining oil and gas, is essentially flat over the 2011 through 2014 period, indicating technological improvements are not improving well productivity.

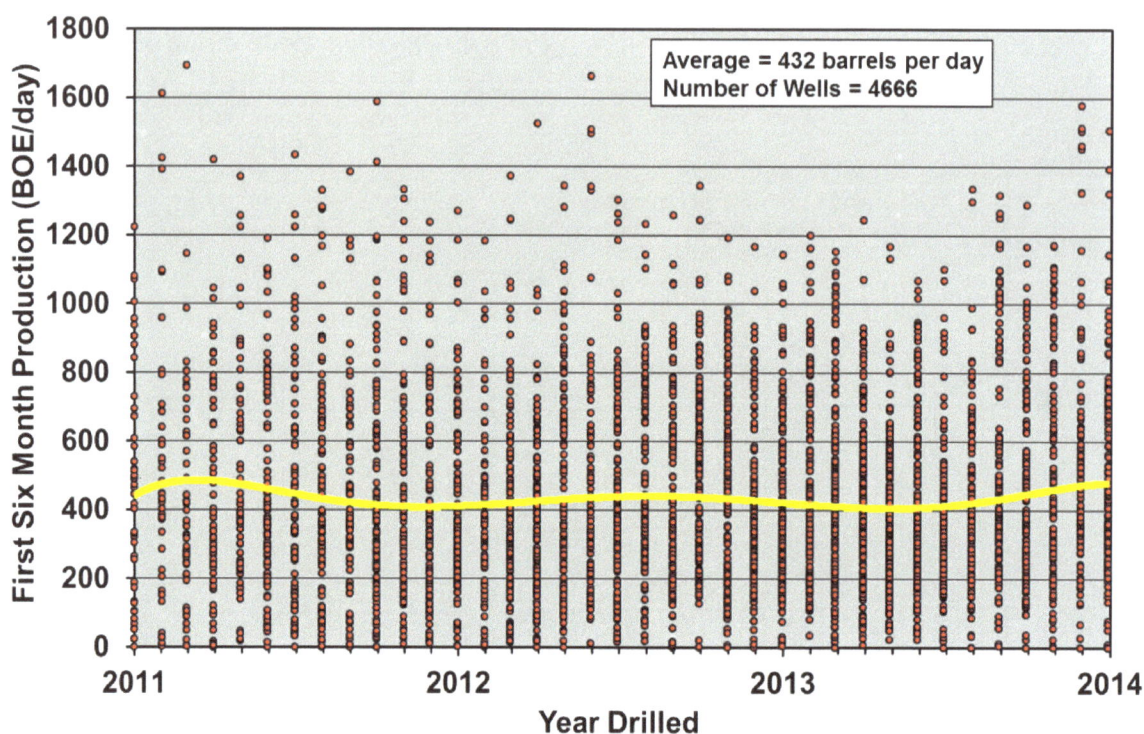

Figure 2-39. Average oil and gas production over the first six months for all wells drilled in the Eagle Ford play on a barrels of oil equivalent basis.[66]

Although there are a few exceptional wells, the average well produced 432 barrels of oil equivalent per day over this period.

---

[66] Data from Drillinginfo retrieved September 2014. Note that only leases with one well and individual wells are included in this figure (Texas has a practice of lumping production from multi-well leases with production from individual wells).

Different counties in the Eagle Ford display markedly different well production rate characteristics which are critical in determining the most likely production profile in the future. Figure 2-40, which illustrates oil production over time by county, shows that the top three counties produce 51% of the total, the top six produce 81% and the remaining 22 counties produce just 19%. Three years of widespread drilling (see Figure 2-41 for number of wells drilled per county) have not resulted in significant production increases outside the top six counties.

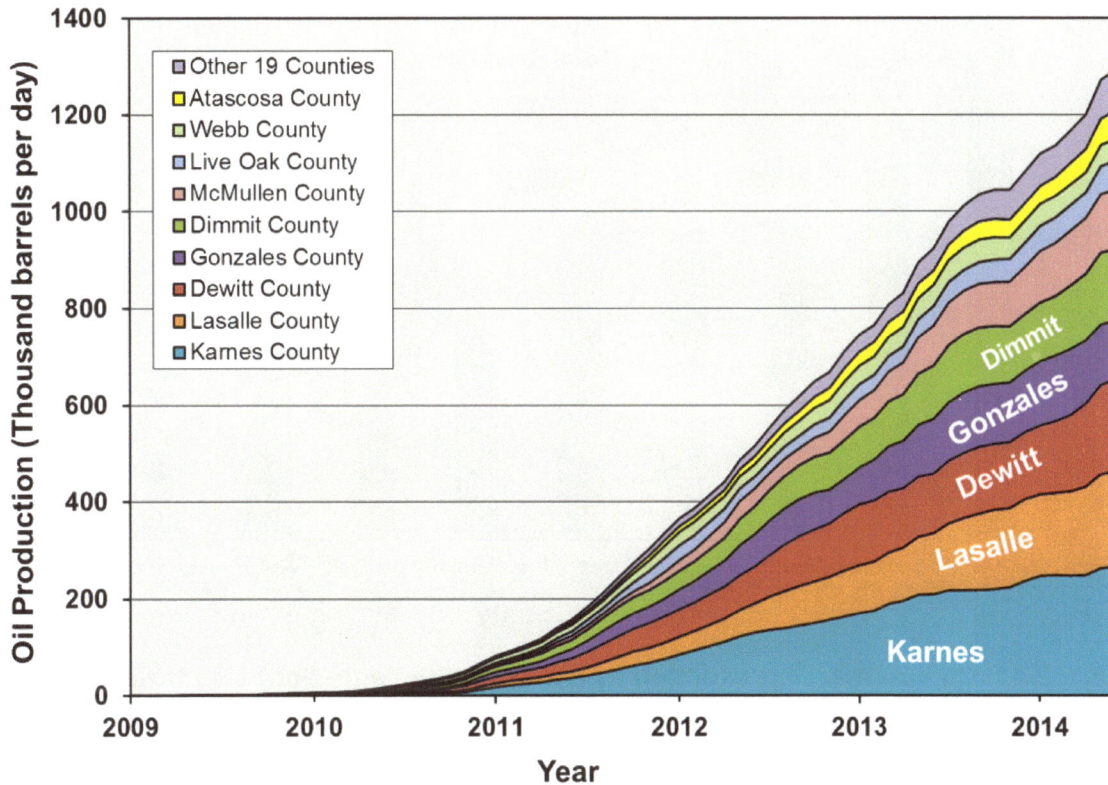

Figure 2-40. Oil production by county in the Eagle Ford play, 2009 through 2014.[67]

Eighty-one percent of production came from just six counties in mid-2014.

---

[67] Data from Drillinginfo retrieved September 2014. Three-month trailing moving average.

The same trend holds in terms of cumulative production since the field commenced. As illustrated in Figure 2-41, the top three counties have produced 51% of the oil and the top six have produced 81%.

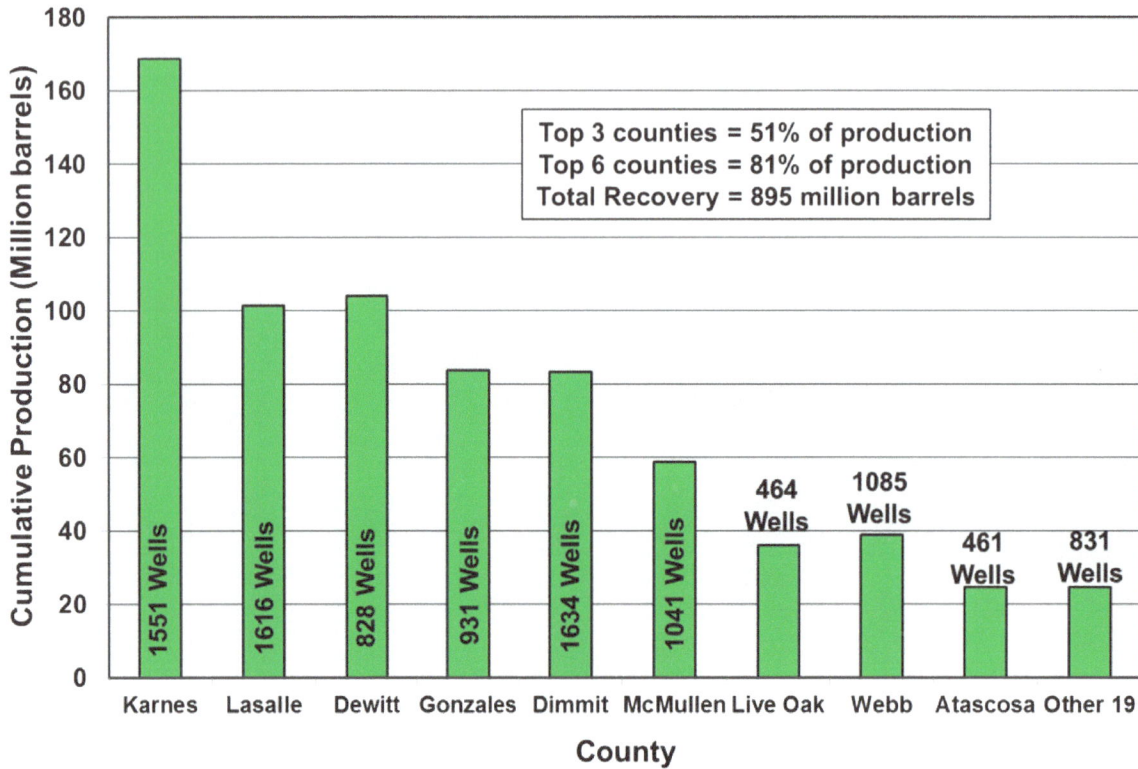

Figure 2-41. Cumulative oil production by county in the Eagle Ford play through 2014.[68]
The top six counties have produced 81% of the 895 million barrels produced to date. Production is growing in all counties.

---

[68] Data from Drillinginfo retrieved September 2014.

Approximately 39% of the energy produced from the Eagle Ford is in the form of natural gas, making the field one of the nation's top five gas fields (see the Eagle Ford section in *Part3: Shale Gas* of this report for a full discussion). The Eagle Ford currently produces 4.9 billion cubic feet per day and has produced nearly four trillion cubic feet since 2009. As with oil, gas production is concentrated in a few counties, but these tend to be different counties than for oil given the segregation of the play into oil and gas windows. Webb County, for example, produces less than 4% of the play's oil but produces 25% of its gas. Figure 2-42 illustrates gas production from the play since 2009 by county. In 2014, the top three counties produced 54% of the gas and the top six produced 87%.

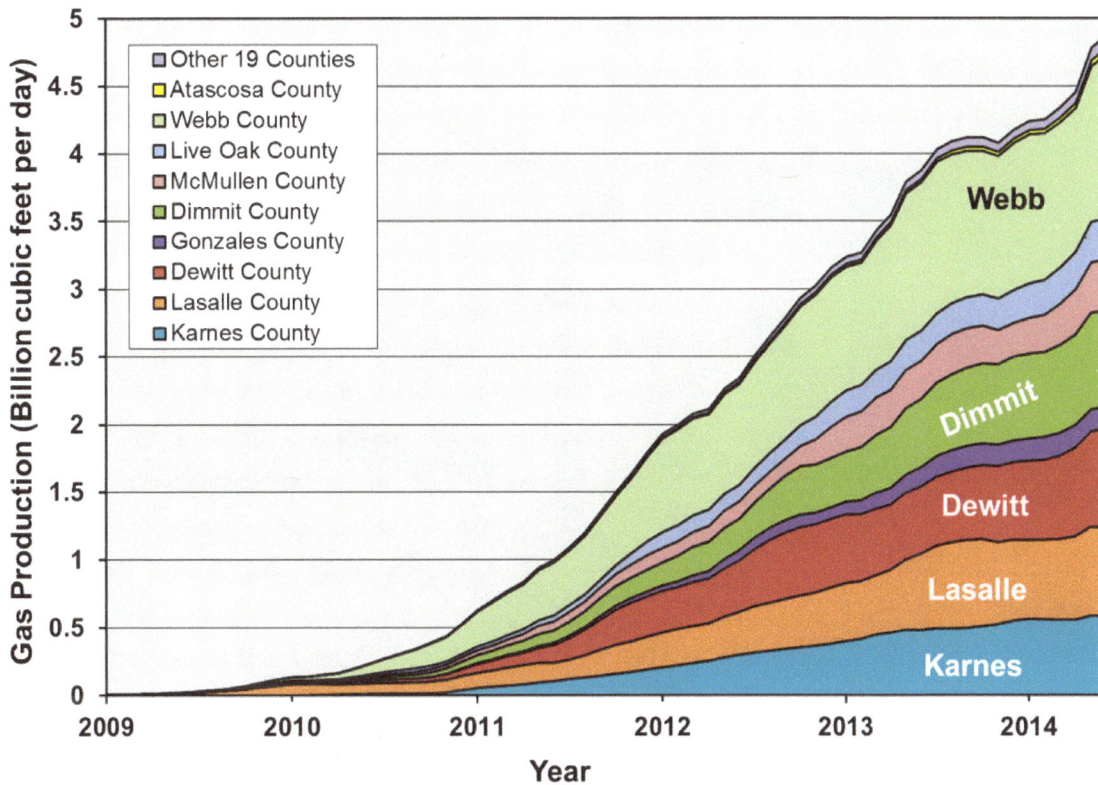

Figure 2-42. Gas production by county in the Eagle Ford play, 2009 through 2014.[69]
Eighty-seven percent of production came from just six counties as of mid-2014. For ease of comparison, the counties in this figure are sorted in the same order as in Figure 2-40, i.e., by *oil* production.

---

[69] Data from Drillinginfo retrieved September 2014. Three-month trailing moving average.

Figure 2-43 illustrates cumulative gas production from the play as of mid-2014. The top three counties have produced 58% of the gas and the top six have produced 89%.

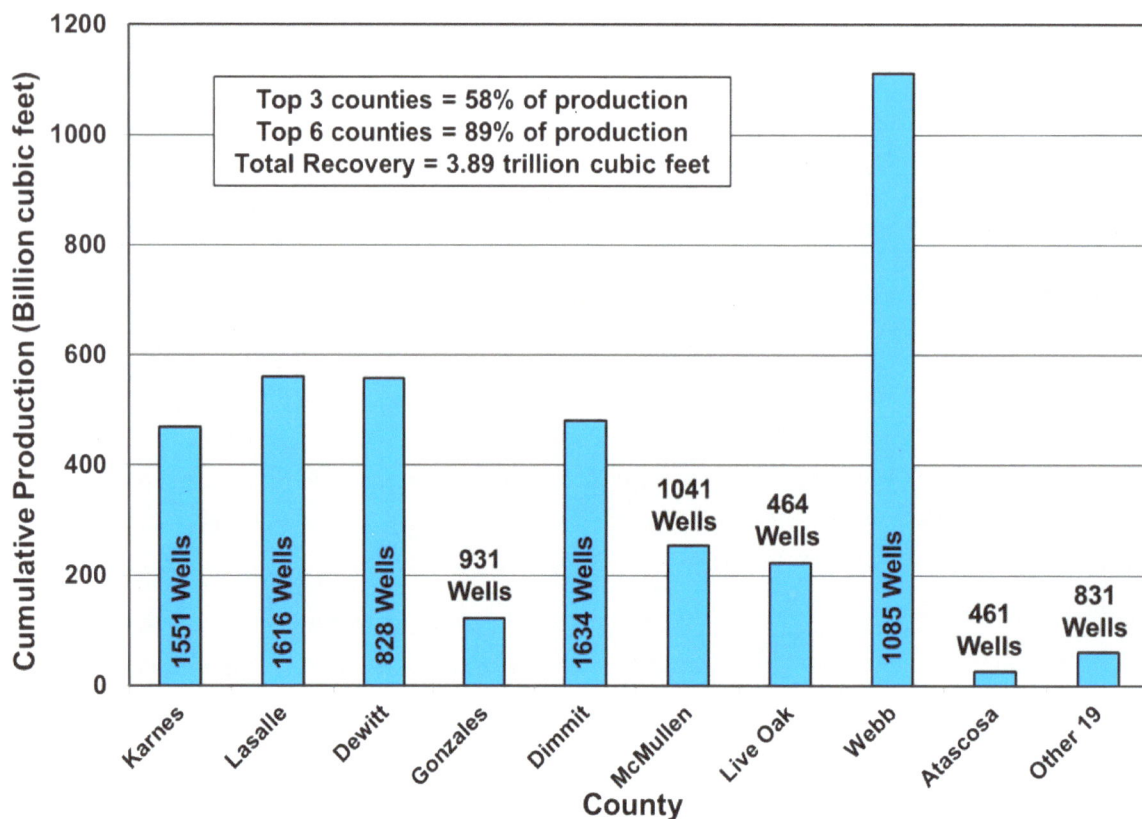

Figure 2-43. Cumulative gas production by county in the Eagle Ford play through 2014.[70]

The top six counties have produced 89% of the 3.89 trillion cubic feet produced to June 2014.

---

[70] Data from Drillinginfo retrieved September 2014.

Operators are highly sensitive to the economic performance of the wells they drill, which typically cost in the order of $8 million each,[71] not including leasing costs and other expenses. The areas of highest quality—the "core" or "sweet spots"—have now been well defined, both for oil and gas. Figure 2-44 illustrates average well decline profiles by county which are a measure of well quality. As can be seen, the decline profiles from the top four counties are all above the Eagle Ford average, hence these counties are attracting the bulk of the drilling and investment.

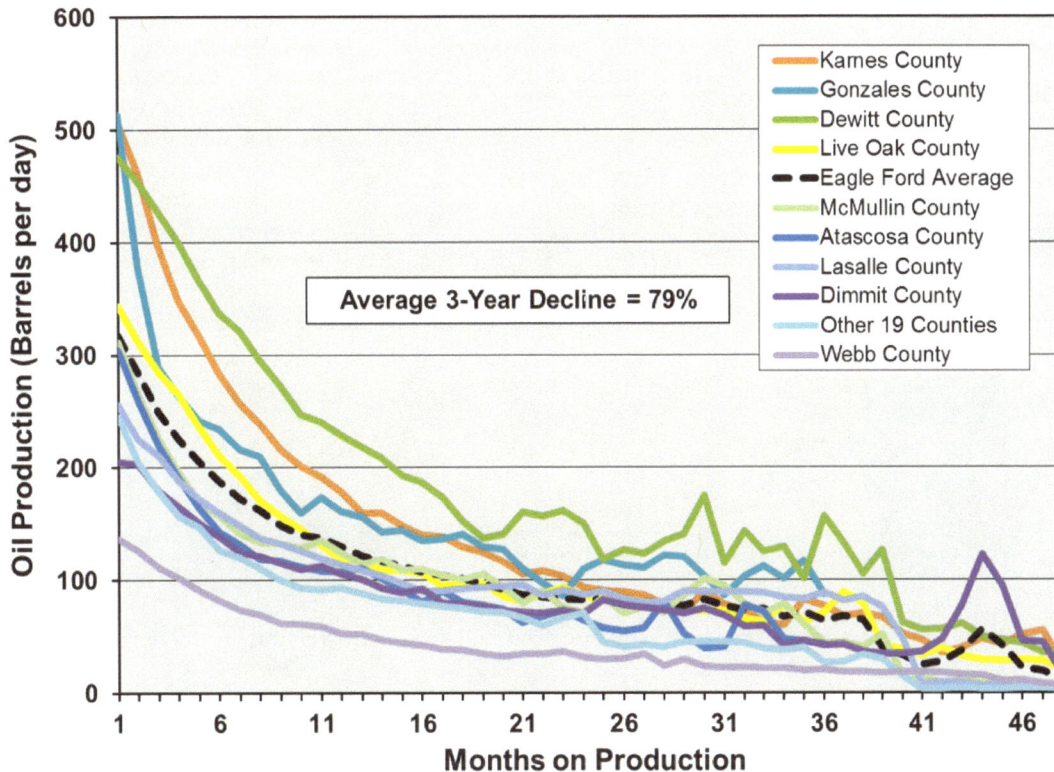

Figure 2-44. Average oil well decline profiles by county for the Eagle Ford play.[72]

The top four counties, which have produced much of the oil in the Eagle Ford, are clearly superior compared the play average and the other 23 counties. Well decline profiles are based on horizontal wells drilled since 2009.

---

[71] Trey Cowan, "Costs for Drilling The Eagle Ford," *Rigzone*, June 20, 2011, https://www.rigzone.com/news/article.asp?a_id=108179.
[72] Data from Drillinginfo retrieved May 2014.

Figure 2-45 illustrates average well decline profiles on a "barrels of oil equivalent" basis which includes the energy value of natural gas. Five counties are above the Eagle Ford average. Although four of these five are also the top four for oil production, Webb County is the second highest county on an energy output basis due to its prolific natural gas output, whereas is it ranks at the bottom for oil output.

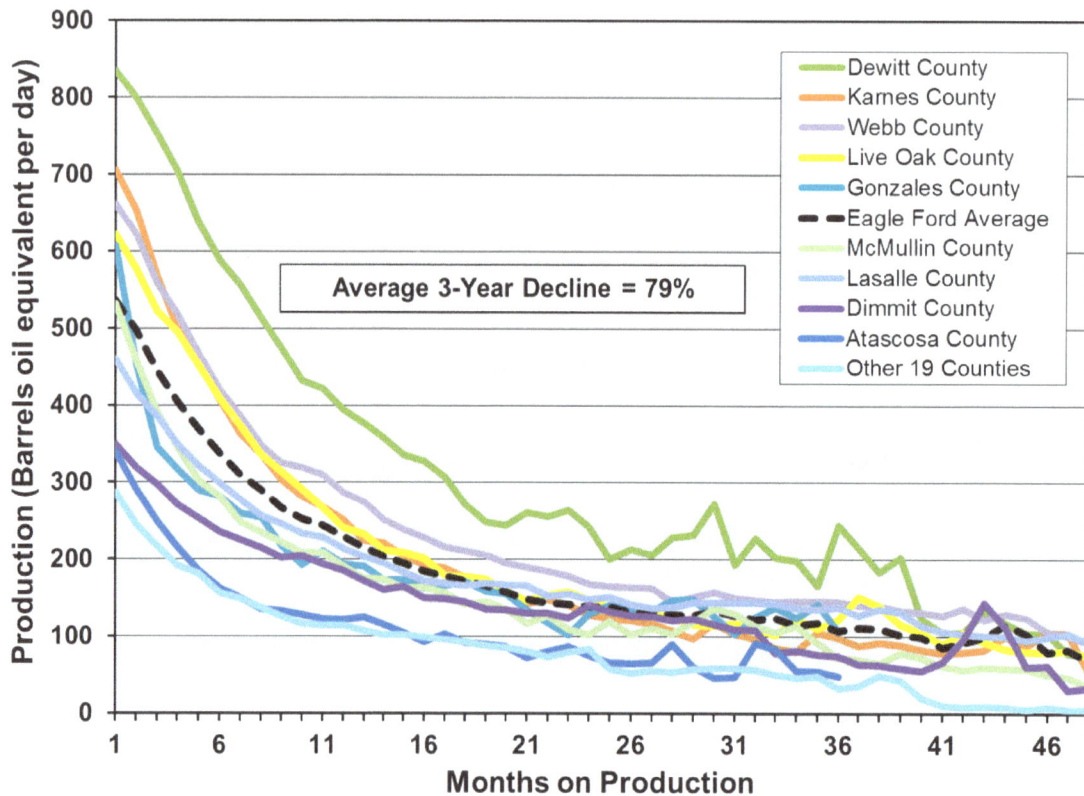

Figure 2-45. Average well decline profiles on a "barrels of oil equivalent" basis including the energy of natural gas produced by county for the Eagle Ford play.[73]

Although the top five counties include the top four for oil, Webb County has moved up to number two on an energy output basis, whereas it ranks at the bottom for oil production. Well decline profiles are based on horizontal wells drilled since 2009.

Another measure of well quality is "estimated ultimate recovery" (EUR), the amount of oil a well will recover over its lifetime. To be clear, no one knows what the lifespan of an Eagle Ford well is, given that few of them are more than five years old. Operators fit hyperbolic and/or exponential curves to data such as presented in Figure 2-44 and Figure 2-45, assuming well life spans of 30-50 years by comparison to conventional wells, but so far this is speculation given the nature of the extremely low permeability reservoirs and the completion technologies used in the Eagle Ford. Nonetheless, for comparative well quality purposes only, one can use the data in Figure 2-44 and Figure 2-45, which show that wells exhibit steep initial decline rates with progressively more gradual decline rates over the first three years, and assume a constant terminal decline rate thereafter to develop a theoretical EUR.

---

[73] Data from Drillinginfo retrieved May 2014.

Figure 2-46 illustrates theoretical EURs per well by county for the Eagle Ford; these range from 101,000 to 531,000 barrels per well. This compares to EURs of 97,000 to 223,000 barrels per well assumed by the EIA (the EIA EURs are not broken down by county and include large areas of limited prospectivity).[74] EURs in the top three counties are nearly 100% higher than in the lowest 22 counties of the play. The steep well production declines mean that well payout, if it is achieved, comes in the first few years of production, as between 46% and 56% of an average well's lifetime production occurs in the first three years.

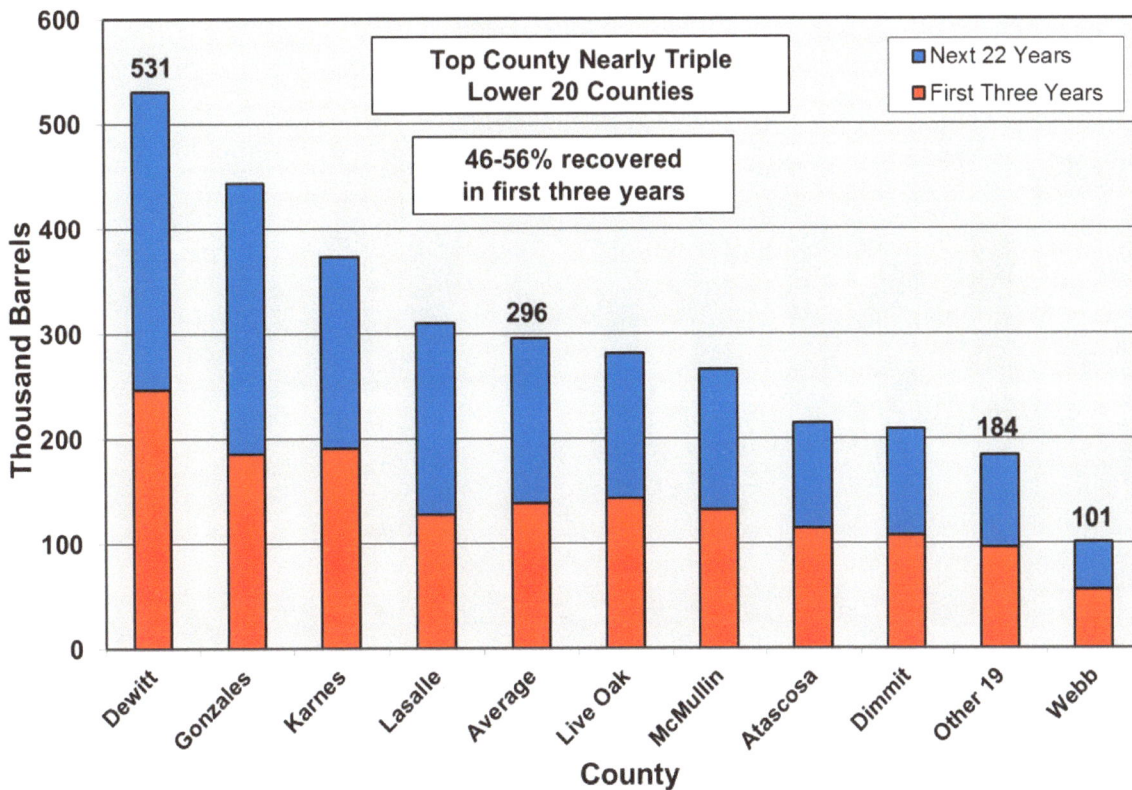

Figure 2-46. Estimated ultimate recovery of oil per horizontal well by county for the Eagle Ford play.[75]

EURs are based on average well decline profiles (Figure 2-44) and a terminal decline rate of 15%. These are for comparative purposes only as it is highly uncertain if wells will last for 25 years, as are the decline rates at the end of well life. The steep decline rates mean that most production occurs early in well life.

[74] EIA, *Assumptions to the Annual Energy Outlook 2014*, http://www.eia.gov/forecasts/aeo/assumptions/pdf/oilgas.pdf.
[75] Data from Drillinginfo retrieved May 2014.

Figure 2-47 illustrates theoretical EURs by county on a "barrels of oil equivalent" basis showing the split between oil and gas by county. The average well has an EUR of nearly 500,000 barrels oil equivalent, with Dewitt, the top county, more than triple the lowest 20 counties.

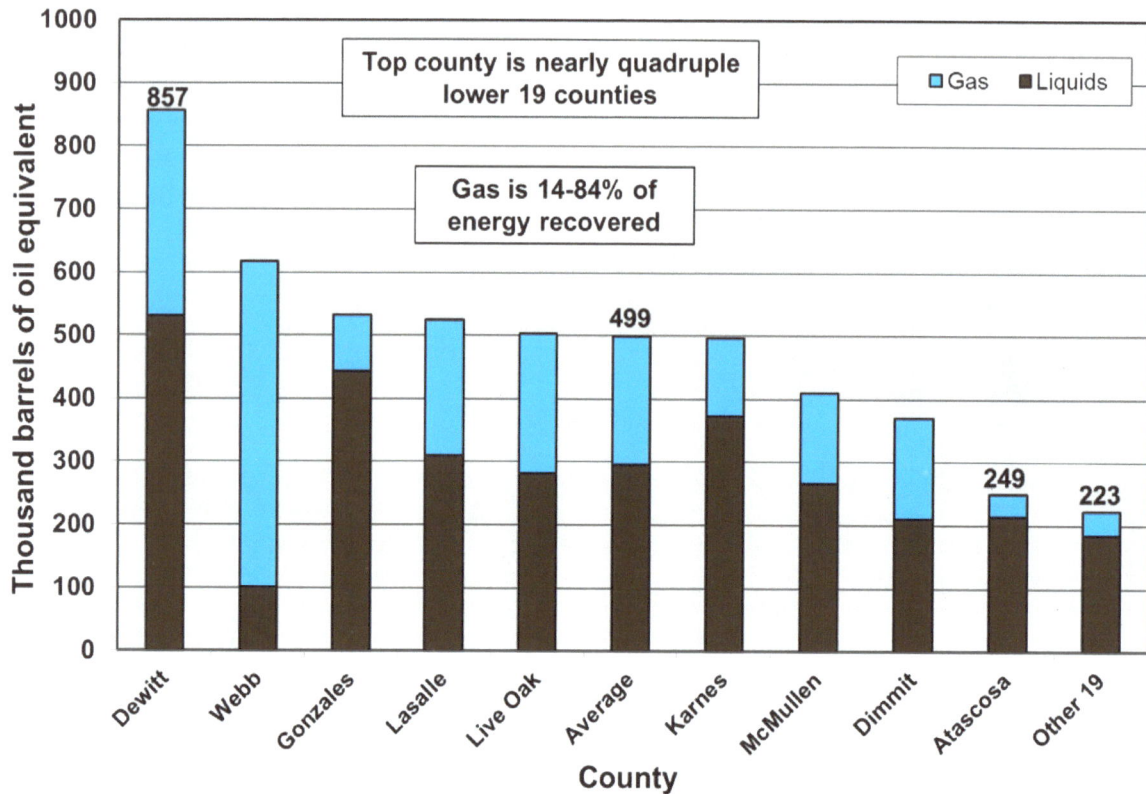

Figure 2-47. Estimated ultimate recovery on a "barrels of oil equivalent" basis, including the energy value of gas, by county for the Eagle Ford play.[76]

EURs are based on average well decline profiles (Figure 2-45) and a terminal decline rate of 15%. These are for comparative purposes only as it is highly uncertain if wells will last for 25 years, as are the decline rates at the end of well life. Gas comprises 14% to 84% of the energy produced with an average of about 39%.

Well quality can also be expressed as the average rate of production over the first year of well life. If we know both the field decline rate and the average well's first-year production rate, we can calculate the number of wells that need to be drilled each year in order to offset field decline and maintain production. Given that drilling is currently focused on the highest quality counties, the average first-year production rate per well will fall as drilling moves into lower quality counties over time as the best locations are drilled off. As average well quality falls, the number of wells that must be drilled to offset field decline must rise, until the drilling rate can no longer offset decline and the field peaks.

---

[76] Data from Drillinginfo retrieved May 2014.

Figure 2-48 illustrates the average first year oil production rate of wells by county over the 2009 to 2013 period. Gains are evident in several counties although Dewitt, the most productive county, is in decline. Only three counties exceed the play average. The average increase in productivity for the play as a whole is just 4% over 2012, suggesting that technological improvements are approaching the limits of diminishing returns. Much of the observed improvement is likely from the shift of drilling from gas prone to oil prone portions of counties. Future technology improvements are unlikely to postpone for long the inevitable decline in average overall well quality as drilling moves into lower quality counties.

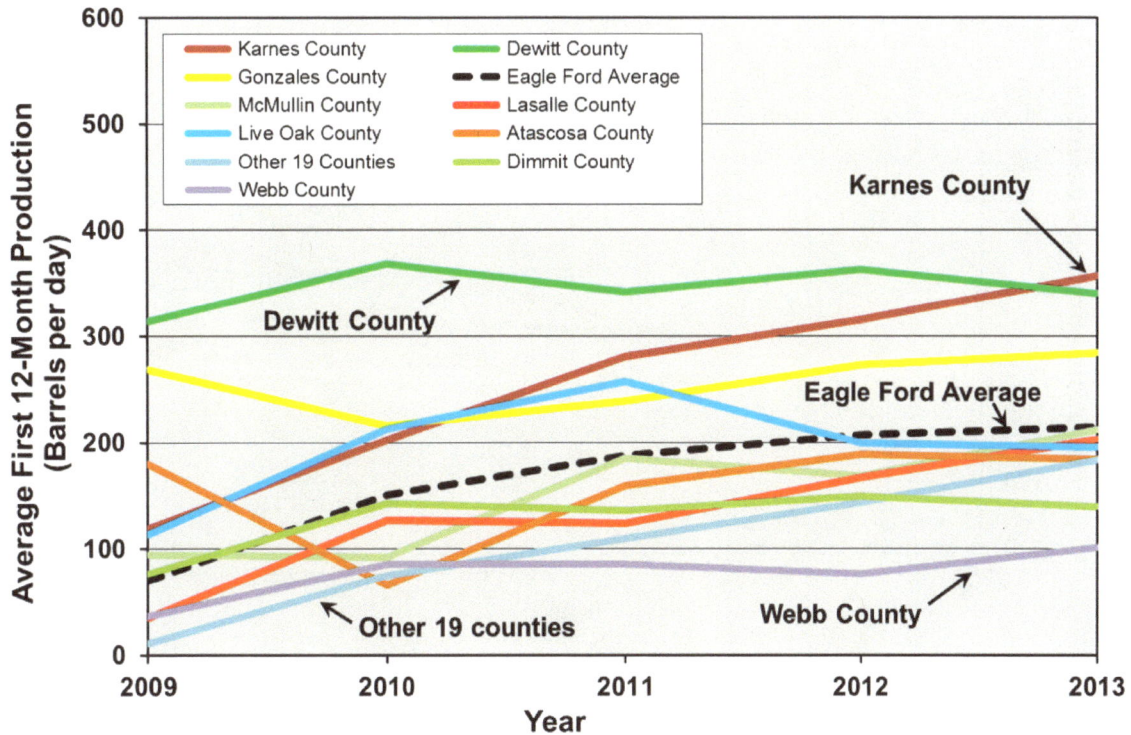

Figure 2-48. Average first-year oil production rates of wells by county for the Eagle Ford play, 2009 to 2013.[77]

Well quality is rising most rapidly in Karnes County, which has the second highest well count. Average first year oil production rates rose 4% over 2012.

---

[77] Data from Drillinginfo retrieved May 2014.

Figure 2-49 illustrates the average first year oil and gas production rate of wells by county on a "barrels of oil equivalent" basis over the 2009 to 2013 period. Gains are evident in several counties although Dewitt, the most productive county, is in decline, and the overall average for the play is unchanged over 2012, suggesting technological improvements are not making much difference overall.

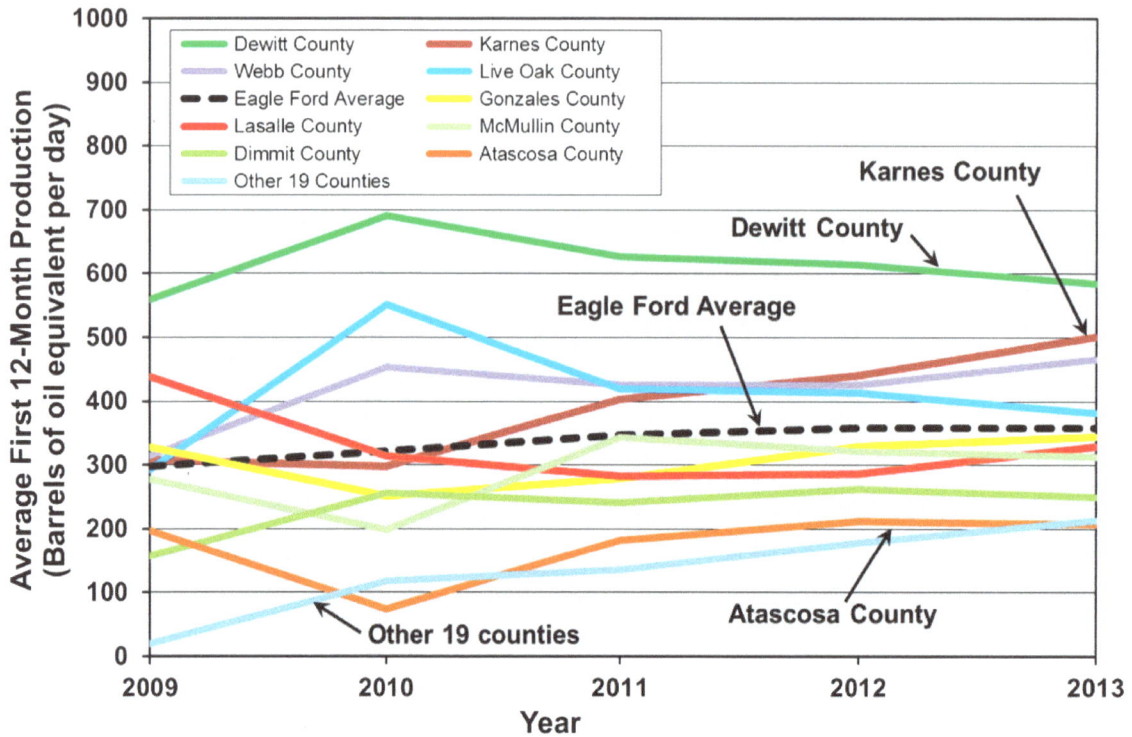

Figure 2-49. Average first-year oil and gas production rates of wells on a "barrels of oil equivalent" basis by county for the Eagle Ford play, 2009 to 2013.[78]

Average first-year production rates were unchanged in 2013 compared to 2012.

[78] Data from Drillinginfo retrieved May 2014.

2.3.2.4   Number of Wells

The fourth key fundamental is the number of wells that can ultimately be drilled in the Eagle Ford, a function of (a) the size of the area worth drilling and (b) the density of drilling that will likely occur. As in the Bakken, this is hotly debated in investor presentations.

Determining the likely density at which operators will drill wells requires consideration of both the geology of the play and the mechanics of hydraulic fracturing. Typical wells in the Eagle Ford have horizontal laterals of 5,000-7,000 feet in length with 20 or more frack stages. The EIA suggests that the area may be drilled at a density of 6 wells per square mile,[79] which would space horizontal laterals at 880 feet from each other. Companies like Marathon claim that spacing in core areas can be reduced to 16 wells per square mile in its pilots (40-acre spacing).[80] This would place horizontal laterals 350 feet apart, implying that frack jobs on wells only effectively drain less than 200 feet from a well.

This seems very optimistic given studies on well interference discussed earlier (section 2.3.1.4) showing that interference may occur with wells separated by less than 2,000 feet in the Bakken.[81] There has been no compelling evidence presented to suggest that 40-acre spacings in the Eagle Ford will not cannibalize production from adjacent wells, meaning that such attempts will not increase ultimate oil and gas recovery, although they may temporarily increase production.

Determining the area actually conducive to drilling is comparatively straightforward. After years of exploration and thousands of wells drilled, operators have delineated the limits of the play and focused their efforts on those areas with proven potential; thus by identifying the farthest-lying wells with little to no production as the likely edge of the play, and estimating the size of the area within that edge which is clearly attracting industry interest, the functional area of the Eagle Ford play can be calculated. By this method, the area likely to be conducive to drilling is approximately 7,200 square miles (see Figure 2-28).

Based on the above parameters, and given the fact that much of the area covered by the Eagle Ford is of considerably lower quality than the top few counties, an estimate of 6 wells per square mile may be reasonable for the whole area, allowing for a higher density in core areas and a lower density in outlying lower quality areas. This translates to approximately 43,200 potential wells if drilled at a density of 6 wells per square mile (compared to EIA's estimated 66,987 locations, determined from the product of the EIA's play area and well density). As more than 10,500 wells have been drilled to date, this means that approximately 32,800 wells remain to be drilled. Of course, these estimates assume that the entire designated area is available, and do not account for parks, towns, rivers, reservoirs, and other areas not conducive to drilling. A more conservative but possibly more realistic calculation would include a "risk" that 20% of the remaining play area will be undrillable. This reduces the remaining number of potential wells to approximately 26,200 which, coupled with wells already drilled, puts the total well count when the play is completely finished at 35,900. Either way, the Eagle Ford play could experience somewhere between three and four times the number of wells drilled to date.

---

[79] EIA, *Assumptions to the Annual Energy Outlook 2014*, http://www.eia.gov/forecasts/aeo/assumptions/pdf/oilgas.pdf.

[80] Marathon Investor Presentation, December 11, 2013, http://files.shareholder.com/downloads/AMDA-DZ30I/2909841818x0x713050/bf6626c0-3865-4e94-a4d4-6d47103dcc6d/Analyst_Day_Final_without_notes_v2.pdf.

[81] Kevin Thuot, "There Will Be Blood: Well Spacing & The Bakken Shale Oil Milkshake," DrillingInfo, November 26, 2013, http://info.drillinginfo.com/well-spacing-bakken-shale-oil.

2.3.2.5   Rate of Drilling

The fifth key fundamental is the *rate of drilling*. As noted earlier, the Eagle Ford play has a field decline of 38% per year (for oil), meaning that 38% of production has to be replaced with new wells each year to keep production flat. As the amount of oil produced from an average well in its first year of production is known from the data, the number of wells needed to offset field production decline each year at a given production level can be easily calculated. For the Eagle Ford at current production levels some 2,285 wells must be drilled each year to keep production flat. Since drilling rates in the Eagle Ford are now at about 3,550 wells per year, production will keep growing as long as these rates are sustained—until drilling locations run out. However, the higher production grows, the more wells are needed to offset the field decline. And as drilling moves into lower quality parts of the play, even more wells will be needed, for as illustrated above (Figure 2-48), well quality in most counties is significantly lower than in the best three.

2.3.2.6   Future Production Scenarios

Based on the five key fundamentals outlined above, several production projections for the Eagle Ford play were developed to illustrate the effects of changing the rate of drilling.

The projections are given in two cases, differentiated by the number of drilling locations:

1.   An "Optimistic Case" of 100% of the play area being drillable, at 6 wells per square mile.

2.   A "Realistic Case" of 80% of the remaining play area being drillable (i.e., the play is "risked" at 80% to account for undrillable areas like parks, towns, rivers, etc.), at 6 wells per square mile.

Each case includes three scenarios, differentiated by the rate of drilling:

1.   MOST LIKELY RATE scenario: Drilling continues at the current rate of 3,550 wells per year and then declines to 2,000 wells per year as drilling moves into the lower quality counties.

2.   EXPANDED RATE scenario: Drilling continues at the current rate of 3,550 wells per year and held constant until locations run out.

3.   FASTEST RATE scenario: Drilling is increased to 4,000 wells per year and held constant until locations run out.

The critical parameters used for determining production rates in these scenarios are given in Table 2-2.

Parameters	Counties										Total
	Atascosa	Dewitt	Dimmit	Gonzales	Karnes	Lasalle	Live Oak	McMullen	Webb	Other 19	
Oil Production Jan 2014 (Kbbl/d)	58.6	190.8	148.3	120.3	267.1	196.0	59.7	123.2	45.2	83.2	1292.2
Gas Production Jan 2014 (Kbbl/d)	7.9	122.3	118.3	26.5	98.3	107.9	50.8	63.0	201.5	16.5	813.1
Gas Production Jan 2014 (Bcf/d)	0.0	0.7	0.7	0.2	0.6	0.6	0.3	0.4	1.2	0.1	4.9
Oil % of Field Production	4.5	14.8	11.5	9.3	20.7	15.2	4.6	9.5	3.5	6.4	100.0
BOE % of Field Production	3.2	14.9	12.7	7.0	17.4	14.4	5.3	8.8	11.7	4.7	100.0
Gas % of Field Production	1.0	15.0	14.6	3.3	12.1	13.3	6.3	7.7	24.8	2.0	100.0
Cumulative Oil (million bbls)	32.4	130.3	102.7	101.6	200.0	122.8	43.1	72.5	43.7	46.4	895.5
Cumulative Gas (Bcf)	25.8	557.5	480.1	122.7	468.9	560.3	223.5	254.6	1112.0	60.6	3866.0
Number of Wells	461	828	1634	931	1551	1616	464	1041	1085	831	10442
Number of Producing Wells	441	797	1576	895	1506	1580	458	990	1023	759	10025
Avg. Oil EUR per well (Kbbls)	215	531	210	443	373	310	281	266	101	184	296
Avg. BOE EUR per well (Kbbls)	249	857	371	532	496	524	503	410	618	223	499
Avg. Gas EUR per well (Bcf)	0.2	2.0	1.0	0.5	0.7	1.3	1.3	0.9	3.1	0.2	1.2
Oil Field Decline (%)	50	41	37	33	40	34	40	30	48	27	38
BOE Field Decline (%)	49	44	36	33	42	37	47	45	46	30	42
Gas Field Decline (%)	43	47	34	26	47	41	54	64	46	43	47
Oil 3-Year Well Decline (%)	88	72	79	81	87	68	74	86	86	89	79
BOE 3-Year Well Decline (%)	88	74	82	81	86	72	76	87	79	88	79
Gas 3-Year Well Decline (%)	86	77	82	80	90	78	78	89	77	81	80
Average First Year Oil Production in 2013 (bbl/d)	184.6	340.9	140.0	284.9	357.5	203.3	195.8	212.8	101.4	183.7	214.9
Average First Year BOE Production in 2013 (bbl/d)	207.6	584.0	249.5	344.6	500.9	328.8	381.5	312.4	466.4	213.6	357.9
Average First Year Gas Production in 2013 (mcf/d)	138.4	1459.1	656.9	358.0	860.8	752.9	1113.9	597.3	2190.2	179.7	858.2
Oil New Wells Needed to Offset Field Decline	159	229	392	139	299	328	122	174	214	122	2285
BOE New Wells Needed to Offset Field Decline	157	236	385	141	306	342	136	268	243	140	2470
Gas New Wells Needed to Offset Field Decline	147	236	368	115	322	353	148	405	254	236	2672
Area in square miles	1232	909	1331	1068	750	1489	1036	1113	3357	20000	32285
% Prospective	50	30	90	40	75	80	25	60	30	5	22
Net square miles	616	273	1198	427	563	1191	259	668	1007	1000	7201
Well Density per square mile	0.75	3.04	1.36	2.18	2.76	1.36	1.79	1.56	1.08	0.83	1.45
Additional locations to 6/sq. Mile	3235	808	5553	1632	1824	5531	1090	2966	4958	5169	11162
Population	44911	20097	9996	19807	14824	6886	11531	707	250304	N/A	N/A
Total Wells 6/sq. Mile	3696	1636	7187	2563	3375	7147	1554	4007	6043	6000	43208
Producing Wells 6/sq. Mile	3676	1605	7129	2527	3330	7111	1548	3956	5981	5928	42791

Table 2-2. Parameters for projecting Eagle Ford tight oil production, by county

Optimistic Case

Figure 2-50 illustrates the production profiles of the three drilling rate scenarios in the "Optimistic Case," where 100% of the play area is drillable at six wells per square mile.

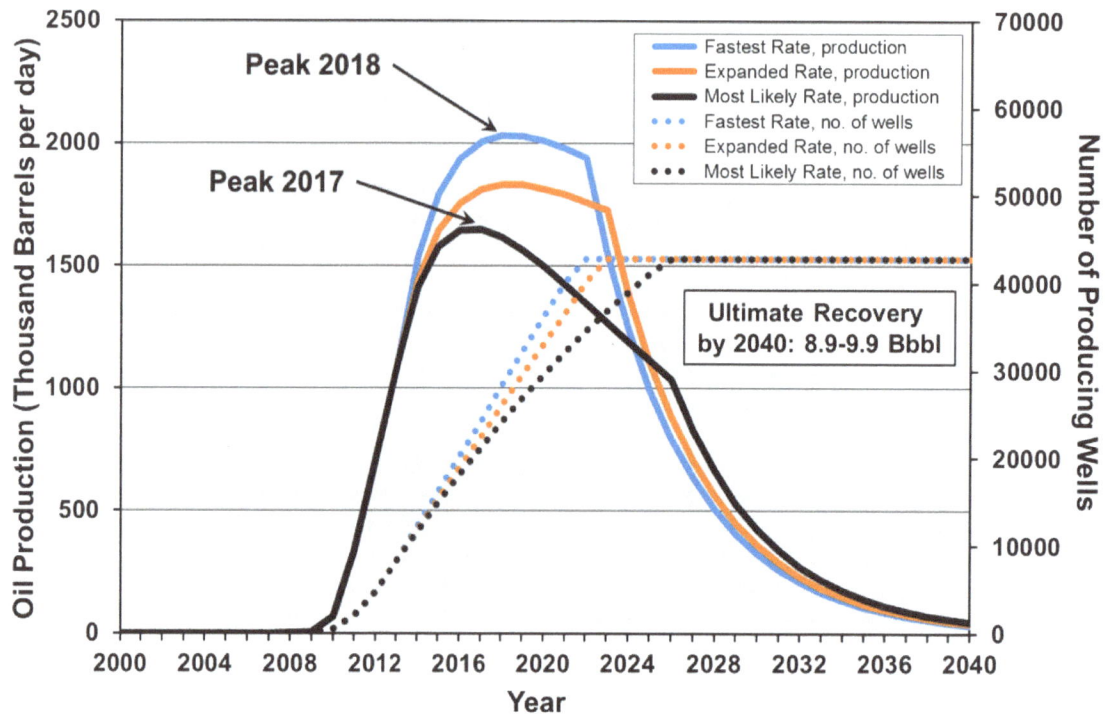

Figure 2-50. Three drilling rate scenarios of Eagle Ford tight oil production, in the "Optimistic Case" (100% of the play area is drillable at six wells per square mile).[82]

"Most Likely Rate" scenario: drilling continues at 3,550 wells/year, declining to 2,000 wells/year.
"Expanded Rate" scenario: drilling continues at 3,550 wells/year, holding constant until locations run out.
"Fastest Rate" scenario: drilling is increased to 4,000 wells/year, holding constant until locations run out.

The drilling rate scenarios in this case have the following results:

1. MOST LIKELY RATE scenario: Peak production occurs in 2017 at 1.65 MMbbl/d. Drilling continues until 2026, and total oil recovery by 2040 is 8.9 billion barrels.

2. EXPANDED RATE scenario: Peak production occurs in 2018 at 1.83 MMbbl/d. Drilling continues until 2023, and total oil recovery by 2040 is 9.6 billion barrels. In this scenario, however, production would be lower after 2026 than in the most likely case; in essence faster drilling recovers the oil sooner but makes future supply more problematic.

3. FASTEST RATE scenario: Peak production occurs in 2018 at 2.03 MMbbl/d. Drilling continues until 2022, and total oil recovery by 2040 is 9.9 billion barrels. In this scenario, however, production would be lower after 2025 than in the most likely case; in essence faster drilling recovers the oil sooner but makes future supply more problematic.

---

[82] Data from Drillinginfo retrieved September 2014.

The following two figures add natural gas, as oil equivalent energy, and differentiate oil from condensate production for the "Most Likely Rate" scenario (the Texas Railroad Commission reports that approximately 20% of liquids production is condensate,[83] which is generally of lower value than oil).

Figure 2-51 illustrates oil, condensate and gas production for the "Most Likely Rate" scenario in the "Optimistic Case" (100% of the prospective area is drillable at six wells per square mile).

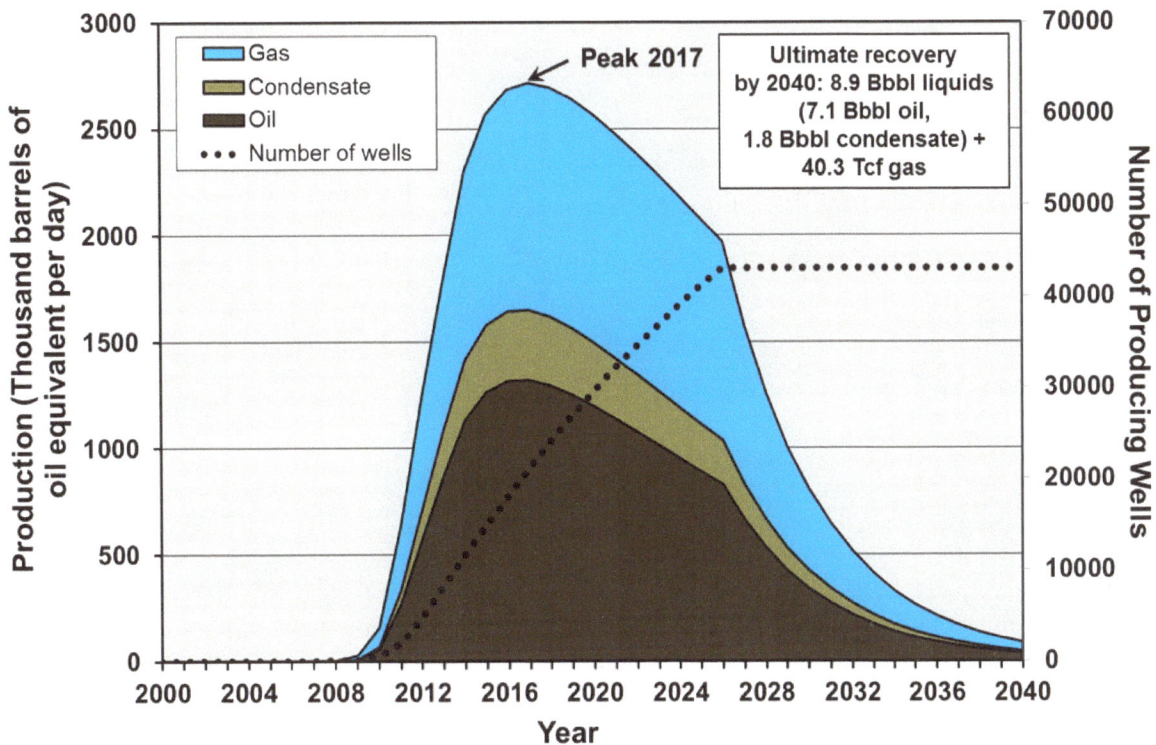

Figure 2-51. "Most Likely Rate" scenario of Eagle Ford production for oil, condensate and gas in the "Optimistic" case (100% of the play area is drillable at six wells per square mile).[84]

In this "Most Likely Rate" scenario, drilling continues at 3,550 wells/year, declining to 2,000 wells/year.

In this case, peak production occurs in 2017 at 2.7 MMbbl/d of oil equivalent. Drilling continues until 2026, total liquids recovery is 8.9 billion barrels (7.1 billion barrels of oil and 1.8 billion barrels of condensate), and total gas recovery is 40.3 trillion cubic feet.

[83] Texas Railroad Commission, "Eagle Ford Shale Information," July 2014, http://www.rrc.state.tx.us/oil-gas/major-oil-gas-formations/eagle-ford-shale.
[84] Data from Drillinginfo retrieved September 2014.

Realistic Case

Figure 2-52 illustrates oil, condensate and gas production in the "Most Likely Rate" scenario in the "Realistic Case" (only 80% of the remaining prospective area is drillable, at six wells per square mile).

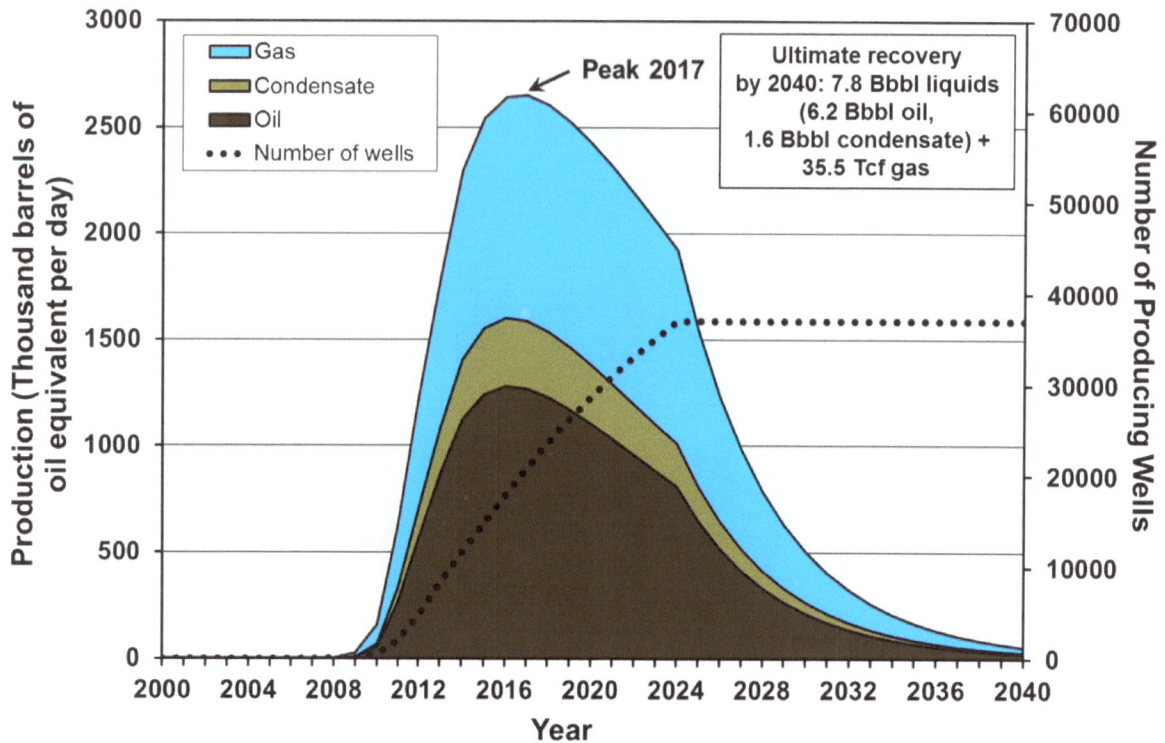

Figure 2-52. "Most Likely Rate" scenario of Eagle Ford production for oil, condensate and gas in the "Realistic Case" (80% of the remaining area is drillable at six wells per square mile).[85]

In this "Most Likely Rate" scenario, drilling continues at 3,550 wells/year, declining to 2,000 wells/year.

In this case, peak production occurs in 2017 at 2.65 MMbbl/d of oil equivalent. Drilling continues until 2024, total liquids recovery by 2040 is 7.8 billion barrels (6.2 billion barrels of oil and 1.6 billion barrels of condensate), and total gas recovery is 35.5 trillion cubic feet.

---

[85] Data from Drillinginfo retrieved September 2014.

2.3.2.7   Comparison to EIA Forecast

Figure 2-53 compares the EIA's reference case projection for Eagle Ford tight oil production to the "Most Likely Rate" scenario of the "Realistic" Case presented above.

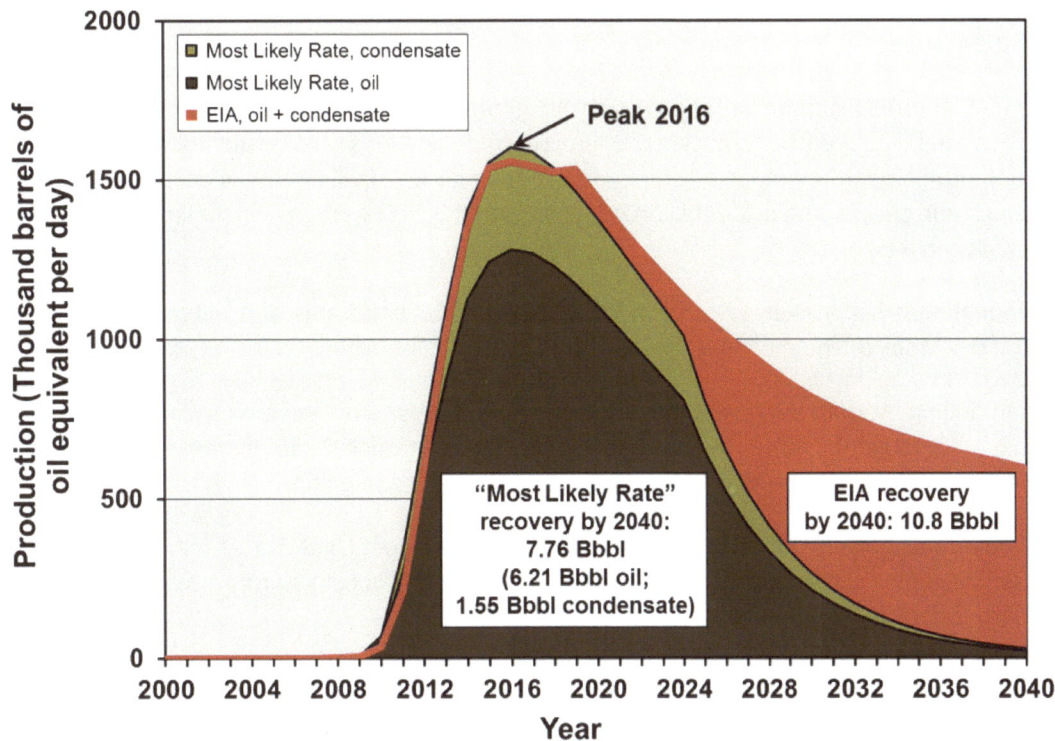

Figure 2-53. "Most Likely Rate" scenario ("Realistic" case) of Eagle Ford tight oil production compared to the EIA reference case, 2000 to 2040.[86]
This "Most Likely Rate" scenario sees 3,550 wells/year, declining to 2,000 wells/year. By 2040, 7.76 billion barrels of liquids would be recovered: 6.21 Bbbls of oil and 1.55 Bbbls of condensate.

This comparison reveals:

- The EIA's forecast of the timing of peak production (2016) in the Eagle Ford is the same as the projection of this report.

- The EIA's forecast of the production rate at peak (1.56 million bpd) is lower than the projection of this report (1.60 million bpd), but only slightly.

- The EIA projects a higher tail of production after peak, with estimated ultimate recovery (EUR) of 10.8 billion barrels by 2040 (10.2 billion for 2014-2040) as opposed this report's projection of 7.8 billion barrels by 2040 (7 billion for 2014-2040).

In short, the EIA is forecasting 3.2 billion additional barrels of future Eagle Ford production than this report finds substantiated. The EIA's assumption that production will be nearly 600,000 barrels per day in 2040 implies that much additional oil will be recovered.

---

[86] EIA, *Annual Energy Outlook 2014*.

2.3.2.8 Eagle Ford Play Analysis Summary

As with the Bakken, several things are clear from this analysis:

1. High well- and field-decline rates mean a continued high rate of drilling is required to maintain, let alone increase, production. Approximately 2,285 wells must be drilled each year to keep production flat at current levels.

2. The production profile is most dependent on drilling rate and to a lesser extent on the number of drilling locations (i.e., greatly increasing the number of drilling locations would not change the production profile nearly as much as changing the drilling rate). Drilling rate is determined by capital input, which currently is about $28 billion per year to drill 3,550 wells, not including leasing and other ancillary costs.

3. Peak production is highly likely to occur in the 2016 to 2018 timeframe and will occur at between 1.6 and 2.0 MMbbl/d. The most likely peak is about 1.6 MMbbl/d in 2016.

4. Increased drilling rates would raise the level of peak production and move it forward a few months but would not appreciably increase cumulative oil recovery through 2040. Increased drilling rates effectively recover the oil sooner making the supply situation worse later.

5. The projected recovery of 7.8 billion barrels by 2040 in the "Most Likely Rate" scenario of the "Realistic" case (i.e., six wells per square mile "risked" at 80%) is considerably less than the 10.8 billion barrels forecast by the EIA to be recovered by 2040.[87]

6. These projections are optimistic in that they assume the capital will be available for the drilling "treadmill" that must be maintained (roughly $210 billion is needed to drill more than 26,200 wells excluding leasing and ancillary costs). This is not a sure thing as drilling in the poorer-quality parts of the play will require much higher oil prices to be economic. Failure to maintain drilling rates will result in a steeper drop off in production.

7. Nearly four times the current number of wells will be required to recover 7.8 billion barrels by 2040 in the "Realistic" case assuming six wells per square mile "risked" at 80%.

8. The concept that the Eagle Ford will maintain a production plateau beyond its peak is unwarranted, even with extremely large capital inputs.

---

[87] EIA, *Annual Energy Outlook 2014*, Unpublished tables from AEO 2014 provided by the EIA.

2.4 THE PERMIAN BASIN PLAYS

The Permian Basin is the third largest source of tight oil production growth in the U.S. after the Bakken and Eagle Ford. The Permian Basin has been a prolific conventional oil and gas producer for nearly 100 years. Some 400,000 wells have been drilled there, producing more than 30 billion barrels of oil and 108 trillion cubic feet of gas.

Figure 2-54 illustrates the distribution of wells drilled since 1970 in the basin in Texas and southeastern New Mexico. The basin contains five major plays and several smaller ones that have collectively allowed oil production to grow by more than 500,000 barrels per day since 2005.[88] Three of these, the Spraberry, Wolfcamp, and Avalon/Bone Spring, are projected by the EIA to be major contributors to future production (see Figure 2-5).

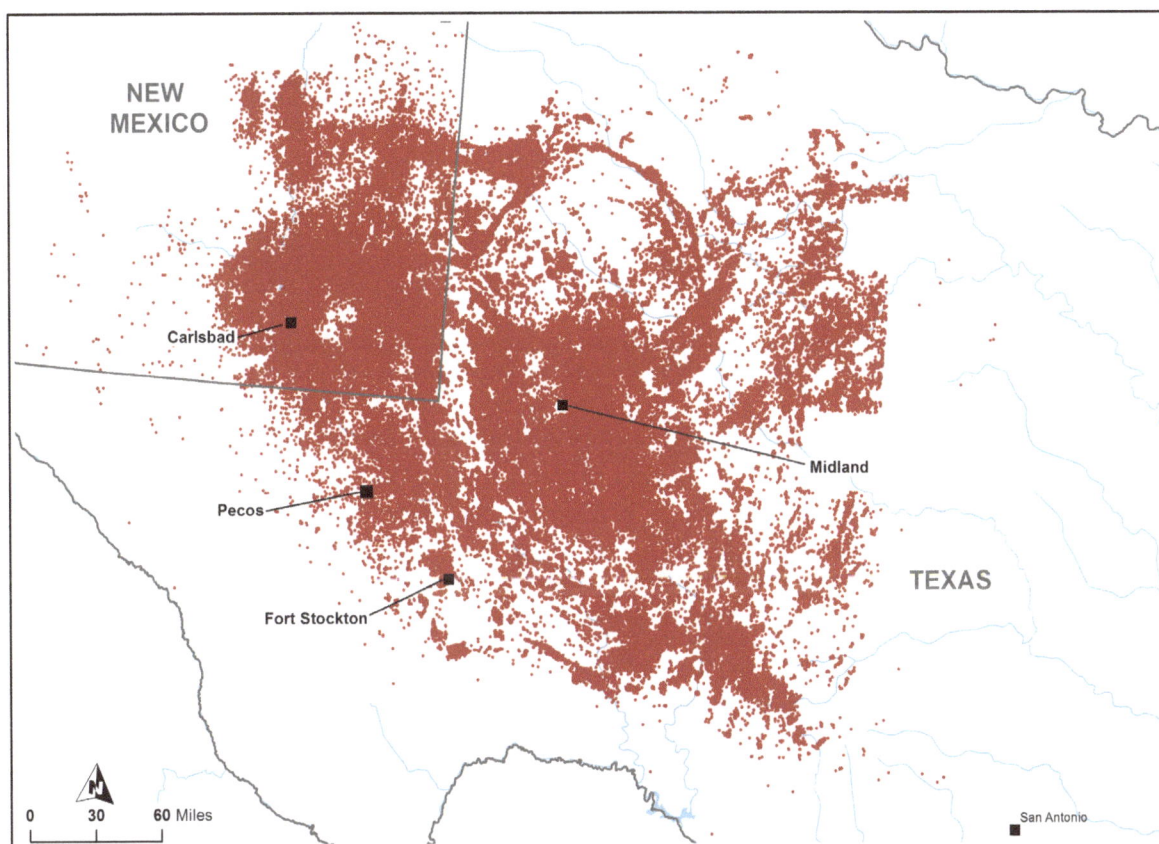

Figure 2-54. Distribution of wells drilled since 1970 in the Permian Basin of Texas and southeastern New Mexico.[89]

---

[88] EIA, "Six formations are responsible for surge in Permian Basin crude oil production," *Today in Energy*, July 9, 2014, http://www.eia.gov/todayinenergy/detail.cfm?id=17031.
[89] Data from Drillinginfo retrieved July 2014.

Production of oil and gas from the Permian Basin peaked in 1973 but has undergone a renaissance since 2010, with the application of new technology to old reservoirs. Figure 2-55 illustrates oil and gas production in the basin since 1960.

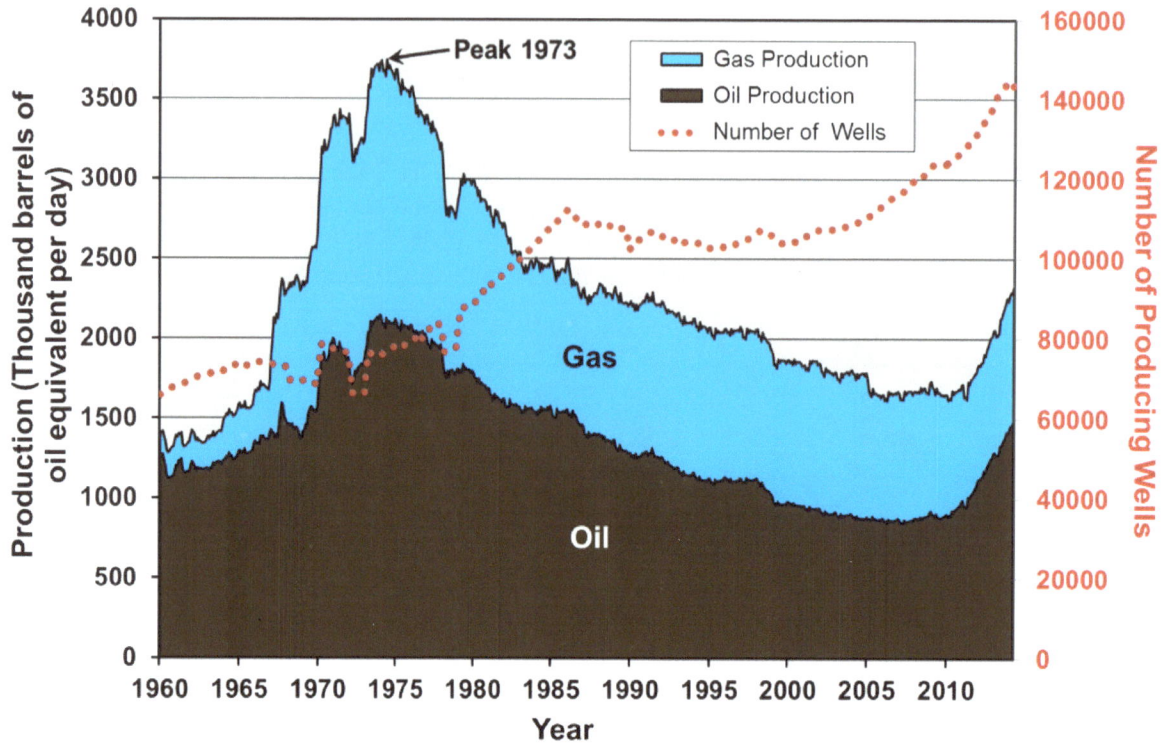

Figure 2-55. Permian Basin oil and gas production and number of producing wells, 1960 to 2014.[90]

Gas production is expressed as "barrels of oil equivalent" (6,000 cubic feet of gas is approximately equivalent to one barrel of oil on an energy basis).

---

[90] Data from Drillinginfo retrieved July 2014. Three-month trailing moving average.

Unlike the Bakken and Eagle Ford shale plays, tight oil production in the Permian Basin is from both horizontal and vertical fracked wells. Coupled with the fact that the Permian Basin has been producing for nearly a century, this makes it difficult to separate truly new "tight oil" production from conventional production. Figure 2-56 illustrates production from vertical and horizontal wells in the Permian Basin. Production growth has occurred from both well types, although horizontal wells appear to contribute a larger proportion of the growth.

As mentioned above, the three plays that the EIA is counting on to meet a significant proportion of its tight oil forecasts from the Permian Basin are the Spraberry, Wolfcamp, and Avalon/Bone Spring. These plays are reviewed below with respect to production characteristics and future growth potential in the light of the EIA projections for them.

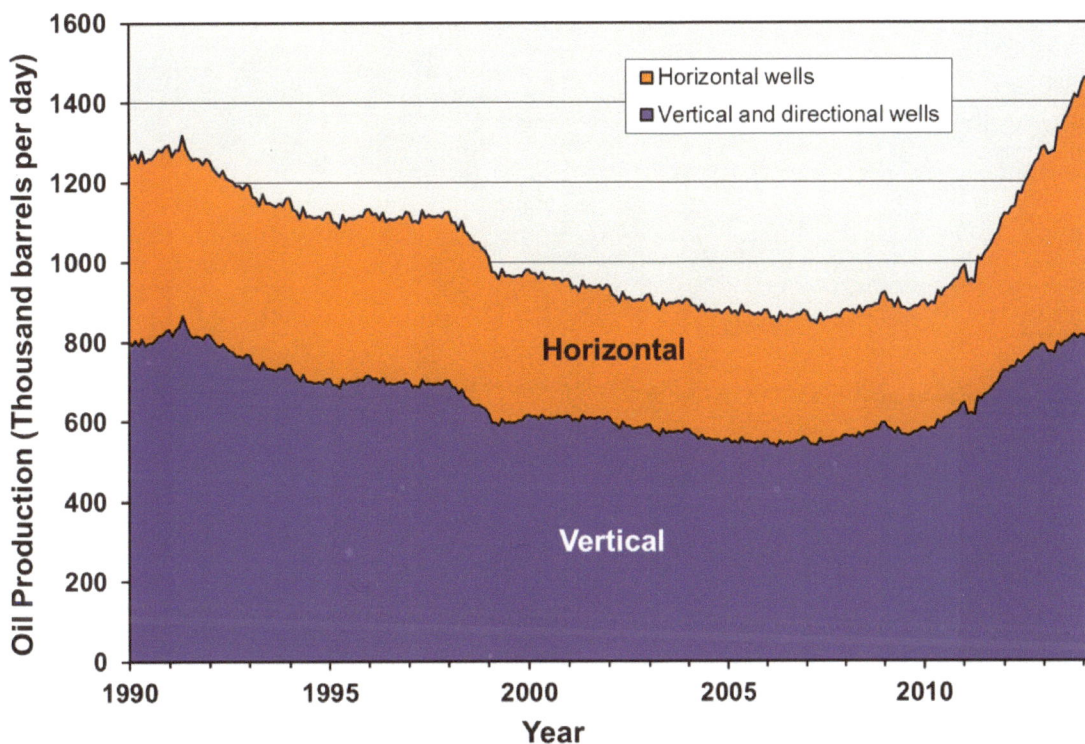

Figure 2-56. Oil production by well type in the Permian Basin, 1990 to 2014.[91]
Recent production growth is a function of both horizontal and vertical wells.

---

[91] Data from Drillinginfo retrieved July 2014. Three-month trailing moving average.

2.4.1   Spraberry Play

The EIA forecasts recovery of 6.5 billion barrels of oil from the Spraberry play between 2012 and 2040. The analysis of actual production data presented below suggests that this forecast is unlikely to be realized.

The Spraberry play has been producing oil and gas for decades. Nearly 37,000 wells have been drilled of which more than 25,000 are currently producing. The play has produced over 1.8 billion barrels of oil and more than 4.2 trillion cubic feet of natural gas over its lifetime. Production comes from the Spraberry reservoir proper, and an equivalent reservoir termed "Trend Area", which together make up the Spraberry play. Figure 2-57 illustrates well distribution within the play.

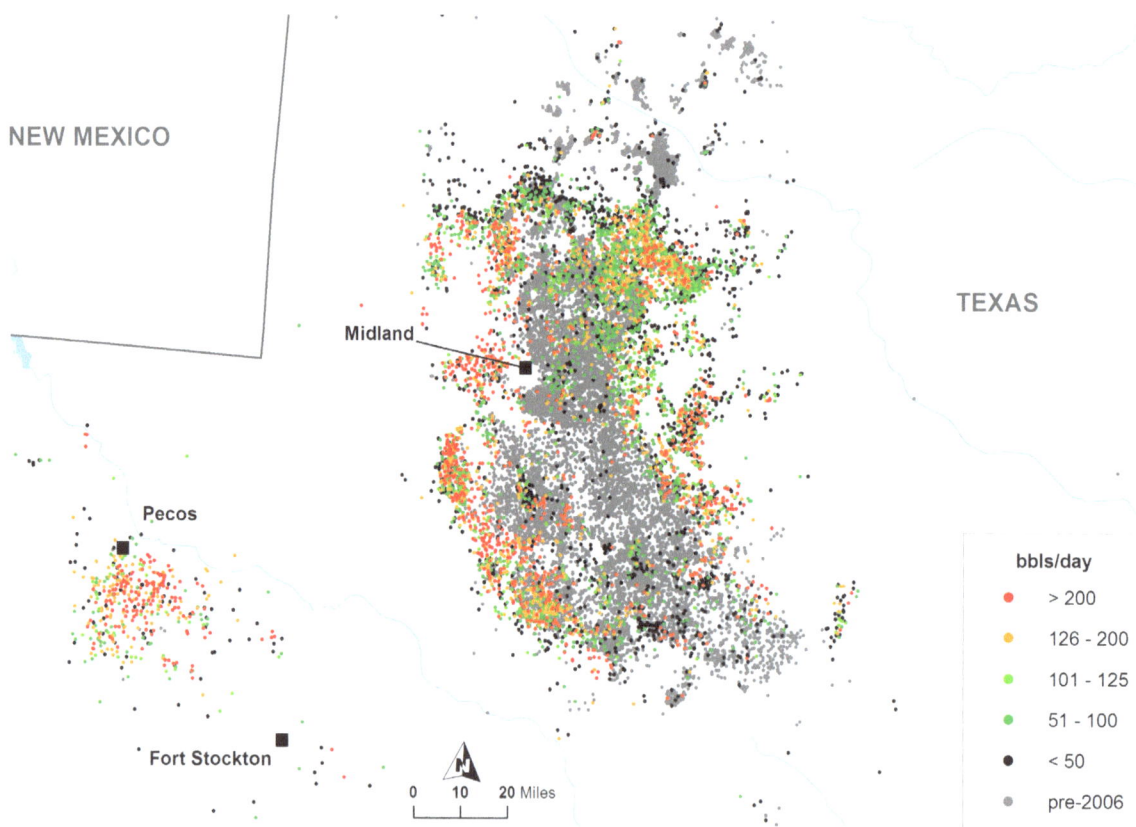

Figure 2-57. Distribution of wells in the Spraberry play as of mid-2014 illustrating highest one-month oil production (initial productivity, IP).[92]

Only wells drilled in 2006 and later are considered as possible "tight oil" production and colored by IP; wells drilled prior to 2006 are predominantly conventional production. Well IPs are categorized approximately by percentile; see Appendix.

---

[92] Data from Drillinginfo retrieved July 2014.

2.4.1.1   Production History

Production of oil in the Spraberry has more than tripled since 2005 and including natural gas (on an energy equivalent basis) is up four-fold as illustrated in Figure 2-58. The number of producing wells has also more than doubled over this period.

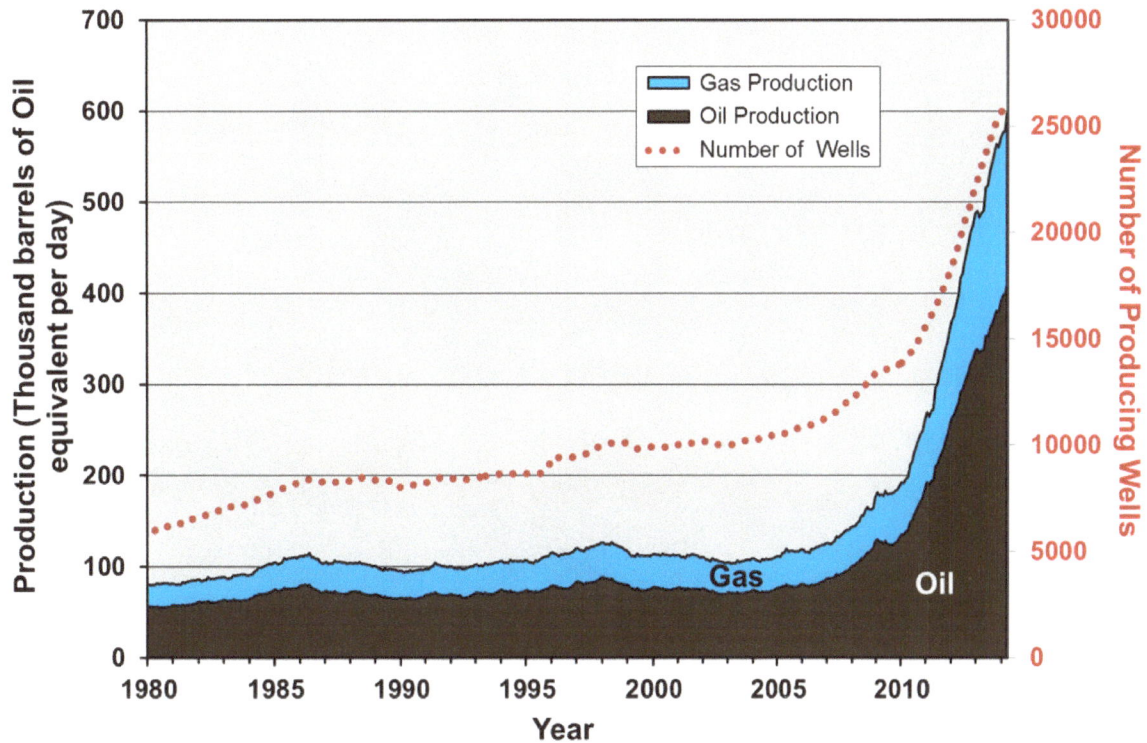

Figure 2-58. Spraberry play oil and gas production and number of producing wells, 1980 to 2014.[93]

Producing well count is now above 25,000.

---

[93] Data from Drillinginfo retrieved July 2014. Three-month trailing moving average.

A look at the split in production by well type reveals that much of this growth is attributable to vertical wells, although horizontal wells are becoming increasingly important (Figure 2-59). New completion technology in both well types is obviously paying dividends.

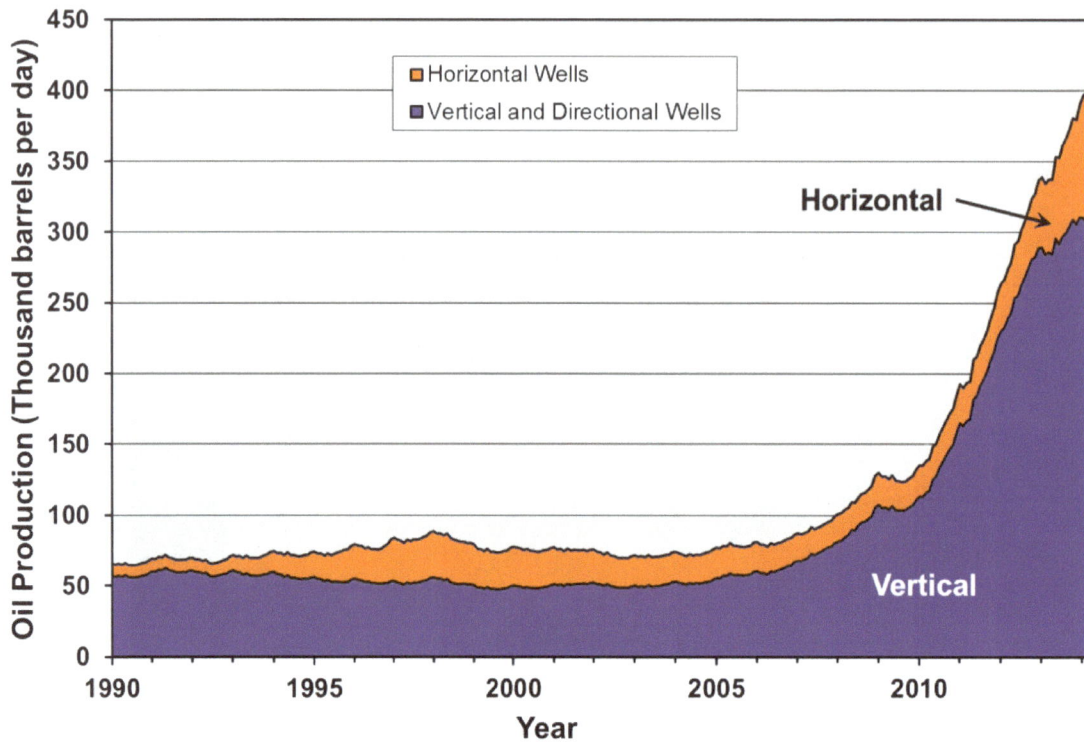

Figure 2-59. Oil production from the Spraberry play by well type.[94]

Although vertical wells have accounted for much of the recent production growth, horizontal wells now appear to be the most important contributors.

---

[94] Data from Drillinginfo retrieved July 2014. Three-month trailing moving average.

2.4.1.2 Well Quality

A look at well quality reveals that the Spraberry is unremarkable by comparison to the Bakken or Eagle Ford. Figure 2-60 illustrates the average well decline profile for all wells; Figure 2-61 illustrates the average well decline profile for horizontal wells only. All wells on an energy equivalent basis are just a tenth of the initial production of an average Bakken well in a top county. Horizontal wells are more than double the initial productivity of the average well but still pale by comparison to a Bakken or Eagle Ford well. The average three-year decline of Spraberry wells is, however, somewhat lower than the Bakken at 60% and 72% for all wells and horizontal wells, respectively.

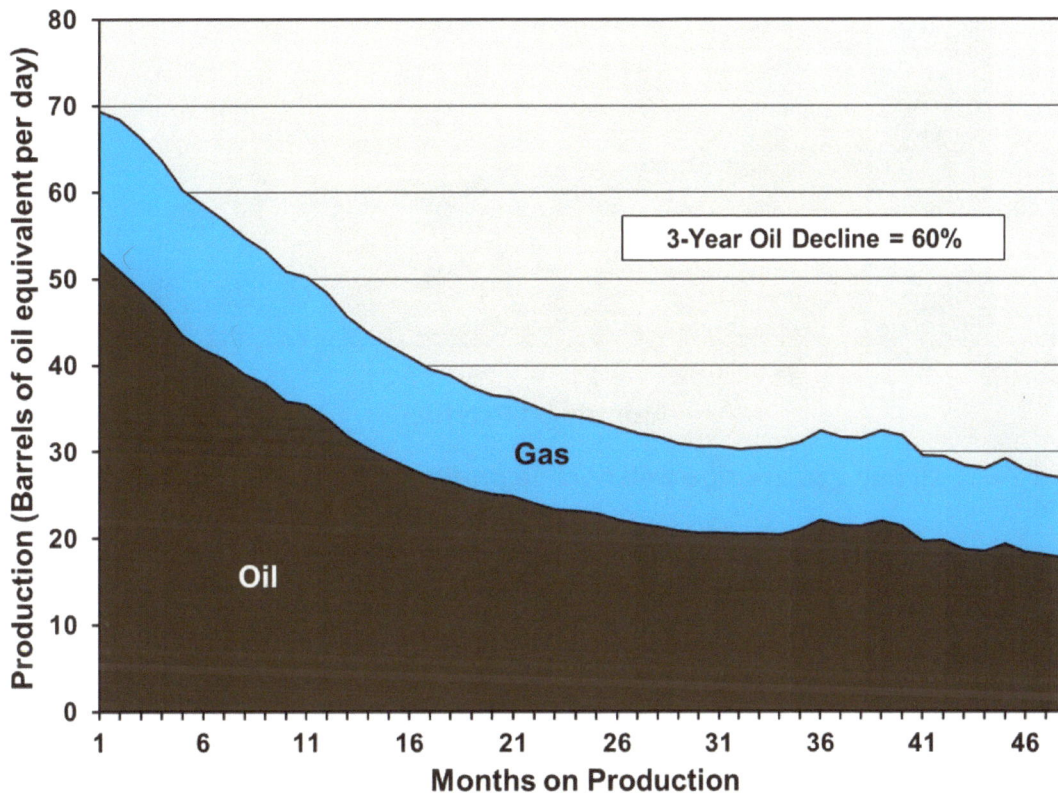

Figure 2-60. Oil and gas average well decline profile for all wells in the Spraberry play.[95]

On an energy equivalent basis these wells have an initial productivity of less than a tenth that of the average well in the top counties of the Bakken play. Decline profile is based on all wells drilled since 2009.

[95] Data from Drillinginfo retrieved July 2014.

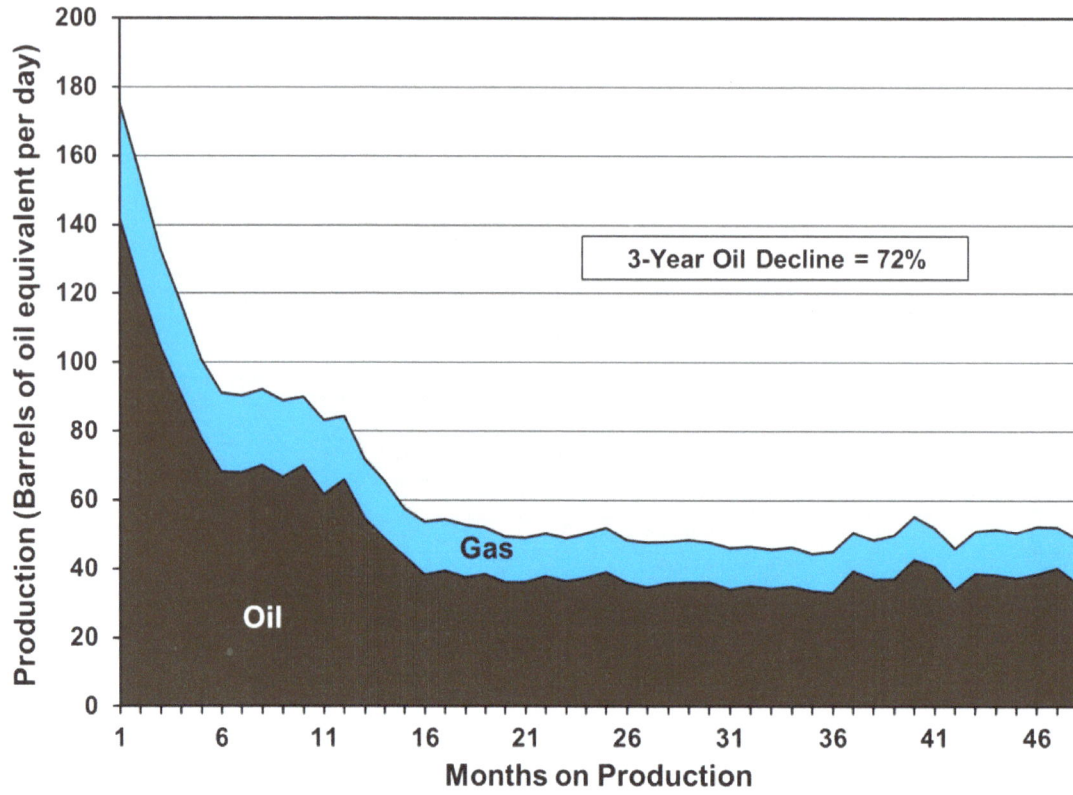

Figure 2-61. Oil and gas average well decline profile for horizontal wells in the Spraberry play.[96]

On an energy equivalent basis these wells have an initial productivity of less than a third that of the average well in the top counties of the Bakken play. Decline profile is based on all horizontal wells drilled since 2009.

[96] Data from Drillinginfo retrieved July 2014.

2.4.1.3   EIA Forecast

The EIA's projection for Spraberry play production through 2040 in its reference case is illustrated in Figure 2-62. Total recovery between 2012 and 2040 is forecast to be 6.5 billion barrels; this amounts to 15% of the EIA's reference case forecast for U.S. tight oil production through 2040. Cumulative production by 2040 amounts to 80% of the "unproved technically recoverable resources" the EIA estimated for the Spraberry as of January 1, 2012.

Given that this is a redevelopment of an old play which is already extensively drilled, the fact that the wells are of relatively low quality, and the nature of likely production profiles from shale plays like the Bakken, this would seem to be a highly optimistic forecast. It is already overestimating actual production by 55% in year one, as actual production for 2013 amounted to 390,000 barrels per day compared to an estimate of 604,000 barrels by the EIA. Furthermore, the EIA is projecting that production will be 505,000 barrels per day in 2040, which is 30% above current levels. High field decline rates make it very likely that production decline after its projected peak in 2021 will be much steeper than projected. Given what is known, this EIA forecast would seem to have a very high optimist bias.

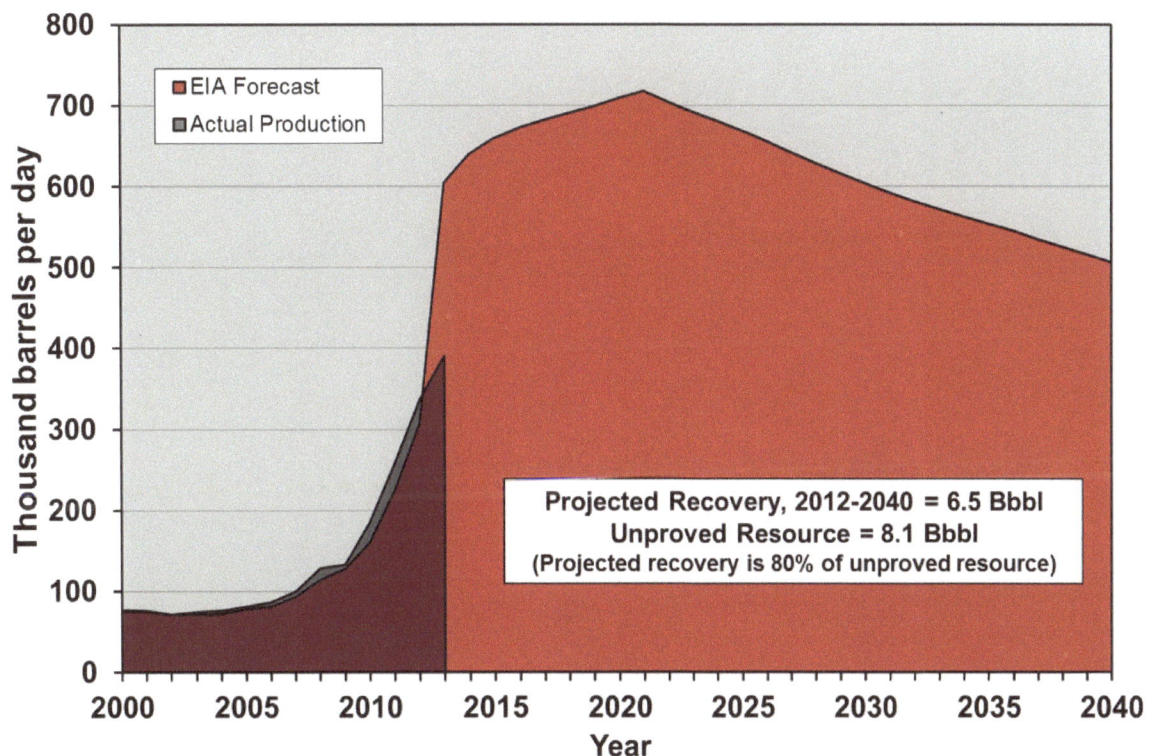

Figure 2-62. EIA reference case projection of oil production from the Spraberry play through 2040, with actual production to 2013.[97]

The forecast total recovery of 6.5 billion barrels over the 2012-2040 period amounts to 80% of the 8.1 billion barrels of "unproved technically recoverable resources as of January 1, 2012".[98]

---

[97] Production data from DrillingInfo, July 2014. Forecast from EIA, *Annual Energy Outlook 2014*, Unpublished tables from AEO 2014 provided by the EIA.
[98] EIA, *Assumptions to the Annual Energy Outlook 2014*, http://www.eia.gov/forecasts/aeo/assumptions/pdf/oilgas.pdf.

2.4.2   Wolfcamp Play

The EIA forecasts recovery of 2.64 billion barrels of oil from the Wolfcamp play between 2012 and 2040. The analysis of actual production data presented below suggests that this forecast is unlikely to be realized.

The Wolfcamp play has also been producing oil and gas for decades. Over 12,800 wells have been drilled of which more than 6,000 are currently producing. The play has produced over 870 million barrels of oil and nearly 4.8 trillion cubic feet of natural gas over its lifetime. Figure 2-63 illustrates well distribution within the Wolfcamp play.

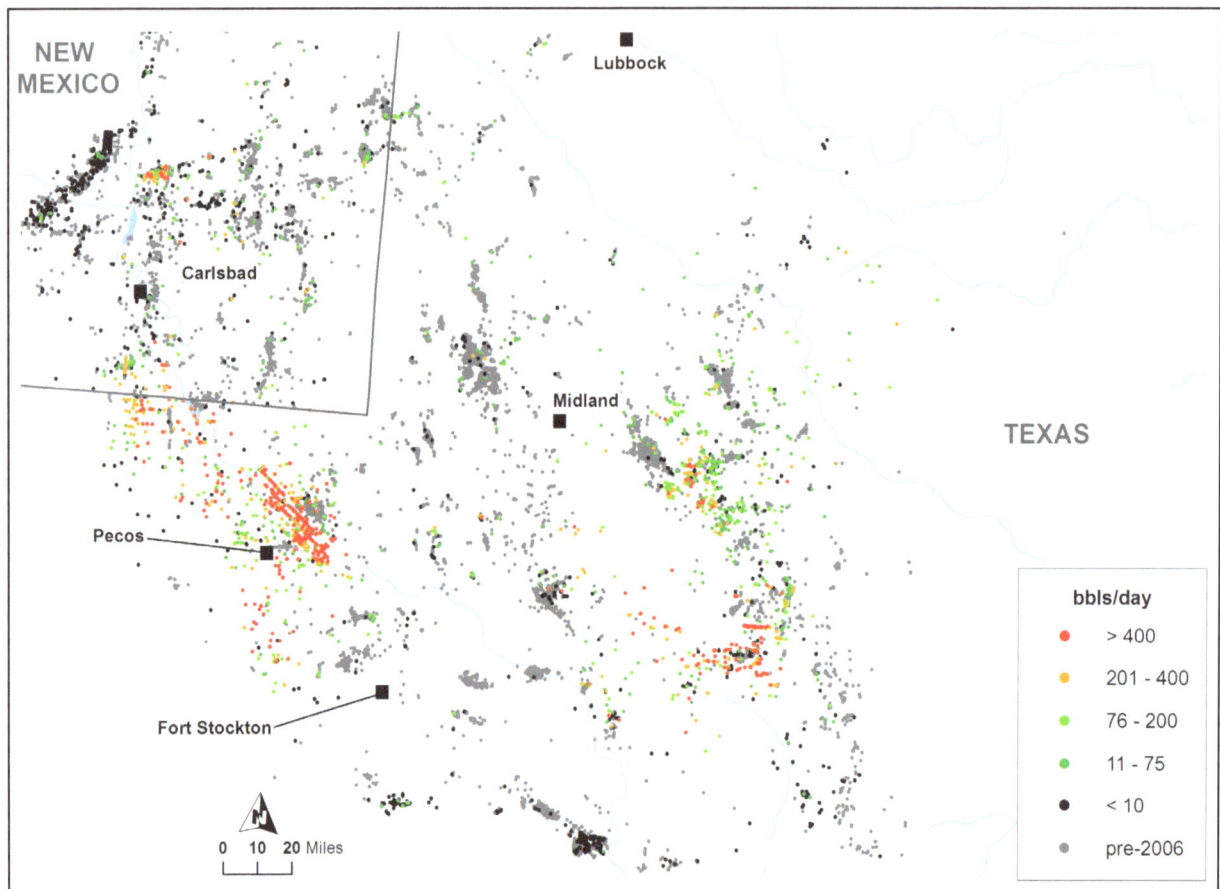

Figure 2-63. Distribution of wells in the Wolfcamp play as of mid-2014 illustrating highest one-month oil production (initial productivity, IP).[99]

Only wells drilled in 2006 and later are considered as possible "tight oil" production and colored by IP; wells drilled prior to 2006 are predominantly conventional production. Well IPs are categorized approximately by percentile; see Appendix.

---

[99] Data from Drillinginfo retrieved July 2014.

2.4.2.1   Production History

Production of oil in the Wolfcamp has quadrupled since 2005, and including natural gas (on an energy equivalent basis) is up three-fold as illustrated in Figure 2-64.

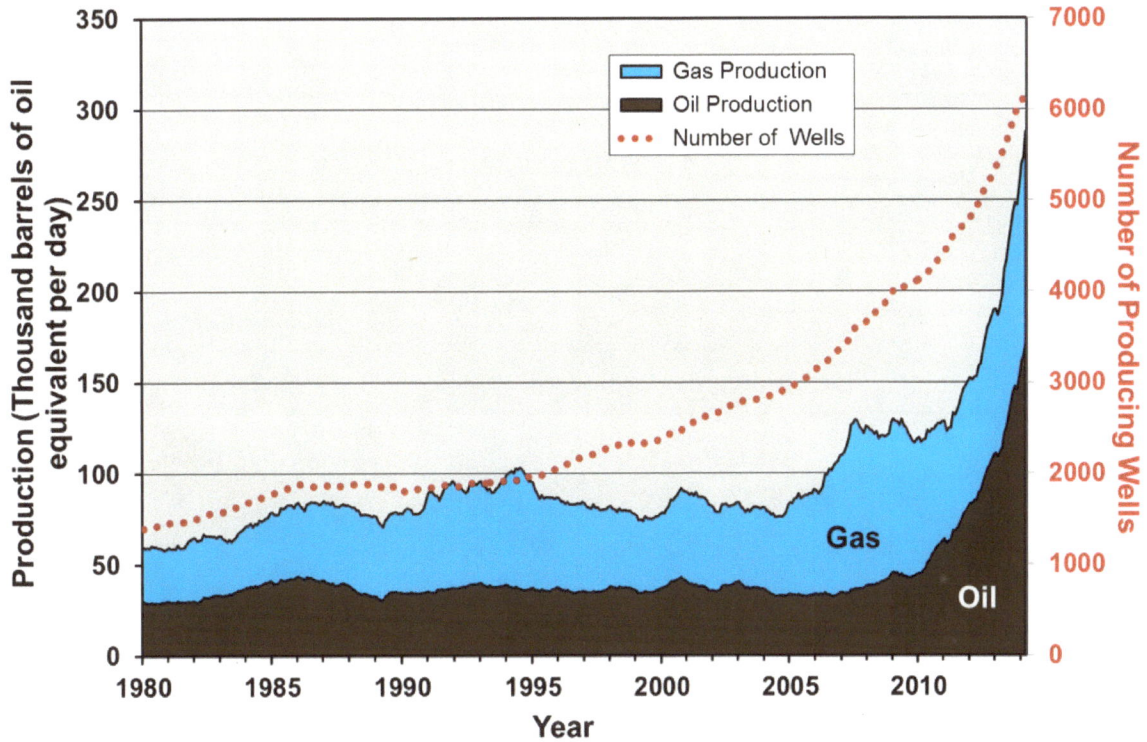

Figure 2-64. Wolfcamp play oil and gas production and number of producing wells, 1980 to 2014.[100]

Producing well count is now over 6,000.

---

[100] Data from Drillinginfo retrieved July 2014. Three-month trailing moving average.

The number of producing wells has also doubled over this period. A look at the split in production by well type reveals that virtually all of this growth is attributable to horizontal wells (Figure 2-65). Horizontal fracking technology is obviously paying dividends.

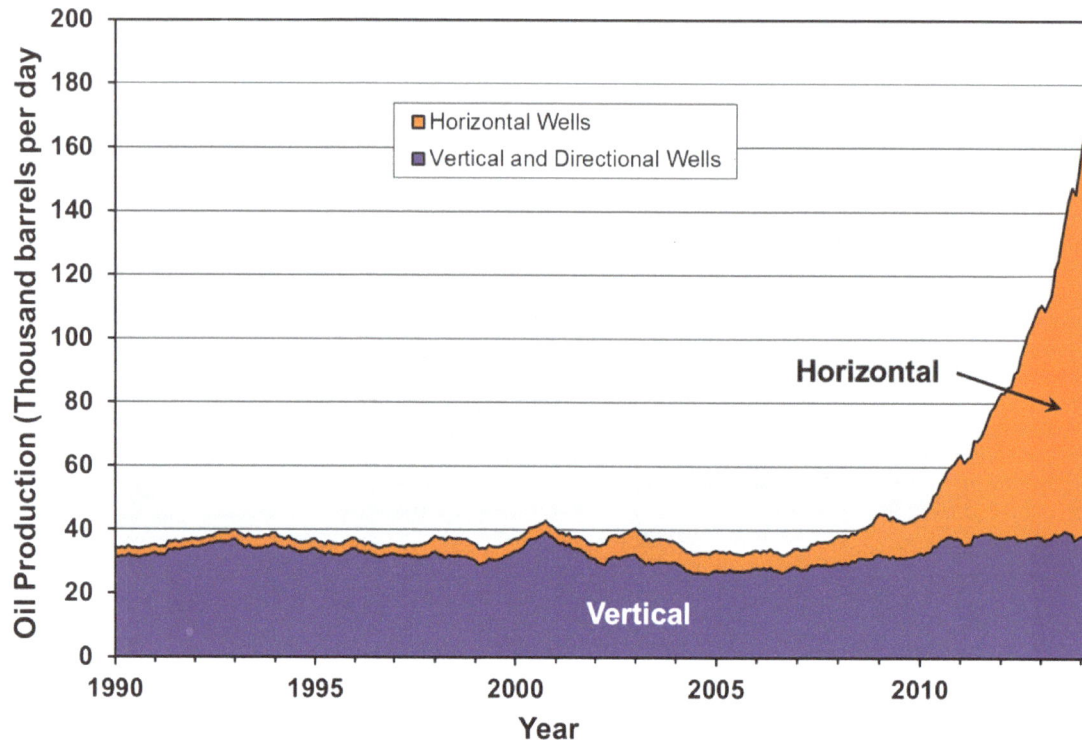

Figure 2-65. Oil production from the Wolfcamp play by well type.[101]
Horizontal wells are now accounting for most of the production growth.

---

[101] Data from Drillinginfo retrieved July 2014. Three-month trailing moving average.

2.4.2.2   Well Quality

A look at well quality reveals that the Wolfcamp, although considerably better than the Spraberry, is unremarkable by comparison to the Bakken or Eagle Ford. Figure 2-66 illustrates the average well decline profile for all wells; Figure 2-67 illustrates the average well decline profile for horizontal wells only. All wells on an energy equivalent basis are just a quarter of the initial production of an average Bakken well in a top county. Horizontal wells are nearly double the initial productivity of the average well but still pale by comparison to a Bakken or Eagle Ford well. The average three-year decline of Wolfcamp wells is comparable to the Bakken at 81% and 85% for all wells and horizontal wells, respectively.

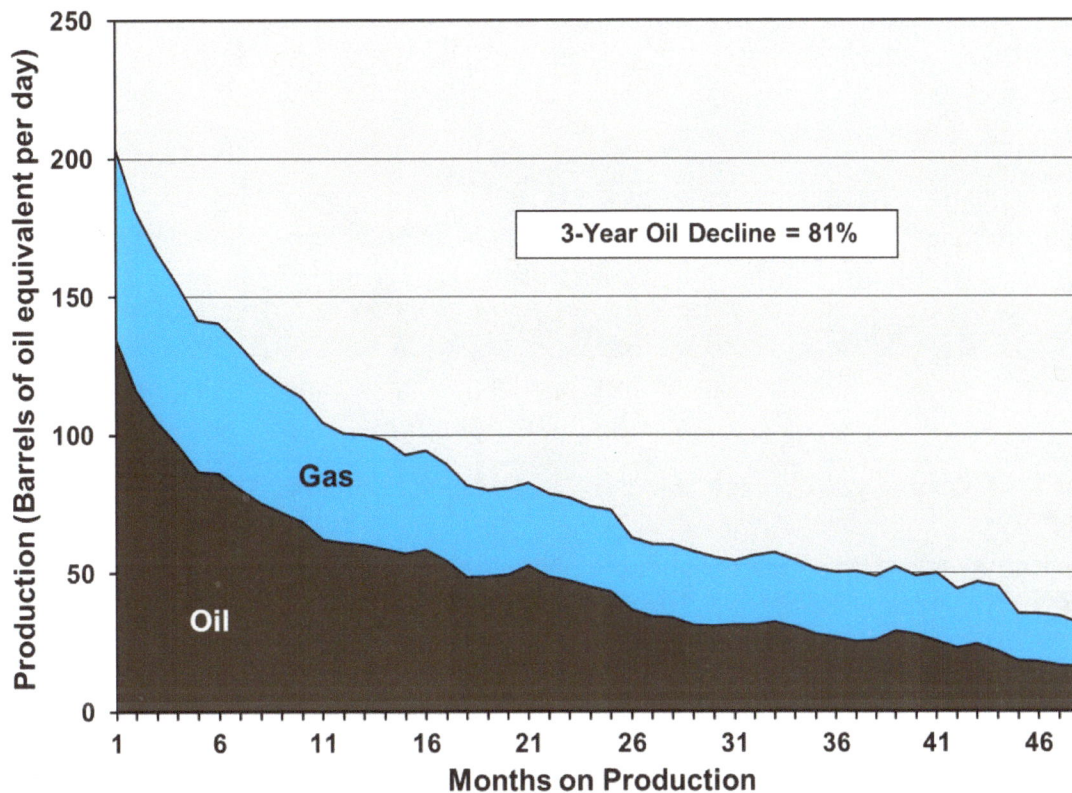

Figure 2-66. Oil and gas average well decline profile for all wells in the Wolfcamp play.[102]

On an energy equivalent basis these wells have an initial productivity of less than a quarter that of the average well in the top counties of the Bakken play. Decline profile is based on all wells drilled since 2009.

[102] Data from Drillinginfo retrieved July 2014.

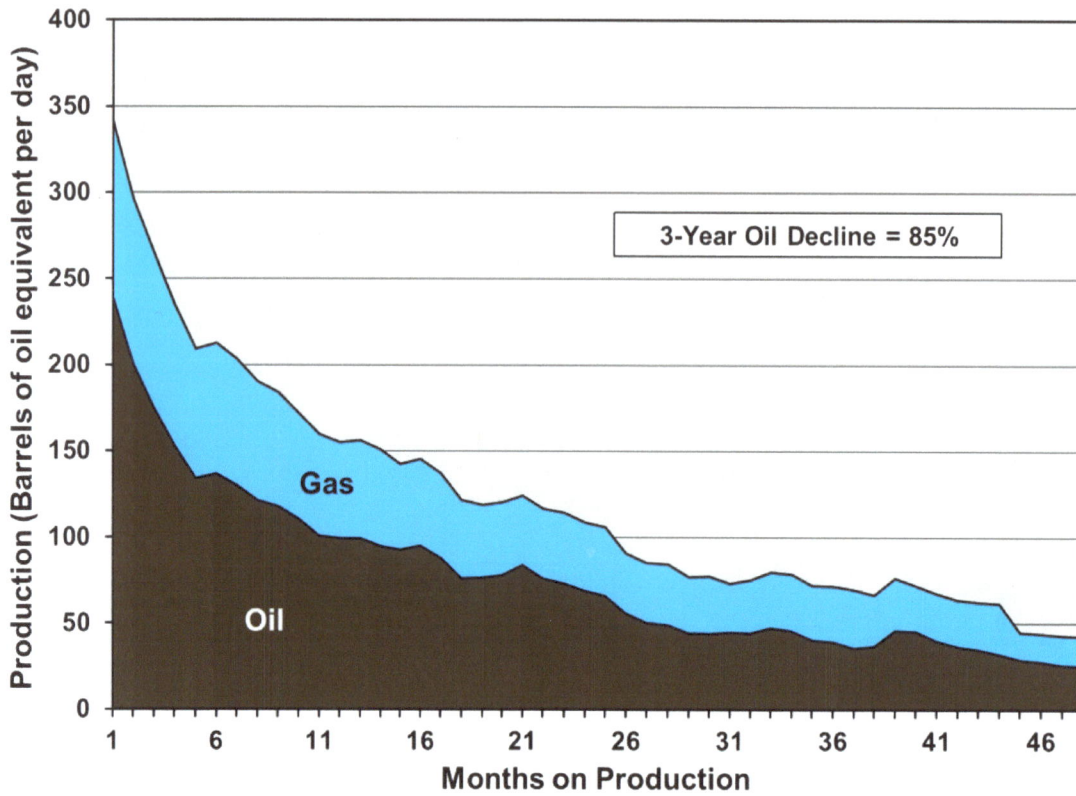

Figure 2-67. Oil and gas average well decline profile for horizontal wells in the Wolfcamp play.[103]

On an energy equivalent basis these wells have an initial productivity of about a third of the average horizontal well in the top counties of the Bakken play. Decline profile is based on all horizontal wells drilled since 2009.

---

[103] Data from Drillinginfo retrieved July 2014.

2.4.2.3   EIA Forecast

The EIA's projection for Wolfcamp play production through 2040 in its reference case is illustrated in Figure 2-68. Total recovery between 2012 and 2040 is forecast to be 2.64 billion barrels. This amounts to 6.1% of its U.S. reference case tight oil production through 2040. Cumulative production by 2040 amounts to 78% of the "unproved technically recoverable resources" the EIA estimated for the Wolfcamp as at January 1, 2012.

Given that this is a redevelopment of an old play which is already extensively drilled, the fact that the wells are of relatively low quality, and the nature of likely production profiles from shale plays like the Bakken, this would seem to be an optimistic forecast. It is already off by 36% on the high side in year one, as actual production for 2013 amounted to 153,000 barrels per day compared to an estimate of 209,000 barrels by the EIA. Furthermore, the EIA is projecting that production will be 220,000 barrels per day in 2040, which is 44% above current levels. High field decline rates make it likely that production decline after its projected peak in 2019 will be much steeper than forecast. Given what is known, this EIA forecast would seem to have a high optimist bias.

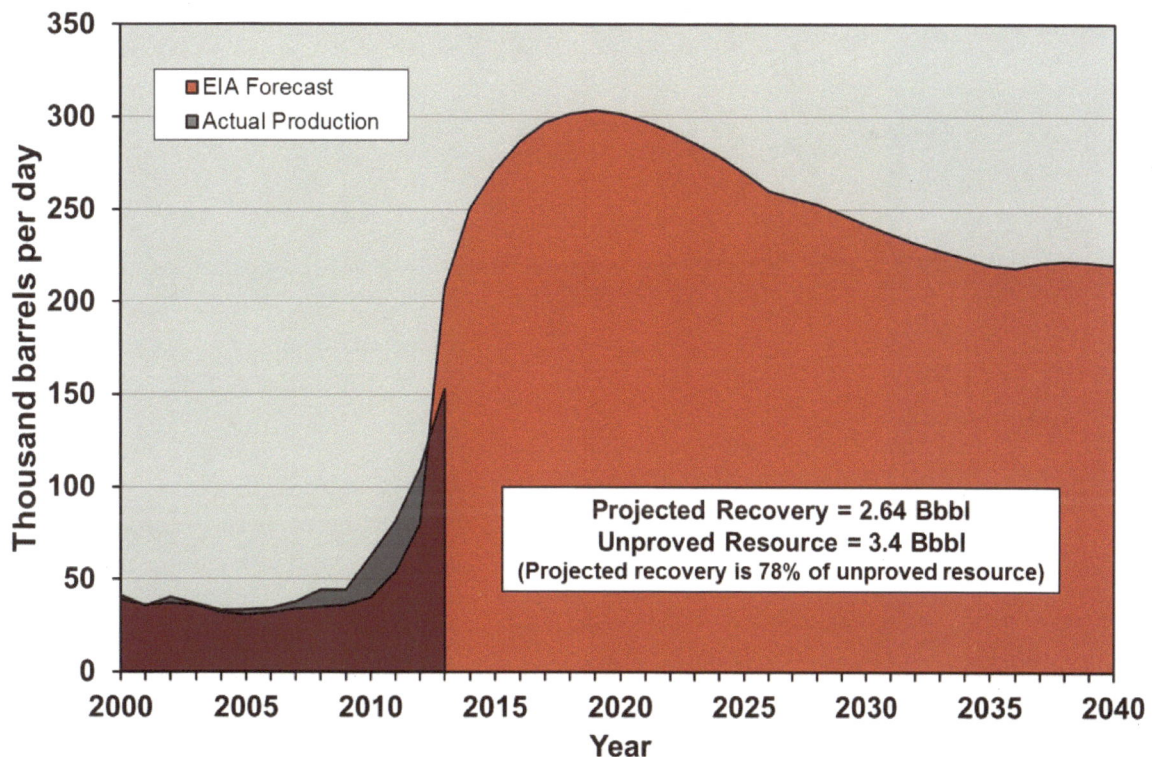

Figure 2-68. EIA reference case projection of oil production from the Wolfcamp play through 2040, with actual production to 2013.[104]

The forecast total recovery of 2.64 billion barrels over the 2012-2040 period amounts to 78% of the 3.4 billion barrels of "unproved technically recoverable resources as of January 1, 2012".[105]

[104] Production data from DrillingInfo, July 2014. Forecast from EIA, *Annual Energy Outlook 2014*, Unpublished tables from AEO 2014 provided by the EIA.
[105] EIA, *Assumptions to the Annual Energy Outlook 2014*, http://www.eia.gov/forecasts/aeo/assumptions/pdf/oilgas.pdf.

2.4.3   Bone Spring Play

The EIA forecasts recovery of 0.68 billion barrels of oil from the Bone Spring play between 2012 and 2040. The analysis of actual production data presented below suggests that this forecast is reasonable and may be on the low end of future production.

The Bone Spring play has, like the Spraberry and Wolfcamp, been producing oil and gas for decades. Over 5,200 wells have been drilled of which 2,500 are currently producing. The play has produced 208 million barrels of oil and 730 billion cubic feet of natural gas over its lifetime. Figure 2-69 illustrates well distribution within the Bone Spring play.

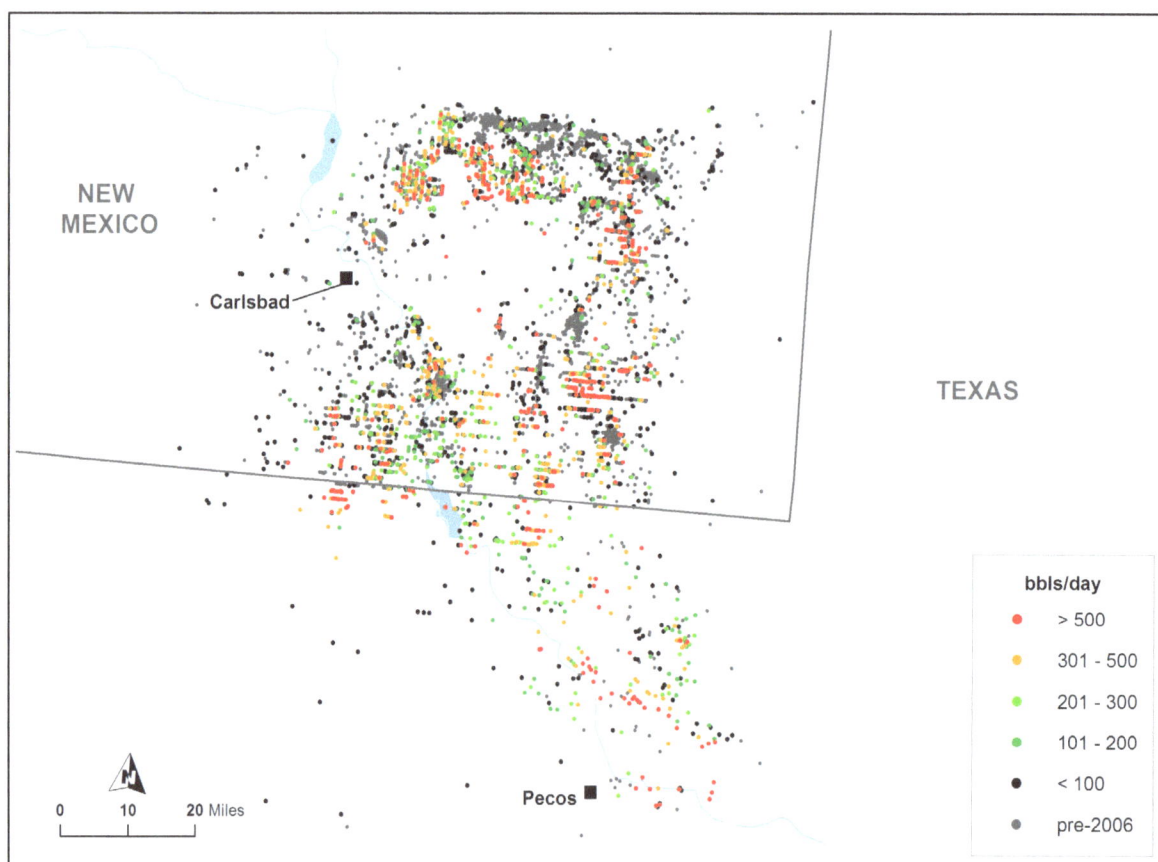

Figure 2-69. Distribution of wells in the Bone Spring play as of mid-2014 illustrating highest one-month oil production (initial productivity, IP).[106]

Only wells drilled in 2006 and later are considered as possible "tight oil" production and colored by IP; wells drilled prior to 2006 are predominantly conventional production. Well IPs are categorized approximately by percentile; see Appendix.

---

[106] Data from Drillinginfo retrieved July 2014.

2.4.3.1   Production History

Production of oil in the Bone Spring has increased more than 10 fold since 2005 and on an energy equivalent basis, including natural gas, is up more than 15-fold as illustrated in Figure 2-70.

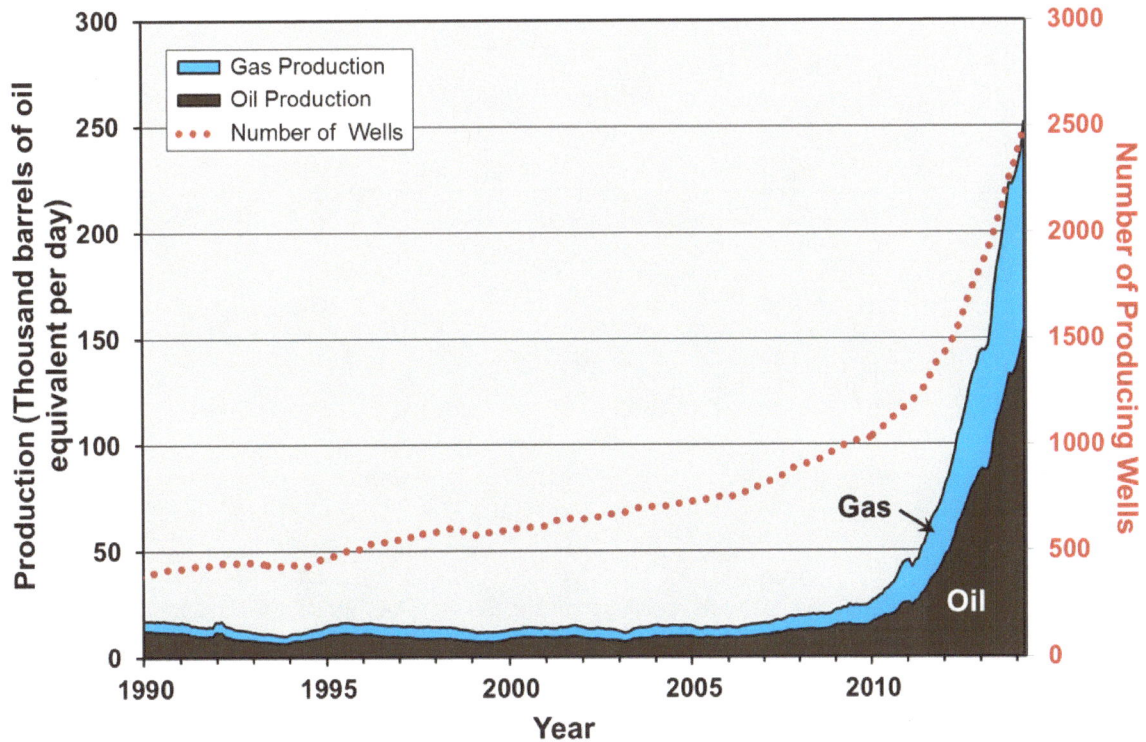

Figure 2-70. Bone Spring play oil and gas production and number of producing wells, 1990 to 2014.[107]

Producing well count is now about 2,500.

---

[107] Data from Drillinginfo retrieved July 2014. Three-month trailing moving average.

The number of producing wells has also more than tripled over this period. A look at the split in production by well type reveals that virtually all of this growth is attributable to horizontal wells (Figure 2-71). Horizontal fracking technology is obviously paying dividends.

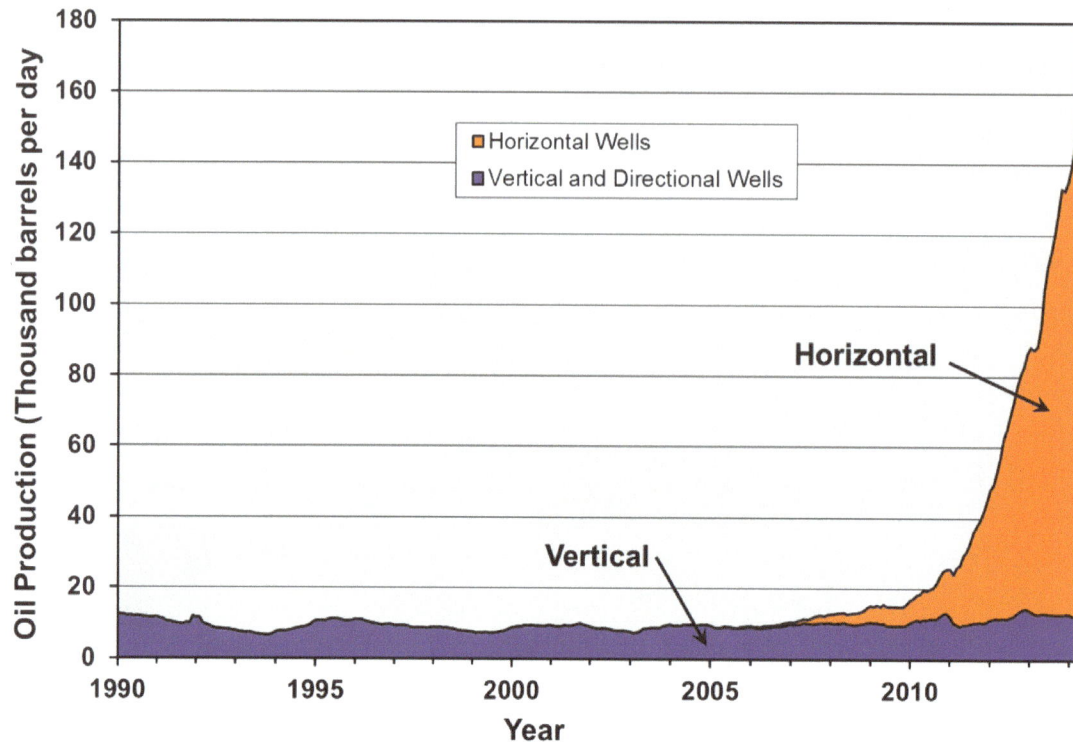

Figure 2-71. Oil production from the Bone Spring play by well type.[108]
Horizontal wells are now accounting for most of the production growth.

[108] Data from Drillinginfo retrieved July 2014. Three-month trailing moving average.

2.4.3.2 Well Quality

A look at well quality reveals that the Bone Spring, although considerably better than either the Spraberry or Wolfcamp, is still unremarkable by comparison to the Bakken or Eagle Ford. Figure 2-72 illustrates the average well decline profile for all wells; Figure 2-73 illustrates the average well decline profile for horizontal wells only. All wells on an energy equivalent basis are about half of the initial production of an average Bakken well in a top county. Horizontal wells are slightly better; the average initial productivity is about two-thirds of an average Bakken well. The average three-year decline of Bone Spring wells is greater that the Bakken at 91% for all wells and for horizontal wells, and is the steepest observed for any shale play.

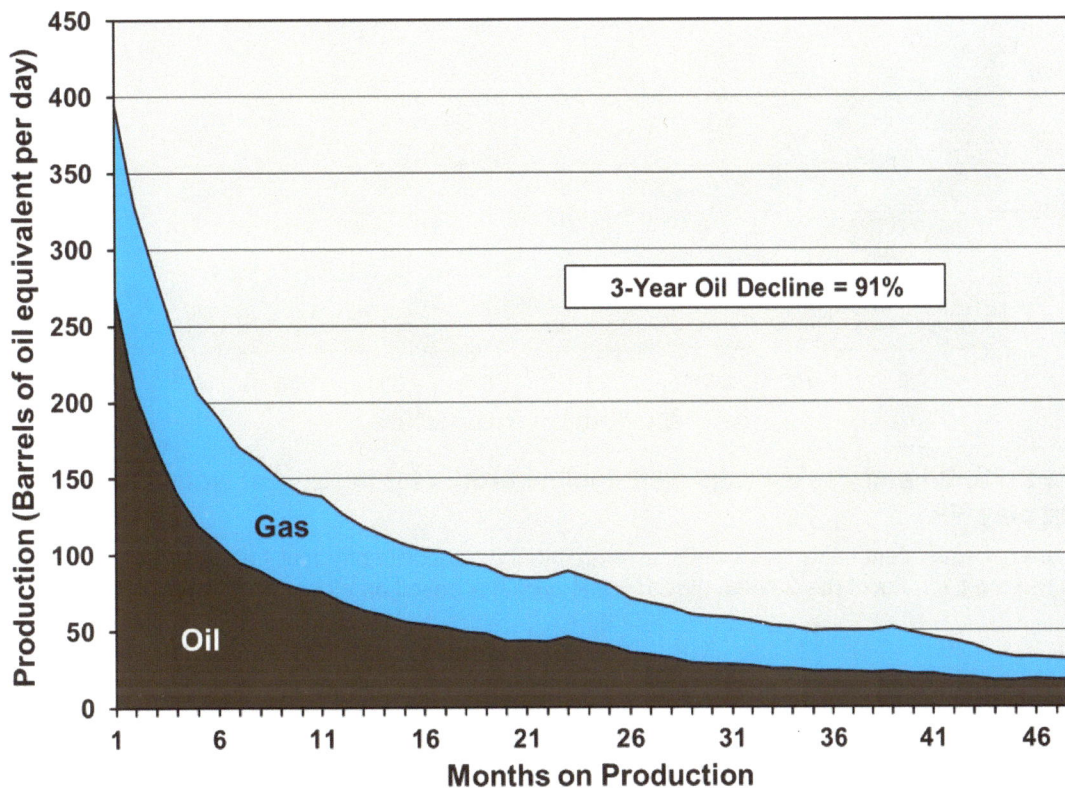

Figure 2-72. Oil and gas average well decline profile for all wells in the Bone Spring play.[109]

On an energy equivalent basis these wells have an initial productivity of about half that of the average well in the top counties of the Bakken play. Decline profile is based on all wells drilled since 2009.

---

[109] Data from Drillinginfo retrieved July 2014.

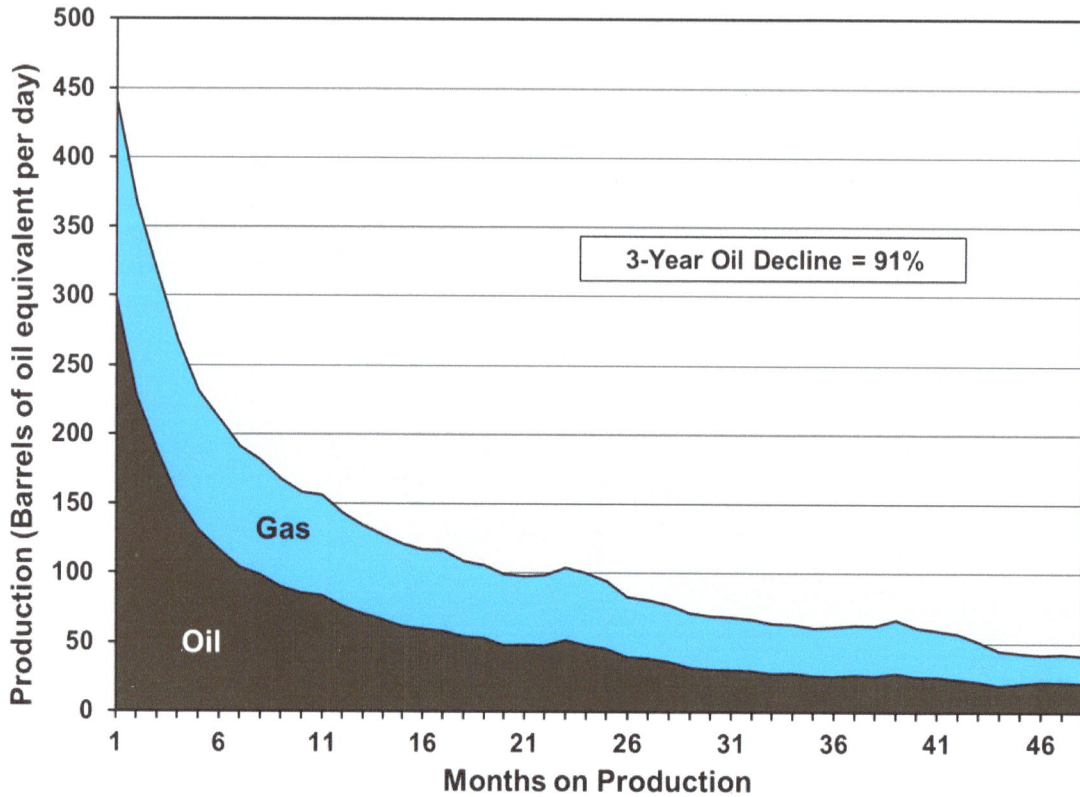

Figure 2-73. Oil and gas average well decline profile for horizontal wells in the Bone Spring play.[110]

On an energy equivalent basis these wells have an initial productivity of about half of the average horizontal well in the top counties of the Bakken play. Decline profile is based on all horizontal wells drilled since 2009.

[110] Data from Drillinginfo retrieved July 2014.

2.4.3.3   EIA Forecast

The EIA's projection for Bone Spring play production through 2040 in its reference case is illustrated in Figure 64. Total recovery between 2012 and 2040 is forecast to be 0.68 billion barrels. This amounts to just 1.6% of its U.S. reference case tight oil production through 2040. Cumulative production by 2040 amounts to 34% of the "unproved technically recoverable resources" the EIA estimated for the Bone Spring as at January 1, 2012.

In this case the EIA's forecast looks conservative, as production is already considerably higher in year one than projected. One could argue with the long extended tail of production but it appears likely that Bone Spring production may rise considerably higher. The very high well- and field-declines noted, which are considerably higher than the other Permian plays examined above, will likely make decline on the far side of peak production much steeper than depicted in the EIA projection. Given what is known, this EIA forecast would seem to have a low optimist bias.

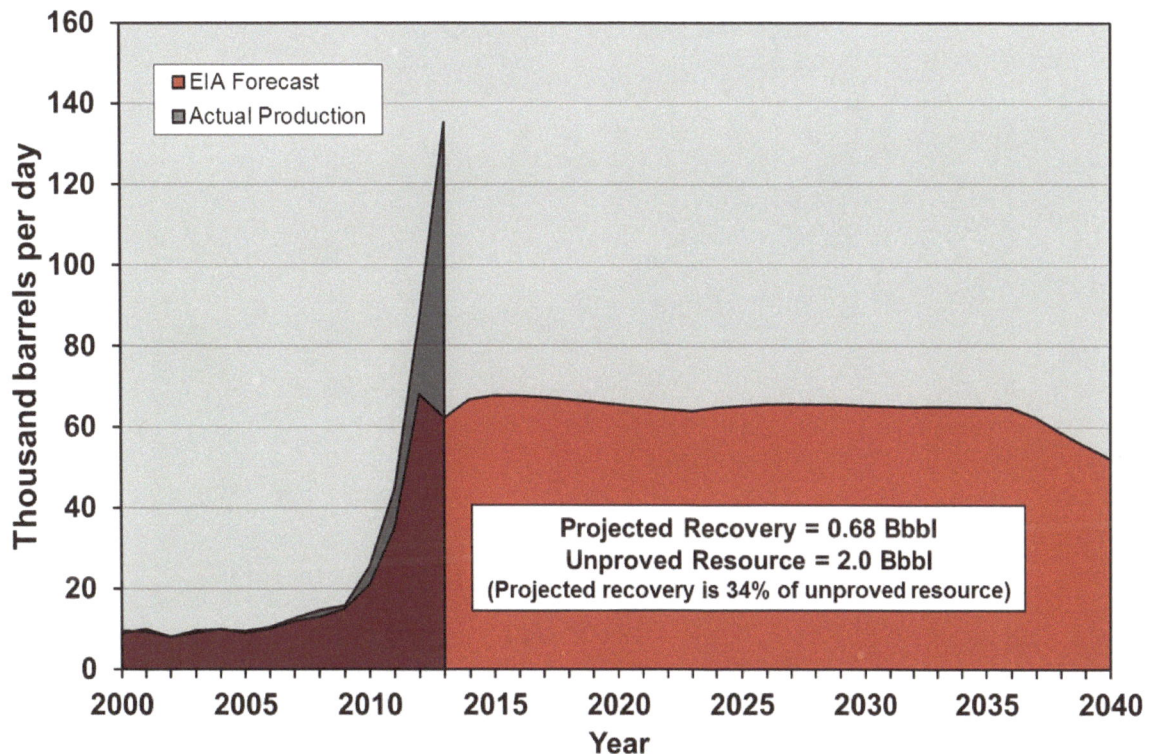

Figure 2-74. EIA reference case projection of oil production from the Bone Spring through 2040, with actual production to 2013.[111]

The forecast total recovery of .68 billion barrels over the 2012-2040 period amounts to 34% of the 2.0 billion barrels of "unproved technically recoverable resources as of January 1, 2012".[112]

---

[111] Production data from DrillingInfo, July 2014. Forecast from EIA, *Annual Energy Outlook 2014*, Unpublished tables from AEO 2014 provided by the EIA.
[112] EIA, *Assumptions to the Annual Energy Outlook 2014*, http://www.eia.gov/forecasts/aeo/assumptions/pdf/oilgas.pdf.

2.4.4   Key Characteristics of the Permian Basin Plays

As mentioned, the Permian Basin is the third largest source of tight oil in the U.S., and the three plays reviewed above constitute 23% of the oil the EIA projects will be recovered by 2040 in its reference tight oil case. In addition to these plays, two smaller Permian plays are listed by the EIA in Figure 2-7 above: the Glorieta-Yeso (actually two separate formations) and the Delaware. These latter two plays display the same characteristics as the first three: they are old plays which have been producing for decades, and although they are increasing somewhat in production, well quality is unremarkable compared to the Bakken and Eagle Ford.

Figure 2-75 illustrates total Permian Basin production, highlighting these five plays which now make up 56% of the total production of the basin.

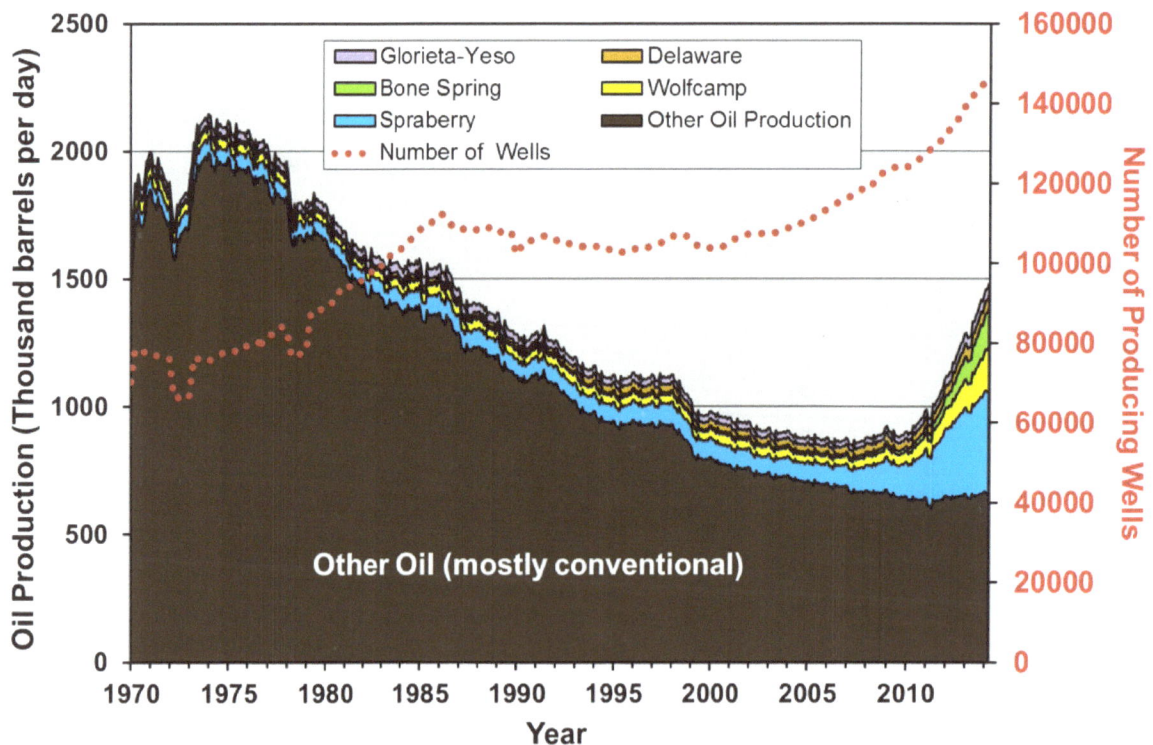

Figure 2-75. Oil production and number of producing wells in the Permian Basin to 2014.[113]

Production from the five tight oil plays the EIA includes in the Permian Basin (see Figure 2-7) is highlighted. As of March 2014, these plays made up 56% of total Permian Basin production.

---

[113] Data from Drillinginfo retrieved July 2014. Three-month trailing moving average.

The EIA only provided projections used in its reference case tight oil forecast for the three Permian plays reviewed in detail above.[114] The aggregate production of these plays compared to the collective forecast of the EIA for them is illustrated in Figure 2-76. The EIA forecast suggests these plays will collectively produce 9.25 billion barrels between 2014 and 2040, which is nearly five times as much oil as they produced in the previous 34 years. Production is projected to rise to a peak in 2021 followed by a gradual decline through 2040, when these plays are forecast to be producing 770,000 barrels per day, or 6% above current levels. This is a very aggressive forecast considering their age and extensive drilling and production history, their relatively low quality wells, and their observed steep well- and field-declines.

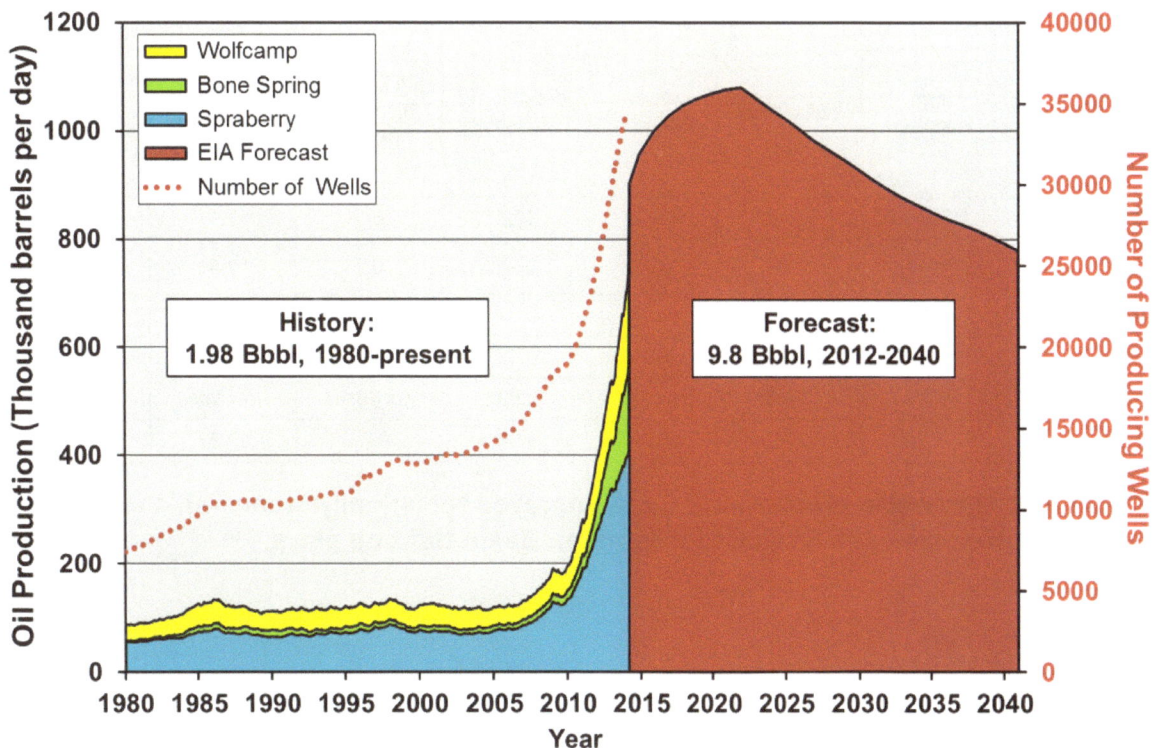

Figure 2-76. Oil production and number of producing wells in the Spraberry, Wolfcamp, and Bone Spring plays to 2014, with EIA reference case projection for these plays through 2040.[115]

The forecast total recovery of 9.8 billion barrels over the 2012-2040 period amounts to nearly five times the 1.98 billion barrels recovered from 1980 to the present, and 73% of the plays "unproved technically recoverable resources as of January 1, 2012.[116]

---

[114] EIA, *Annual Energy Outlook 2014*, Unpublished tables from AEO 2014 provided by the EIA.

[115] Production data from DrillingInfo, July 2014. Forecast from EIA, *Annual Energy Outlook 2014*, Unpublished tables from AEO 2014 provided by the EIA.

[116] EIA, *Assumptions to the Annual Energy Outlook 2014*, http://www.eia.gov/forecasts/aeo/assumptions/pdf/oilgas.pdf.

Growth in the Permian Basin plays is largely a result of redevelopment of long-established plays with better technology, including horizontal drilling and fracking, rather than the new discoveries represented by the Bakken and Eagle Ford. Most of the Permian plays first began to produce significant amounts of oil and gas back in the 1950s. More than 70,000 wells have been drilled of which 43,000 are currently producing. As such they are not analogues to the Bakken and Eagle Ford, from which significant production is just twelve and six years old, respectively. The Bakken and Eagle Ford currently produce 62% of all U.S. tight oil (Figure 5), compared to 25% for the Permian plays. At least some of the oil produced from these so-called Permian "tight oil" plays is conventional, as is most of the rest of Permian Basin production. Table 2-3 summarizes the long history of development of these Permian Basin plays and contrasts that with the EIA's tight oil forecast.

Play	Years Produced	Wells Drilled	Wells Producing	Production to Date (Bbbls)	EIA Recovery 2012-2040 (Bbbls)	EIA Unproved Resources as of January 1, 2012 (Bbbls)	EIA Production in 2040 (MMbbl/d)
Spraberry	60+	36756	25939	1.83	6.5	8.1	0.51
Avalon / Bone Spring	40+	5287	2473	0.21	0.7	2.0	0.05
Wolfcamp	60+	12837	6124	0.87	2.6	3.4	0.22
Delaware	60+	8468	3995	0.43	Not Stated	Not Stated	Not Stated
Glorieta-Yeso	60+	9365	4492	0.59	Not Stated	Not Stated	Not Stated
Total		72713	43023	3.93	9.8+	13.5+	0.78+

Table 2-3. Age, wells, production[117], EIA unproved technically recoverable resources[118] and EIA reference case forecast for Permian Basin tight oil plays.[119]

[117] Data from Drillinginfo retrieved July 2014.
[118] EIA, *Assumptions to the Annual Energy Outlook 2014*, http://www.eia.gov/forecasts/aeo/assumptions/pdf/oilgas.pdf.
[119] EIA, *Annual Energy Outlook 2014*, Unpublished tables from AEO 2014 provided by the EIA.

2.4.5 Permian Basin Plays Analysis Summary

Several conclusions can be made from the foregoing analysis of the Permian Basin plays:

1. Growth in Permian Basin production is largely a result of application of new technologies to old plays, rather than significant new discoveries such as represented by the Bakken and Eagle Ford, although there are some emerging Permian plays lumped by the EIA into "other" in its reference case tight oil forecast.[120]

2. Productivity of wells in Permian tight oil plays is generally much lower on average than in the Bakken and Eagle Ford. Well costs are also lower with both vertical and horizontal development possible, and extensive infrastructure is in place, hence improving the economics of drilling despite the lower well productivity.

3. These plays exhibit steep well- and field-declines mandating continuous high levels of drilling and capital input to maintain production, although in the Spraberry declines are somewhat lower than in the other Permian plays.

4. The EIA is projecting aggressive continued growth in production from these plays with a peak in 2021 followed by a gradual decline, and the recovery of nearly five times as much oil by 2040 as these plays have produced in the past 34 years. This forecast is highly optimistic given the number of wells that would have to be drilled and the amount of capital required.

5. Although these plays were not reviewed on a detailed county-by-county basis, they are highly likely to exhibit "sweet spots" or "core areas" which are being targeted first, hence the number of wells and capital input will need to increase later in the EIA's forecast to moderate production decline.

---

[120] EIA, *Annual Energy Outlook 2014*, Unpublished tables from AEO 2014 provided by the EIA. Note that the EIA did not provide play specific projections for the Glorieta-Yeso and Delaware plays.

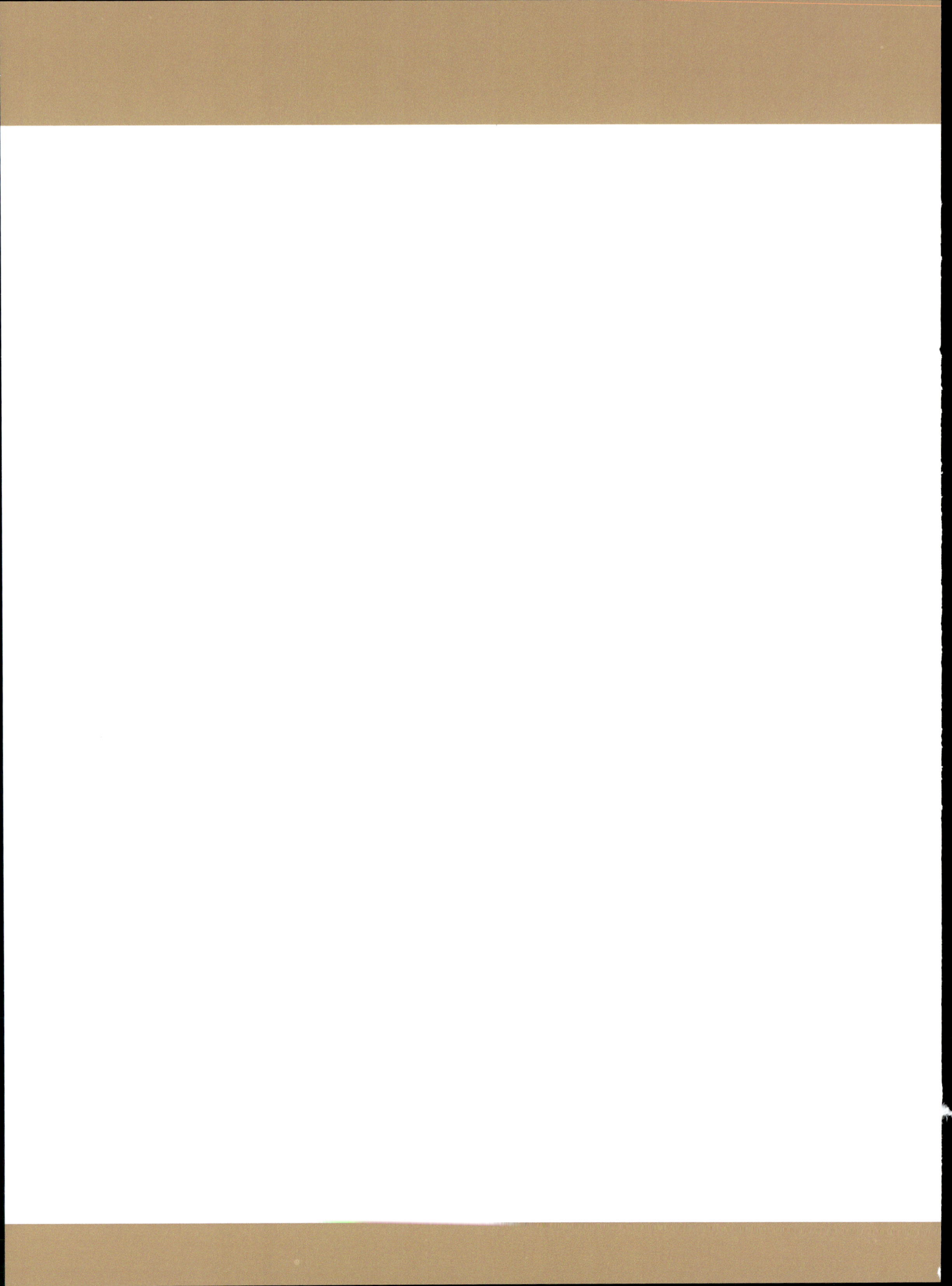

2.5 OTHER MAJOR PLAYS

Two other plays with significant production were singled out by the EIA[121] in its reference case tight oil forecast: the Austin Chalk in the Gulf Coast region and the Niobrara-Codell, in Colorado and Wyoming (a projection for the Monterey was also provided by the EIA but has been dealt with in a previous report[122], and the Woodford Shale, which was also provided, has relatively insignificant oil production). These are reviewed below.

---

[121] EIA, *Annual Energy Outlook 2014*, Unpublished tables from AEO 2014 provided by the EIA.
[122] J. David Hughes, *Drilling California: A Reality Check on the Monterey Shale*, Post Carbon Institute, 2013, http://www.postcarbon.org/publications/drilling-california.

2.5.1　Austin Chalk Play

The EIA forecasts recovery of 4.9 billion barrels of oil from the Austin Chalk play between 2012 and 2040. The analysis of actual production data presented below suggests that this forecast is highly unlikely to be realized.

The Austin Chalk play has, like the Permian plays, been producing oil and gas for decades. Over 15,000 wells have been drilled of which 5,000 are currently producing. The play has produced 1.17 billion barrels of oil and 6.1 trillion cubic feet of natural gas over its lifetime. Figure 2-77 illustrates well distribution within the Austin Chalk play. The play has seen the application of horizontal drilling for many years. Figure 2-78 illustrates the distribution of horizontal wells in the play which tend to be concentrated within certain areas.

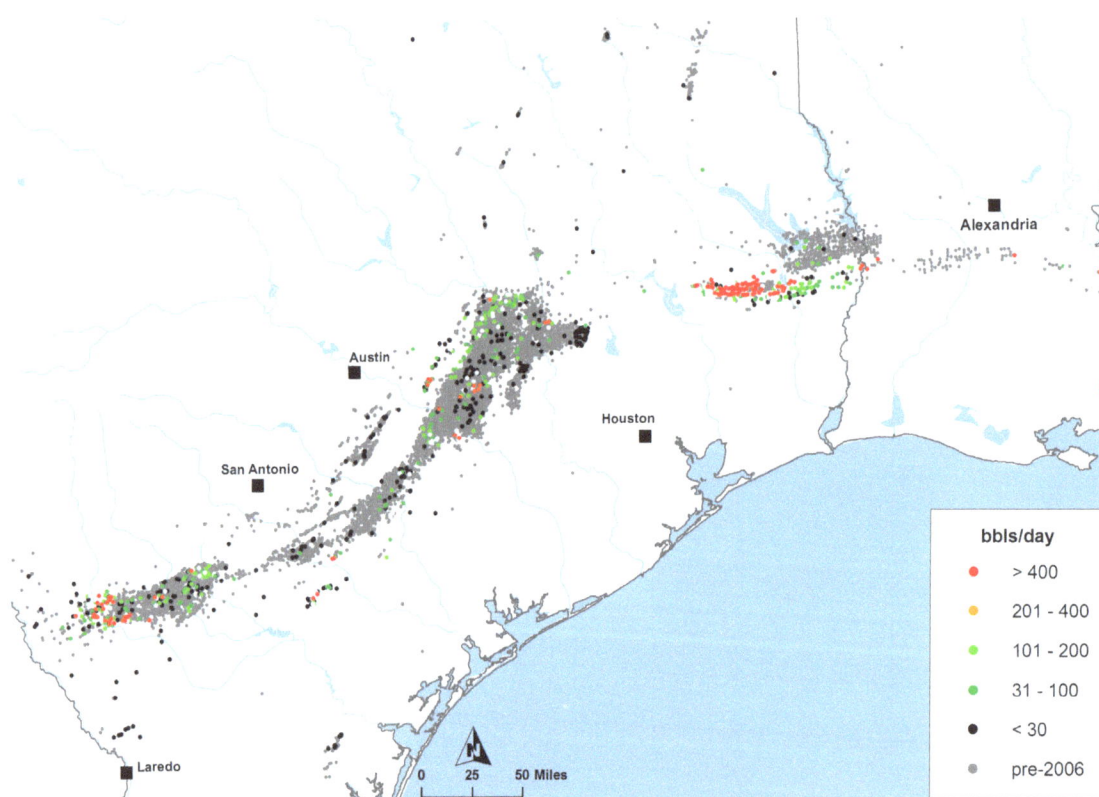

Figure 2-77. Distribution of wells in the Austin Chalk play as of mid-2014 illustrating highest one-month oil production (initial productivity, IP).[123]

Only wells drilled in 2006 and later are considered as possible "tight oil" production and colored by IP; wells drilled prior to 2006 are predominantly conventional production. Well IPs are categorized approximately by percentile; see Appendix.

---

[123] Data from Drillinginfo retrieved July 2014.

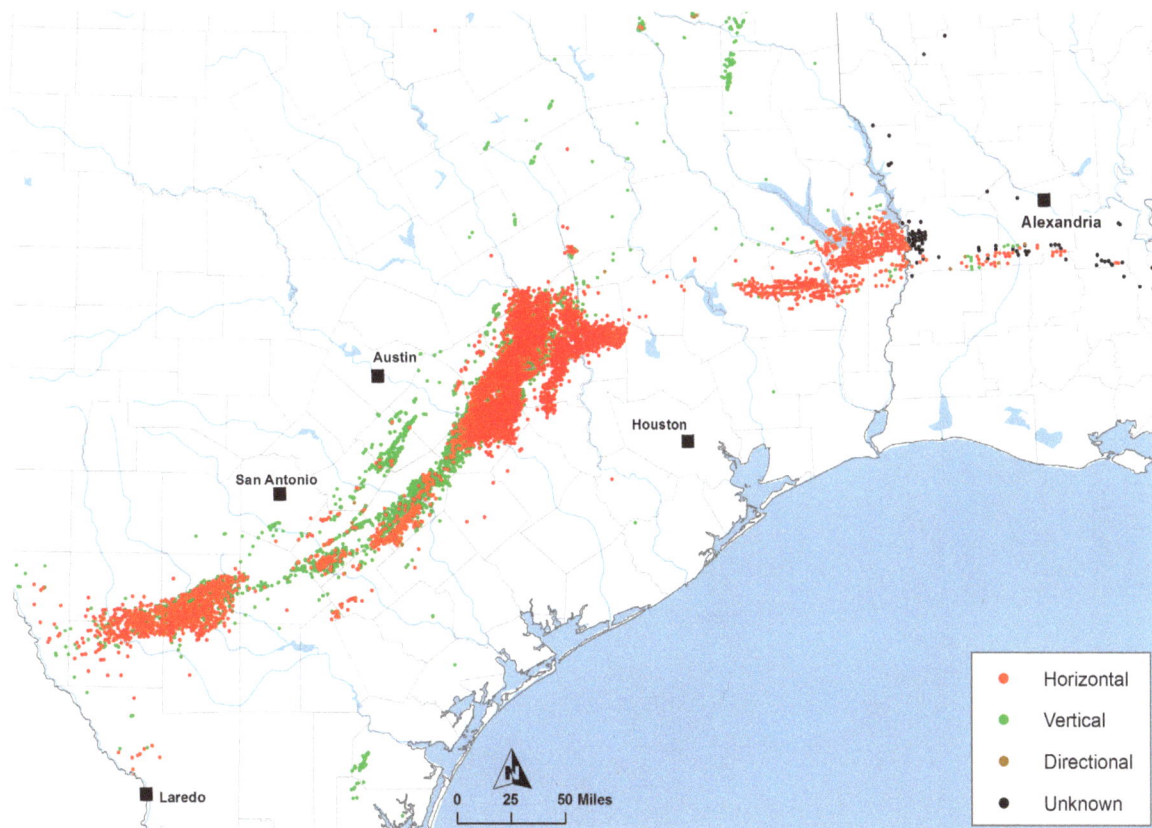

Figure 2-78. Distribution of wells in the Austin Chalk play categorized by drilling type, as of early 2014.[124]

[124] Data from Drillinginfo retrieved July 2014.

2.5.1.1   Production History

Production of oil in the Austin Chalk has been declining and the number of producing wells is also falling as illustrated in Figure 2-79. Oil production has declined by 83% since its peak in 1991.

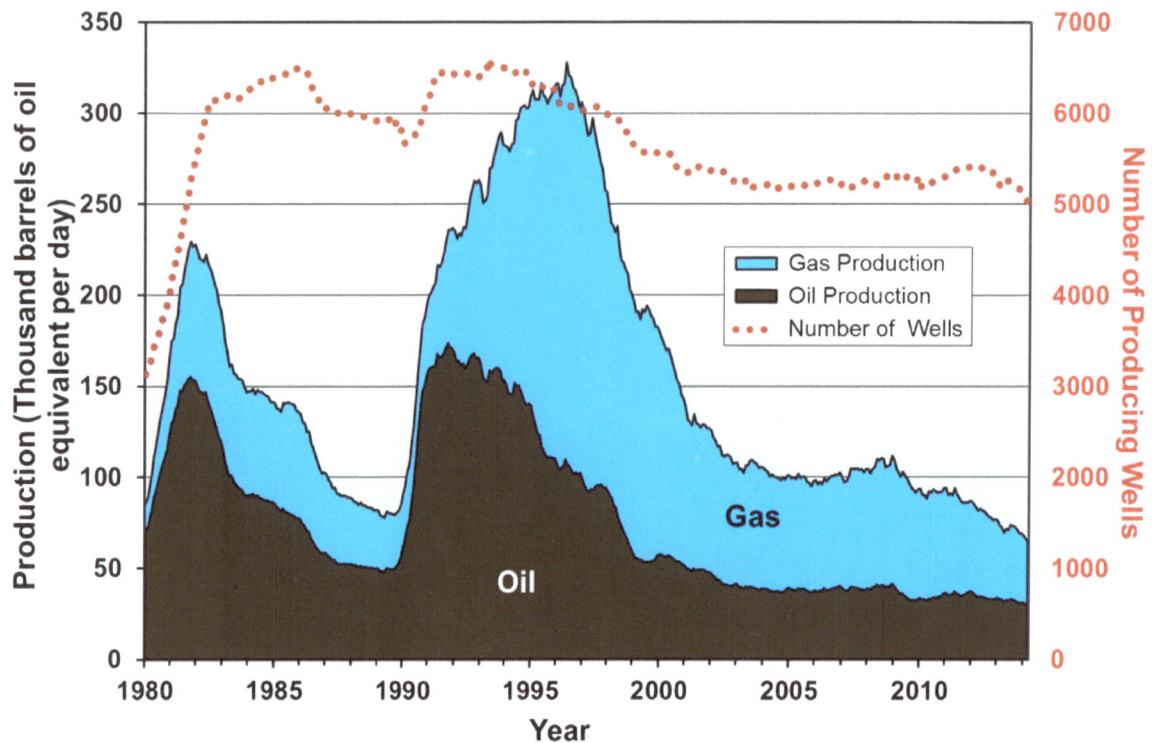

Figure 2-79. Austin Chalk play oil and gas production and number of producing wells, 1980 to 2014.[125]

Producing well count is now about 5,000.

---

[125] Data from Drillinginfo retrieved July 2014. Three-month trailing moving average.

A look at the split in production by well type reveals that horizontal wells have contributed the bulk of oil production over the past 25 years and currently provide 90% of production (Figure 2-80).

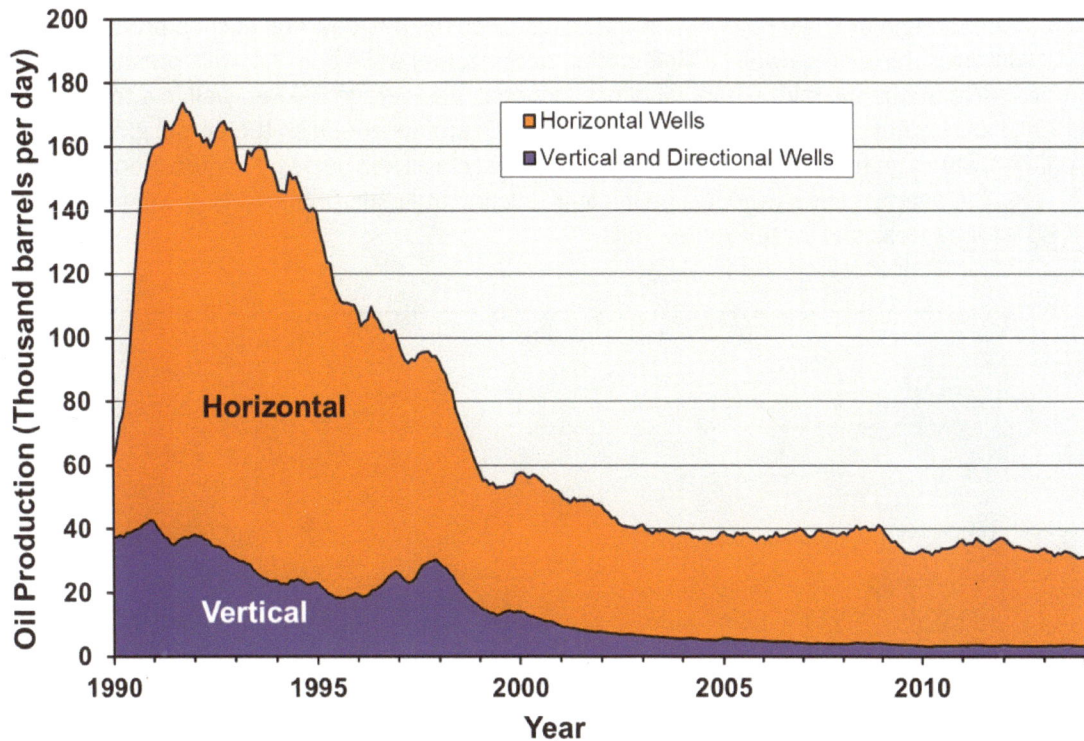

Figure 2-80. Oil production from the Austin Chalk play by well type.[126]

Horizontal wells have been the major contributors since the early 1990s.

---

[126] Data from Drillinginfo retrieved July 2014. Three-month trailing moving average.

2.5.1.2 Well Quality

A look at well quality reveals that the Austin Chalk is, like the Permian Basin plays, unremarkable by comparison to the Bakken or Eagle Ford. Figure 2-81 illustrates the average well decline profile for all wells; Figure 2-82 illustrates the average well decline profile for horizontal wells only. All wells on an energy equivalent basis are about one third of the initial production of an average Bakken well in a top county. Horizontal wells are slightly better (although 90% of "all" wells are horizontal so the only slight improvement is not surprising), although the initial productivity of the average well still pales by comparison to a Bakken or Eagle Ford well. The average three-year decline in oil production of Austin Chalk wells is comparable to the Bakken at 85% for all wells and for horizontal wells.

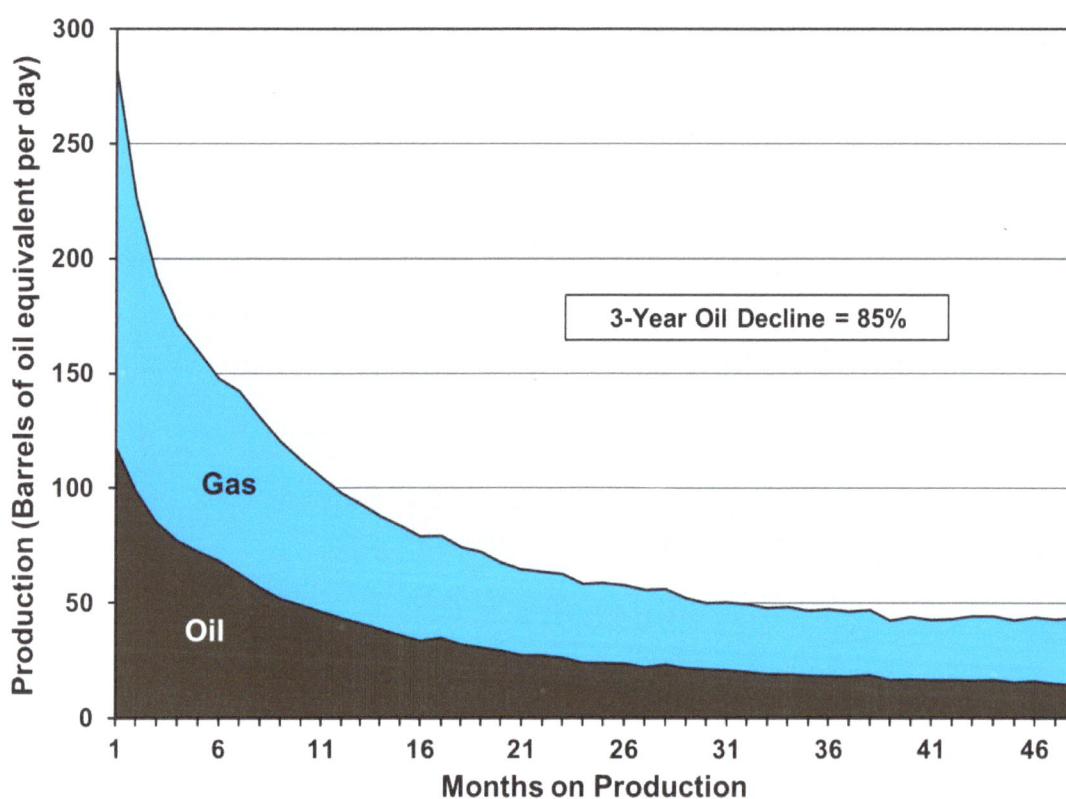

Figure 2-81. Oil and gas average well decline profile for all wells in the Austin Chalk play.[127]

On an energy equivalent basis these wells have an initial productivity of about one third that of the average well in the top counties of the Bakken play. Decline profile is based on all wells drilled since 2009.

---

[127] Data from Drillinginfo retrieved July 2014.

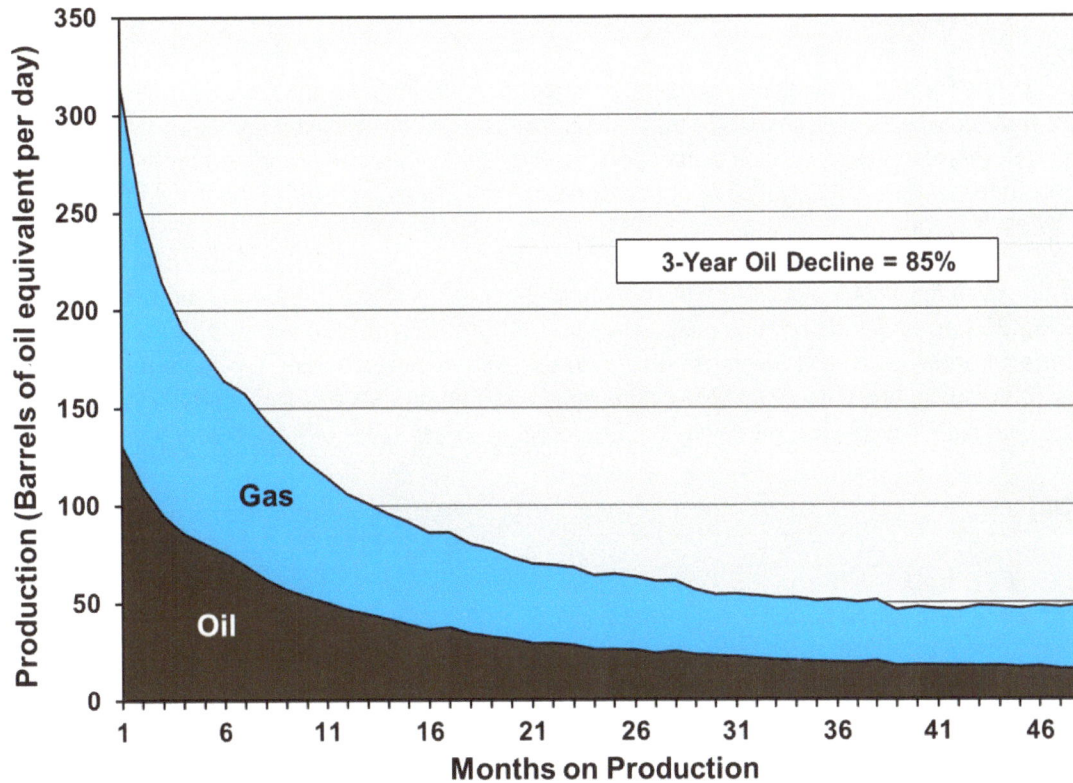

Figure 2-82. Oil and gas average well decline profile for horizontal wells in the Austin Chalk play.[128]

On an energy equivalent basis these wells have an initial productivity of about one third of the average horizontal well in the top counties of the Bakken play. Decline profile is based on all horizontal wells drilled since 2009.

[128] Data from Drillinginfo retrieved July 2014.

2.5.1.3   EIA Forecast

The EIA's projection for Austin Chalk play production through 2040 in its reference case is illustrated in Figure 2-83. Total recovery between 2012 and 2040 is forecast to be 4.9 billion barrels. This amounts to 11.3% of its U.S. reference case tight oil production through 2040. Cumulative production by 2040 amounts to 65% of the "unproved technically recoverable resources" the EIA estimated for the Austin Chalk as at January 1, 2012.

In this case the EIA's forecast looks extremely optimistic. They are projecting a production rise to a peak in 2031, at 656,830 barrels per day, which is 20 times current production, followed by a gradual decline to 513,000 barrels per day in 2040—16 times current production. As noted earlier, production in this play along with well count is falling, and well- and field-decline rates are high. In year one this forecast is already off by 145% on the high side. Given what is known, this EIA forecast would seem to have a very high optimist bias.

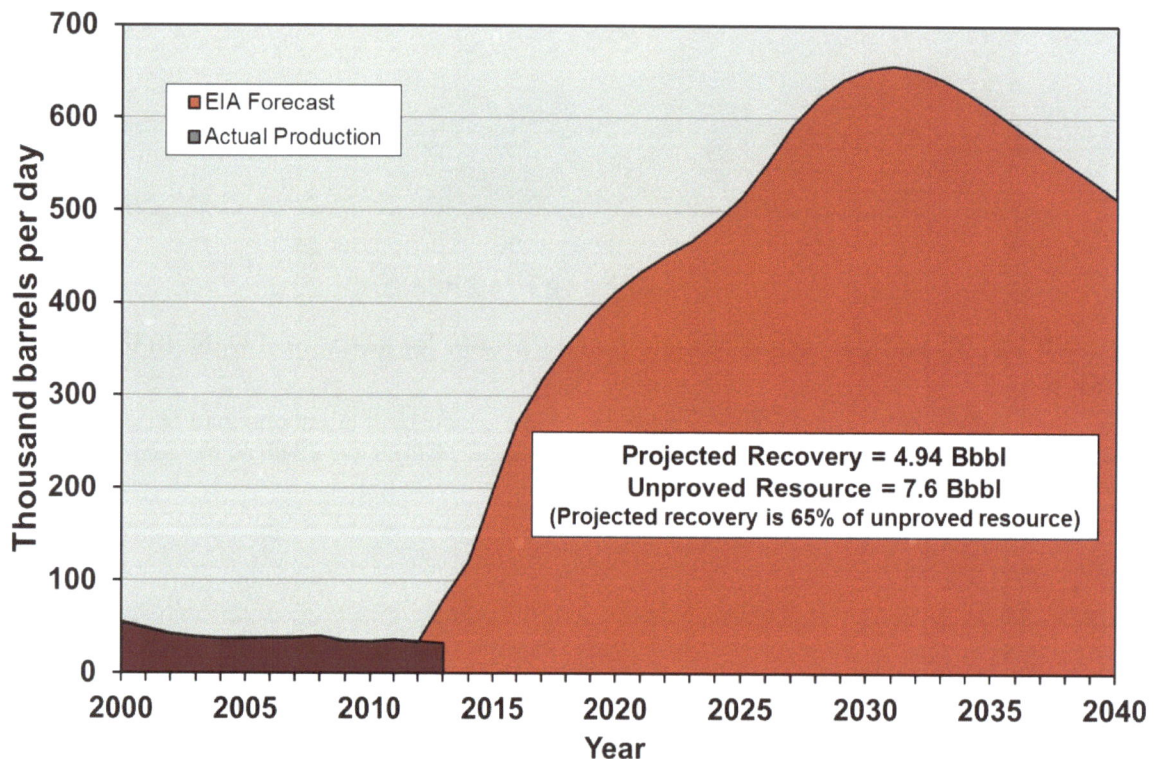

Figure 2-83. EIA reference case projection of oil production from the Austin Chalk through 2040, with actual production to 2013.[129]

The forecast total recovery of 4.94 billion barrels over the 2012-2040 period amounts to 65% of the 7.6 billion barrels of the EIA's "unproved technically recoverable resources as of January 1, 2012.[130]

---

[129] Production data from DrillingInfo, July 2014. Forecast from EIA, *Annual Energy Outlook 2014*, Unpublished tables from AEO 2014 provided by the EIA.
[130] EIA, *Assumptions to the Annual Energy Outlook 2014*, http://www.eia.gov/forecasts/aeo/assumptions/pdf/oilgas.pdf.

2.5.2   Niobrara-Codell Play

The EIA forecasts recovery of 4.9 billion barrels of oil from the Niobrara-Codell play between 2012 and 2040. The analysis of actual production data presented below suggests that this forecast is unlikely to be realized.

The Niobrara-Codell play, like the Permian Basin plays and the Austin Chalk play, has been producing oil and gas for decades. Over 30,800 wells have been drilled of which 13,900 are currently producing. The play has produced 357 million barrels of oil and 3.8 trillion cubic feet of natural gas over its lifetime. Figure 2-84 illustrates well distribution within the Niobrara-Codell play. Figure 2-85 illustrates the distribution of wells in the Wattenberg Field located mainly in Weld County of Colorado, where much of the drilling has occurred.

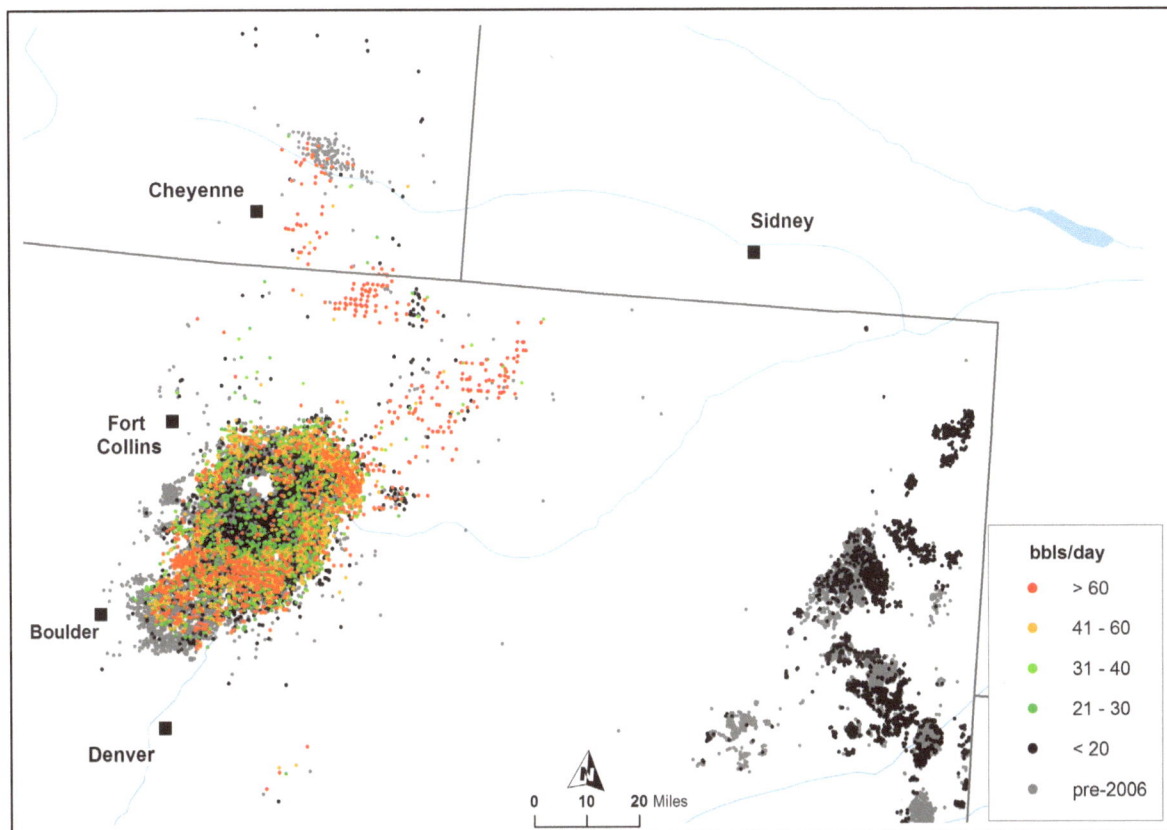

Figure 2-84. Distribution of wells in the Niobrara-Codell play as of mid-2014 illustrating highest one-month oil production (initial productivity, IP).[131]

Only wells drilled in 2006 and later are considered as possible "tight oil" production and colored by IP; wells drilled prior to 2006 are predominantly conventional production. Well IPs are categorized approximately by percentile; see Appendix.

[131] Data from Drillinginfo retrieved July 2014.

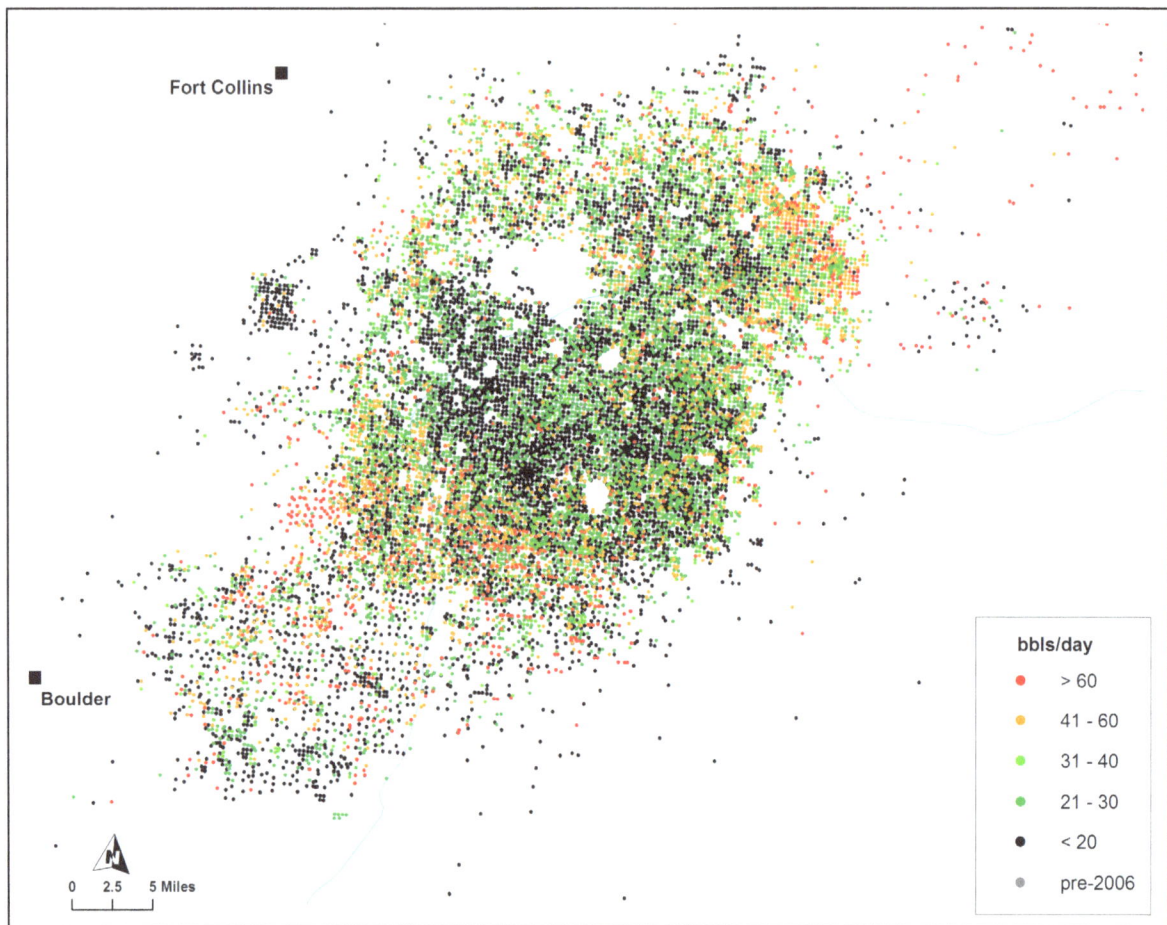

Figure 2-85. Detail of Niobrara-Codell play showing distribution of wells as of mid-2014, illustrating highest one-month oil production (initial productivity, IP).[132]

Map shows the Wattenberg Field of Weld County, Colorado, where much of the drilling has occurred. Only wells drilled in 2006 and later are considered as possible "tight oil" production and colored by IP; wells drilled prior to 2006 are predominantly conventional production. Well IPs are categorized approximately by percentile; see Appendix.

---

[132] Data from Drillinginfo retrieved July 2014.

2.5.2.1 Production History

Production of oil in the Niobrara-Codell has been growing although the number of producing wells has been falling recently as illustrated in Figure 2-86 (this may in part be related to flooding that occurred in Colorado in late 2013). Oil production hit an all-time high in December 2013, but has declined by 18% since then (again possibly related to the flooding).

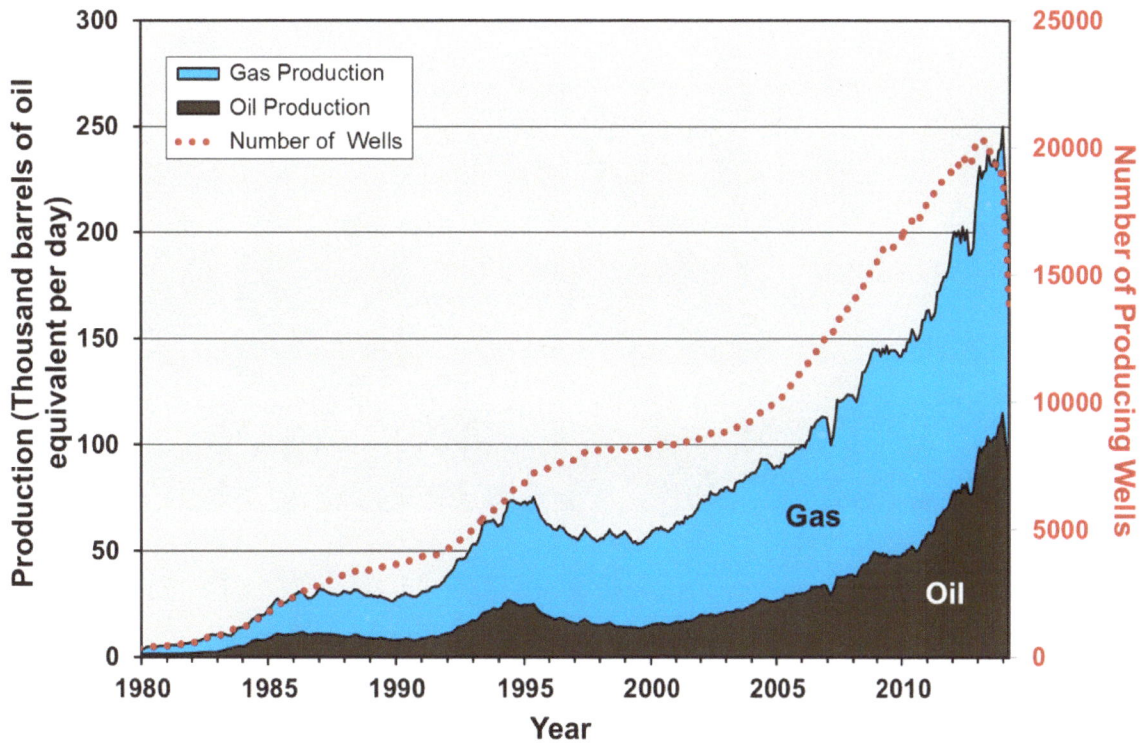

Figure 2-86. Niobrara-Codell play oil and gas production and number of producing wells, 1980 to 2014.[133]

Producing well count is now about 13,900.

---

[133] Data from Drillinginfo retrieved July 2014. Three-month trailing moving average.

A look at the split in production by well type reveals that horizontal wells now account for 77% of oil production (Figure 2-87).

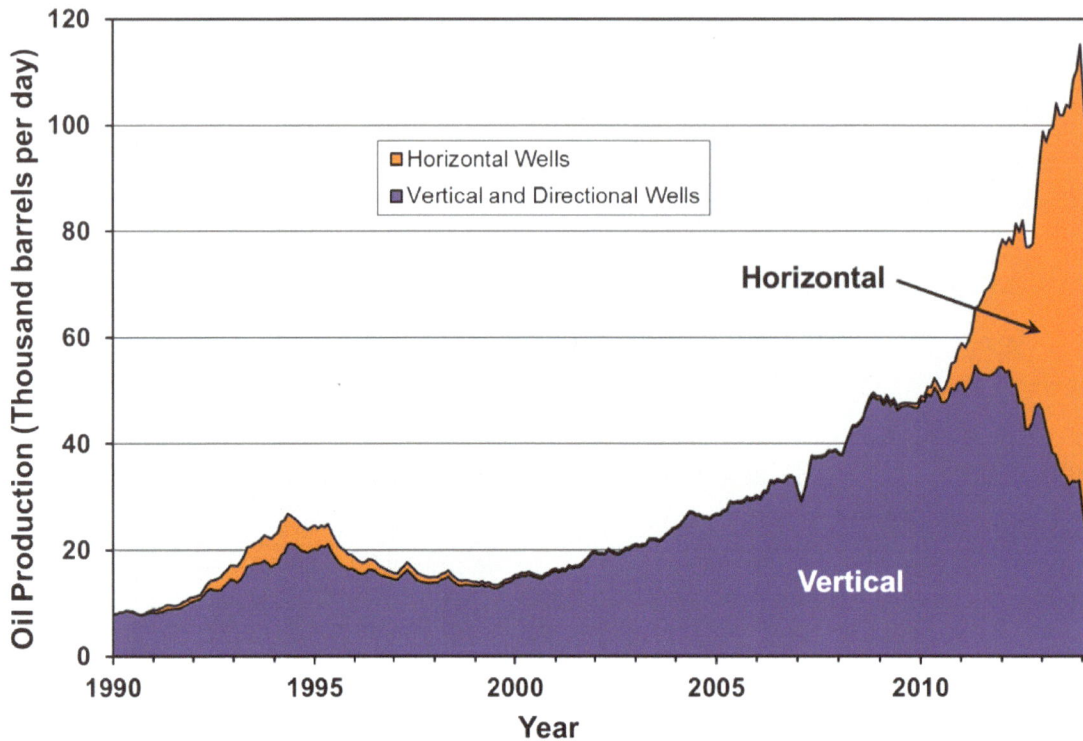

Figure 2-87. Oil production from the Niobrara-Codell play by well type.

Horizontal wells now produce 77% of the oil.[134]

---

[134] Data from Drillinginfo retrieved July 2014. Three-month trailing moving average.

2.5.2.2 Well Quality

A look at well quality reveals that the Niobrara-Codell is unremarkable by comparison to the Bakken or Eagle Ford. Figure 2-88 illustrates the average well decline profile for all wells; Figure 2-89 illustrates the average well decline profile for horizontal wells only. All wells on an energy equivalent basis are about a tenth of the initial production of an average Bakken well in a top county. Horizontal wells are much better (hence the fact that they now make up 77% of production), although the initial productivity of the average well still pales by comparison to a Bakken or Eagle Ford well. The average three-year decline of Niobrara-Codell wells is higher than that of the Bakken at 93% for all wells and 90% for horizontal wells.

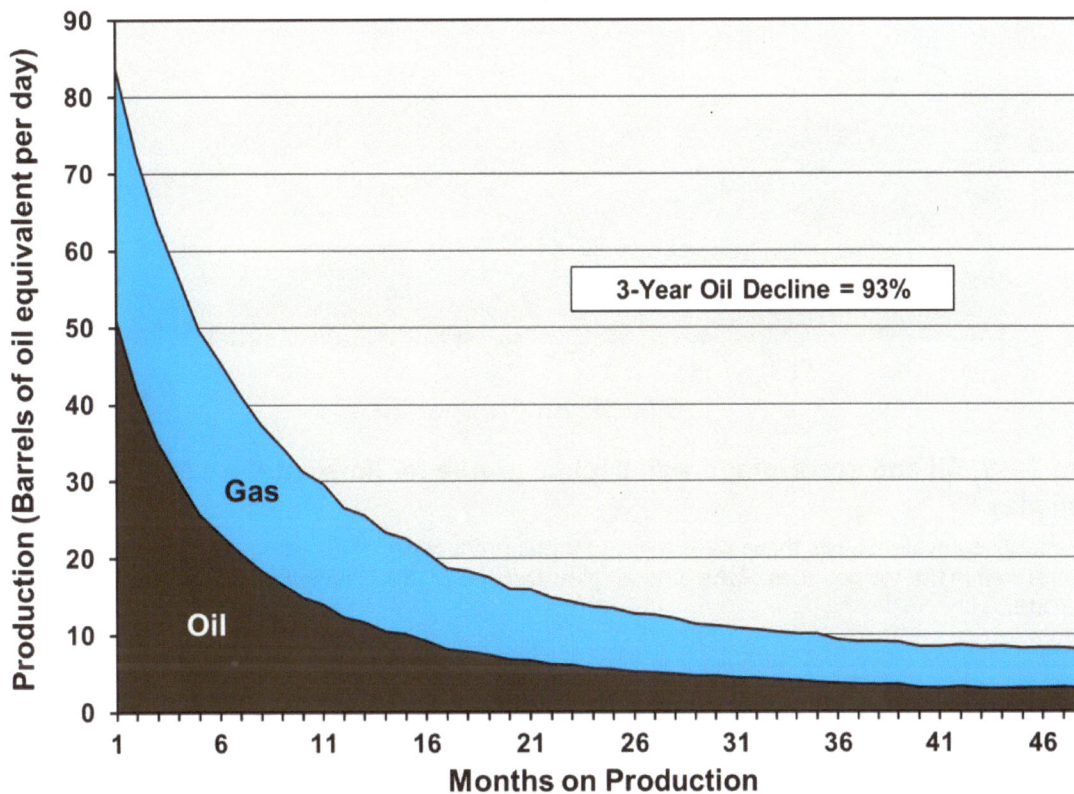

Figure 2-88. Oil and gas average well decline profile for all wells in the Niobrara-Codell play.[135]

On an energy equivalent basis these wells have an initial productivity of about a tenth that of the average well in the top counties of the Bakken play. Decline profile is based on all wells drilled since 2009.

[135] Data from Drillinginfo retrieved July 2014.

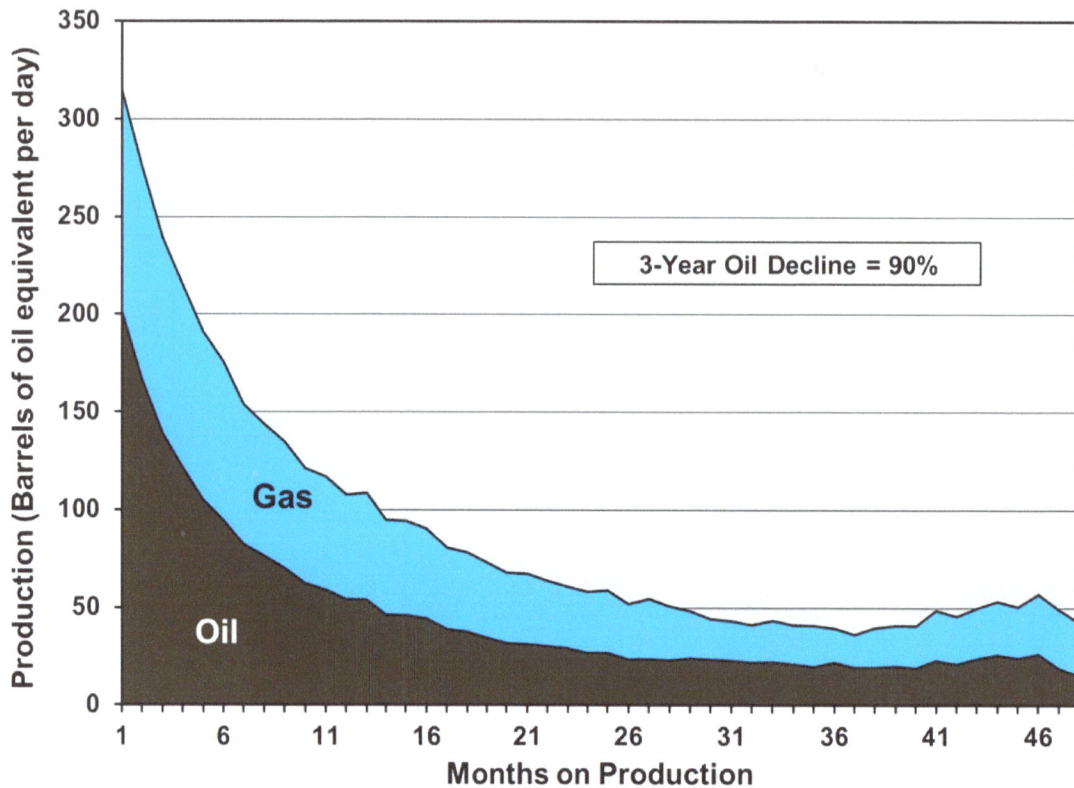

Figure 2-89. Oil and gas average well decline profile for horizontal wells in the Niobrara-Codell play.[136]

On an energy equivalent basis these wells have an initial productivity of about one third of the average horizontal well in the top counties of the Bakken play. Decline profile is based on all horizontal wells drilled since 2009.

[136] Data from Drillinginfo retrieved July 2014.

2.5.2.3 EIA Forecast

The EIA's projection for Niobrara-Codell play production through 2040 in its reference case is illustrated in Figure 2-90. Total recovery between 2012 and 2040 is forecast to be 4.9 billion barrels. This amounts to 4% of its U.S. reference case tight oil production through 2040. Cumulative production by 2040 is much higher than the resource estimate, amounting to 423% of the "unproved technically recoverable resources" the EIA estimated for the Niobrara-Codell as at January 1, 2012.

Notwithstanding the apparent overestimate of the EIA's production forecast compared to resources, the forecast has already been exceeded by production in year one. Nonetheless, the EIA projects that production will be double current levels in 2031 followed by a gradual decline to 76% above current levels in 2040. Given the very high well and field declines, among the highest of any play examined to date, this EIA forecast would seem to have a high optimist bias.

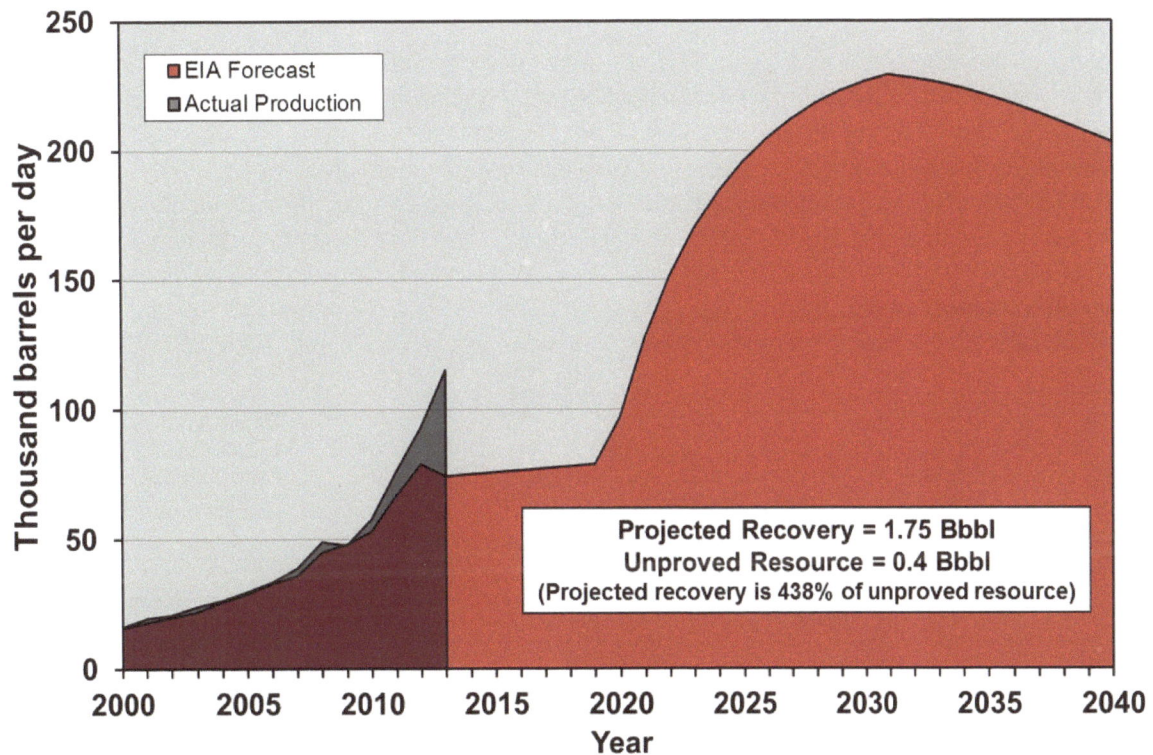

Figure 2-90. EIA reference case projection of oil production from the Niobrara-Codell through 2040, with actual production to 2013.[137]

The forecast total recovery of 1.75 billion barrels over the 2012-2040 period amounts to 438% of the 0.4 billion barrels of the EIA's "unproved technically recoverable resources as of January 1, 2012."[138]

---

[137] Production data from DrillingInfo, July 2014. Forecast from EIA, *Annual Energy Outlook 2014*, Unpublished tables from AEO 2014 provided by the EIA.
[138] EIA, *Assumptions to the Annual Energy Outlook 2014*, http://www.eia.gov/forecasts/aeo/assumptions/pdf/oilgas.pdf.

2.5.3 Key Characteristics of the Austin Chalk and Niobrara-Codell Plays

The Austin Chalk and Niobrara-Codell plays are together projected to account for 15.3% of the tight oil production in the EIA's reference case tight oil forecast[139] (the two other plays for which the EIA provided individual play projections, the Woodford and Monterey, contribute only 2.4%). The EIA suggests these plays will collectively produce 6.6 billion barrels between 2014 and 2040, which is more than four times as much oil as they produced since their discoveries more than 40 years ago. Production is projected to rise to a peak in 2031 at 890,000 barrels per day followed by a gradual decline through 2040, when these plays are forecast to still be producing 720,000 barrels per day, which is nearly five times current levels (current combined production is 147,000 barrels per day[140]). This is a very aggressive forecast considering their age and extensive drilling and production history, their relatively low quality wells, and their observed steep well- and field-declines.

Production growth in the Austin Chalk and Niobrara-Codell plays is largely a result of redevelopment of long established plays with better technology, including horizontal drilling and fracking, rather than the new discoveries represented by the Bakken and Eagle Ford. The Austin Chalk began production in the 1950s and the Niobrara-Codell in the 1970s. More than 46,000 wells have been drilled of which 18,800 are currently producing. As such they are not analogues to the Bakken and Eagle Ford, from which significant production is just twelve and six years old, respectively. The Bakken and Eagle Ford currently produce 62% of all U.S. tight oil (Figure 5), compared to 5.3% for the Austin Chalk and Niobrara-Codell. At least some of the oil produced from these so-called "tight oil" plays is conventional. Table 2-4 summarizes the long history of development of these plays and contrasts that with the expectations for them in EIA's tight oil forecast.

Play	Years Produced	Wells Drilled	Wells Producing	Production to Date (Bbbls)	EIA Recovery 2012-2040 (Bbbls)	EIA Unproved Resources as of January 1, 2012 (Bbbls)	EIA Production in 2040 (MMbbl/d)
Austin Chalk	60+	15308	4988	1.17	4.9	7.6	0.51
Niobrara-Codell	40+	30871	13888	0.36	1.8	0.4	0.20
Total		46179	18876	1.53	6.7	8.0	0.72

Table 2-4. Age, wells, production[141], EIA unproved technically recoverable resources[142] and EIA reference case forecast for the Austin Chalk and Niobrara-Codell plays.[143]
Numbers may not add due to rounding.

[139] EIA, *Annual Energy Outlook 2014*, Unpublished tables from AEO 2014 provided by the EIA.
[140] Data from Drillinginfo retrieved July 2014.
[141] Data from Drillinginfo retrieved July 2014.
[142] EIA, *Assumptions to the Annual Energy Outlook 2014*, http://www.eia.gov/forecasts/aeo/assumptions/pdf/oilgas.pdf.
[143] EIA, *Annual Energy Outlook 2014*, Unpublished tables from AEO 2014 provided by the EIA.

2.5.4   Austin Chalk and Niobrara-Codell Plays Analysis Summary

Several conclusions can be made from the foregoing analysis of the Austin Chalk and Niobrara-Codell plays:

1. Oil production in the Austin Chalk and Niobrara-Codell plays is largely a result of application of new technologies to old plays, rather than significant new discoveries such as represented by the Bakken and Eagle Ford. Despite the application of new technology oil production in the Austin Chalk is falling, and the Niobrara-Codell may have peaked.

2. Productivity of wells in the Austin Chalk and Niobrara-Codell plays is generally much lower on average than in the Bakken and Eagle Ford. Well costs may also be somewhat lower, although most new production utilizes horizontal drilling, and extensive infrastructure is in place, hence improving the economics of drilling despite the lower well productivity.

3. These plays exhibit steep well- and field-declines mandating continuous high levels of drilling and capital input to maintain production.

4. The EIA is projecting aggressive growth in production from these plays with a peak in 2031 followed by a gradual decline, and the recovery of more than four times as much oil by 2040 as they have produced since their discoveries more than 40 years ago. This forecast is extremely optimistic given the number of wells that would have to be drilled and the amount of capital required.

5. Although these plays were not reviewed on a detailed county-by-county basis, they are highly likely to exhibit "sweet spots" or "core areas" which are being targeted first, hence the number of wells and capital input will need to increase later in the EIA's forecast to moderate production decline.

2.6   ALL-PLAYS ANALYSIS

The foregoing analysis has reviewed—on a play-by-play basis—82% of the projected U.S. tight oil production in the EIA reference case forecast through 2040. Eighty percent of this projected production has a "high" or "very high" optimism bias, suggesting that actual production is likely to be far less than that projected by the EIA over the long term. Moreover, the analysis suggests that the Bakken and Eagle Ford plays will remain the foundation of the U.S. tight oil "shale revolution." The plays outside of the Bakken and Eagle Ford are mainly redevelopments of old plays, with tens of thousands of wells drilled over the preceding 40 to 60 years. Despite the EIA's assertion, for example, that Permian Basin plays such as the Spraberry, Wolfcamp, and Bone Spring "have initial well production rates comparable to those found in the Bakken and Eagle Ford shale formations"[144], this is belied by the actual data. Average initial oil well productivities of these plays are a half or less of the average initial production of a high quality county in the Bakken or Eagle Ford.

This section will further explore the outlook for overall U.S. tight oil production with a summary analysis of the plays' EIA forecasts, estimated ultimate recovery per well, associated natural gas production, and production prospects to 2040.

---

[144] EIA, "Six formations are responsible for surge in Permian Basin crude oil production," *Today in Energy*, July 9, 2014, http://www.eia.gov/todayinenergy/detail.cfm?id=17031.

2.6.1 Summary of EIA Forecasts

Table 2-5 summarizes the salient details of the EIA's tight oil production projections and estimates of "unproved technically recoverable resources" and "proved reserves"; it also includes historical production for context, and an "optimism bias" rating.

Play	EIA Projected Recovery 2012-2040 (Bbbls)	Production to Date (Bbbls)[145]	EIA Unproved Resources as of January 1, 2012 (Bbbls)	EIA Proved Reserves as of 2012 (Bbbls)	EIA Total Proved and Unproved Technically Recoverable (Bbbls)	Percent of Unproved Resources and Proved Reserves Recovered by 2040 in EIA Forecast	Play's Share of Total Recovery (%)	EIA Production in 2040 (MMbbl/d)	Optimism Bias
Bakken	8.4	1.16	9.2	3.12	12.32	68.3	19.3	0.45	High
Eagle Ford	10.7	0.90	9.3	3.37	12.67	84.8	24.6	0.59	High
Woodford	0.4	0.03	0.2	--	0.20	207.4	1.0	0.03	Very High
Austin Chalk	4.9	1.17	7.6	--	7.60	65.0	11.3	0.51	Very High
Spraberry	6.5	1.83	8.1	--	8.10	80.0	14.9	0.51	Very High
Niobrara	1.8	0.36	0.4	0.01	0.41	423.8	4.0	0.20	High
Avalon/Bone Spring	0.7	0.21	2.0	--	2.00	34.1	1.6	0.05	Low
Monterey	0.6	--	0.6	--	0.60	102.3	1.4	0.06	High
Wolfcamp	2.6	0.87	3.4	--	3.40	77.6	6.1	0.22	High
Other	6.9	1.50	18.4	0.65	19.05	36.3	15.8	0.58	Unknown
Total	43.6	8.03	59.2	7.15	66.35	65.7	100.0	3.20	High to Very High

Table 2-5. Summary of EIA reference case tight oil forecast and assumptions[146] and stated unproved technically recoverable resources[147] and proved reserves[148], with historical production and "optimism bias" rating.[149]

The "optimism bias" rating is based on the analysis in this report.

[145] "Other" category estimated: Delaware and Glorieta-Yeso plays have cumulative production of 1.02 Bbbls over the last 40+ years and 0.48 Bbbl is estimated for other plays, which include the Utica, Tuscaloosa Marine Shale, Albany and others including liquids produced from shale gas plays.
[146] EIA, *Annual Energy Outlook 2014*, Unpublished tables from AEO 2014 provided by the EIA.
[147] EIA, *Assumptions to the Annual Energy Outlook 2014*, http://www.eia.gov/forecasts/aeo/assumptions/pdf/oilgas.pdf.
[148] EIA, http://www.eia.gov/naturalgas/crudeoilreserves/index.cfm.
[149] Data from Drillinginfo retrieved May-July 2014.

2.6.2 Estimated Ultimate Recovery per Well

Average per-well estimated ultimate recovery (EUR) for each of the analyzed plays is illustrated in Figure 2-91. These EURs are offered for comparative purposes only; each play is treated the same, with the average well decline data used in the first three years followed by an exponential decline at a terminal decline rate (the jury is out on the actual long term oil recovery of tight oil wells). This comparison highlights that the Bakken's and Eagle Ford's per-well EURs are two to more than four times higher than that of the other plays. For all plays, high decline rates of tight oil wells mean that 43% to 64% of the EUR is recovered in the first three years.

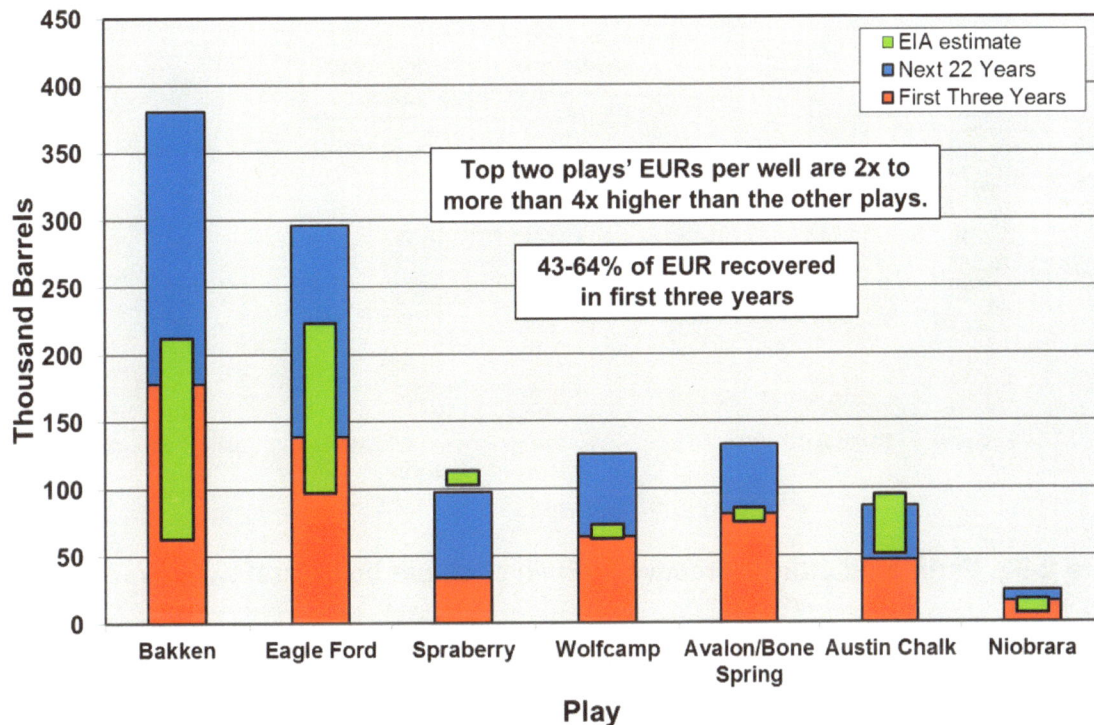

Figure 2-91. Estimated ultimate recovery (EUR) of oil per well of reviewed plays.[150]

Roughly half of the EUR is recovered in the first three years due to steep decline rates. These estimates of EUR per well are generally higher than those provided by the EIA[151] which are (in Kbbls): Bakken, 63-212; Eagle Ford, 97-223; Spraberry, 108; Wolfcamp, 68; Avalon/Bone Spring, 80; Austin Chalk, 51-95; Niobrara, 12.

---

[150] Based on data from Drillinginfo retrieved May-July 2014.
[151] EIA, *Assumptions to the Annual Energy Outlook 2014*, http://www.eia.gov/forecasts/aeo/assumptions/pdf/oilgas.pdf.

Horizontal wells generally improve the per-well EURs somewhat. Figure 2-92 illustrates the same comparison for horizontal wells only. Although looking at only horizontal wells markedly improves plays like the Niobrara-Codell, illustrating the difference that new technology is making, the Bakken's and Eagle Ford's EURs per well are still nearly double to triple the average well performance of the other plays.

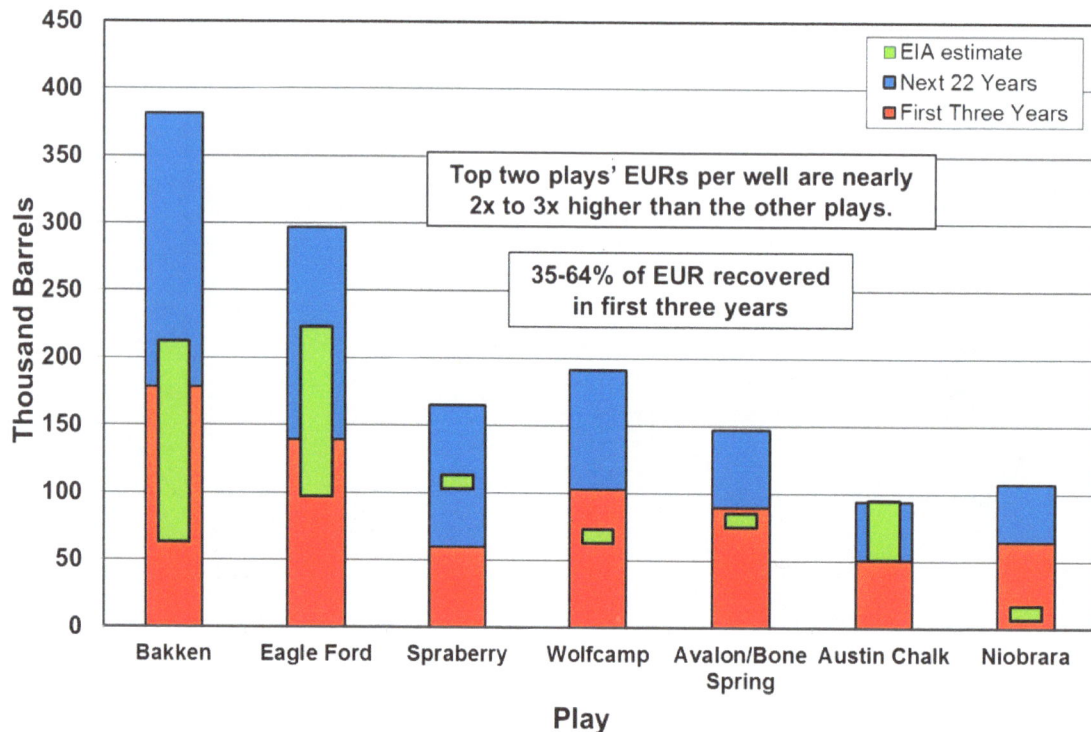

Figure 2-92. Estimated ultimate recovery (EUR) of oil per horizontal well for reviewed plays.[152]

Roughly half of the EUR is recovered in the first three years due to steep decline rates. These estimates of EUR per well are generally higher than those provided by the EIA[153] which are (in kbbls): Bakken, 63-212; Eagle Ford, 97-223; Spraberry, 108; Wolfcamp, 68; Avalon/Bone Spring, 80; Austin Chalk, 51-95; Niobrara, 12.

[152] Based on data from Drillinginfo retrieved May-July 2014.
[153] EIA, *Assumptions to the Annual Energy Outlook 2014*, http://www.eia.gov/forecasts/aeo/assumptions/pdf/oilgas.pdf.

2.6.3   Natural Gas Production Component

The natural gas production component of many of these plays is also an important contributor to energy production and economics (all these plays produce both oil and gas). Natural gas can be converted to its oil energy equivalent at a ratio of 6,000 cubic feet of gas to one barrel of oil. On a price basis, however, oil is far more valuable, so whereas 1,000 cubic feet of gas is equivalent to one sixth of a barrel of oil on an energy equivalent basis, it is only equivalent to one twentieth or less of the value of a barrel of oil at current prices. Figure 2-93 illustrates the EUR comparison between plays on a "barrels of oil equivalent" basis. The same pattern holds: the Bakken's and Eagle Ford's EURs per well are two to more than six times higher than the EURs per well of the other plays.

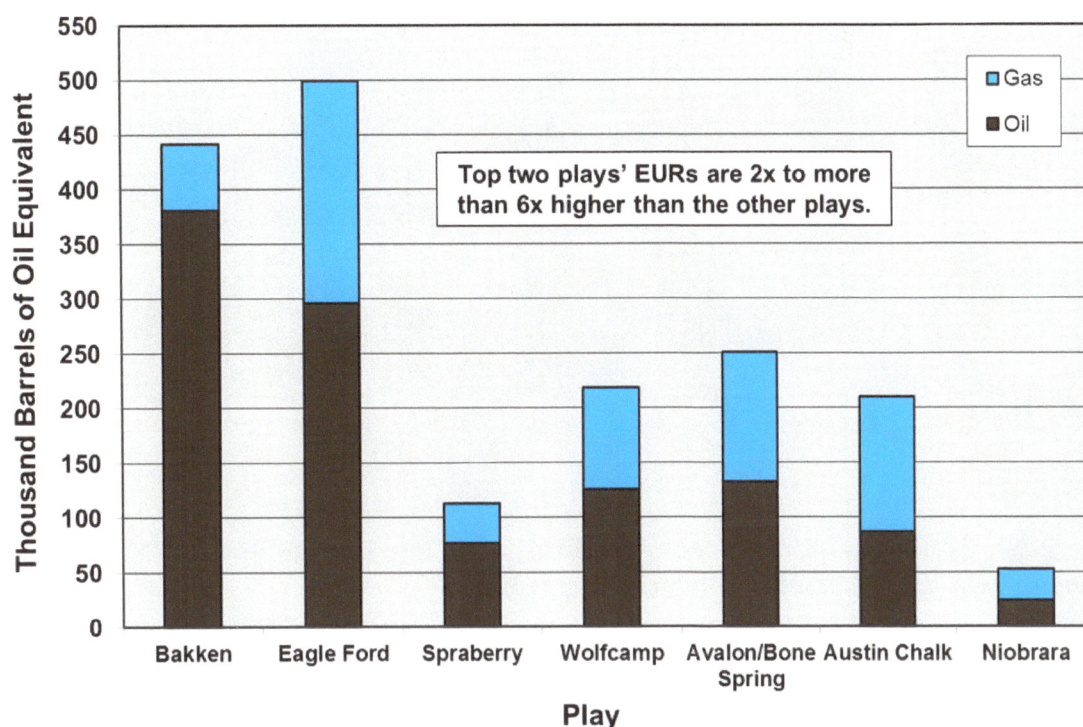

Figure 2-93. Estimated ultimate recovery (EUR) of oil and gas per well of reviewed plays, on a "barrels of oil equivalent" basis.[154]

The Bakken's and Eagle Ford's EURs per well are two to more than six times the EURs per well of the other five plays.

---

[154] Based on data from Drillinginfo retrieved May-July 2014.

Looking at horizontal wells only on an oil and gas EUR energy equivalency basis, production from some of these plays is considerably higher—and in plays like the Austin Chalk, Bone Spring, and Niobrara-Codell, natural gas is half or more of total energy production. Nonetheless, the Bakken's and Eagle Ford's EURs per well remain 39% to 141% higher than the other plays on an energy equivalency basis.

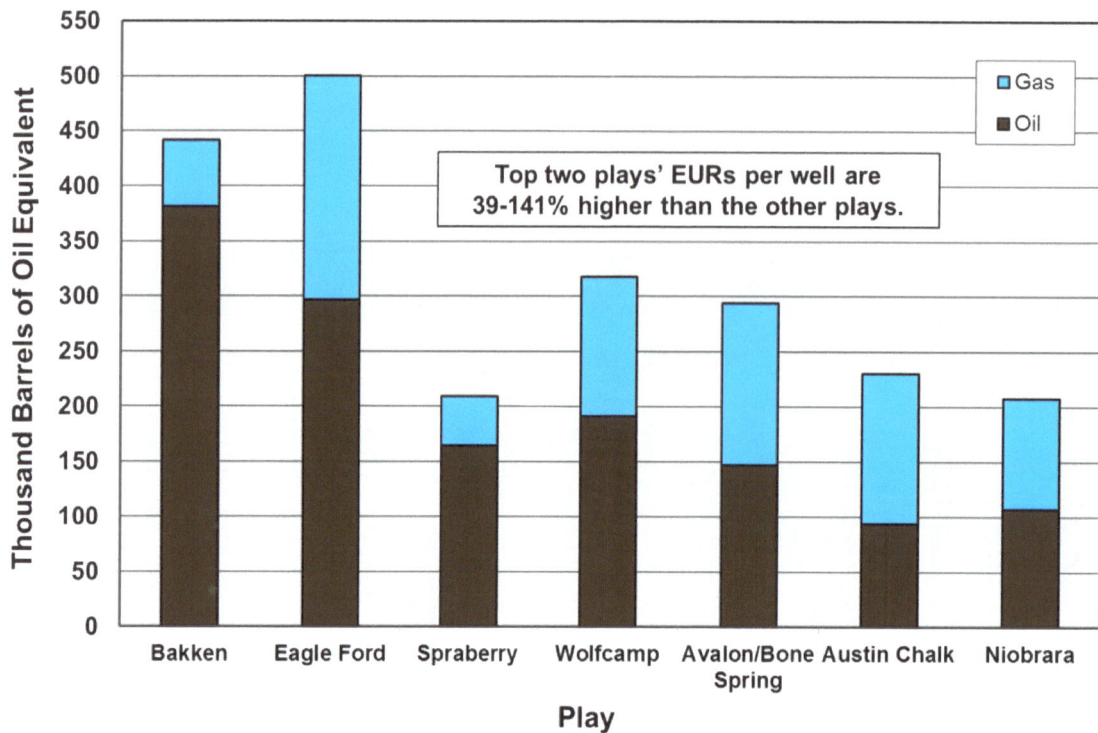

Figure 2-94. Estimated ultimate recovery (EUR) of oil and gas per horizontal well of reviewed plays, on a "barrels of oil equivalent" basis.[155]

The Bakken's and Eagle Ford's EURs per well are 34% to 141% higher than the other plays.

---

[155] Based on data from Drillinginfo retrieved May-July 2014.

2.6.4   Production Through 2040

This report provides tight oil production projections for the Bakken and Eagle Ford plays—which account for 62% of current production—and production history, well quality and other factors controlling future production for additional major plays which comprise a further 27% of tight oil production. The Bakken and Eagle Ford are particularly important as they are projected to account for over half of total production well into the next decade. This analysis reveals that more than two times the projected production from the Bakken and Eagle Ford will have to be produced from other plays to meet the EIA reference case forecast by 2040: a tall order which is unlikely to be realized given the fundamentals of these plays as outlined in this report.

Figure 2-95 compares the EIA's reference case projection through 2040 for tight oil production[156] to the most likely of the Bakken and Eagle Ford scenarios presented in sections 2.3.1.6 and 2.3.2.6, respectively (the "Most Likely Rate" scenarios of the "Realistic" cases of the respective plays).

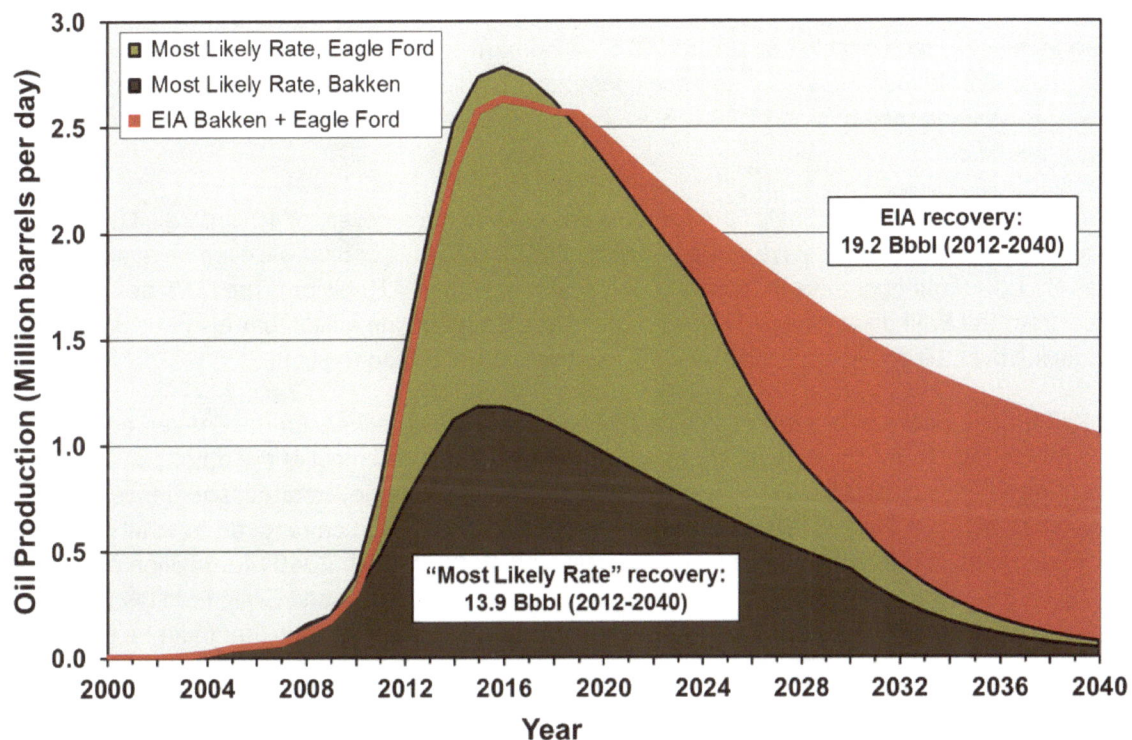

Figure 2-95. "Most Likely Rate" scenarios ("Realistic" cases) of Bakken and Eagle Ford tight oil production compared to the EIA reference case, 2000 to 2040.[157]

Total oil recovery forecast by the EIA from these plays is 19.2 billion barrels from 2012-2040 versus 13.7 billion barrels in this report.

---

[156] EIA, *Annual Energy Outlook 2014*, Unpublished tables from AEO 2014 provided by the EIA.
[157] EIA, *Annual Energy Outlook 2014*, Unpublished tables from AEO 2014 provided by the EIA.

This comparison reveals:

- The EIA's forecast of the timing of peak production in the Bakken and Eagle Ford is similar to this report.

- The EIA's forecast of the rate at peak production is lower than this report, but only slightly.

- The EIA projects a much higher tail after peak production, with recovery of 19.2 billion barrels between 2012 and 2040, as opposed to 13.9 billion barrels forecast in this report.

- The EIA forecasts collective production from these plays to be 1 million barrels per day in 2040, suggesting considerably more oil will be recovered after that date; in contrast, the "Most Likely" drilling rate scenario presented in this report forecasts that production will fall to about 73,000 barrels per day by 2040.

The EIA's reference case projections for the Bakken and Eagle Ford require the recovery of 19.2 billion barrels by 2040. This amounts to 77% of the sum of proved reserves (6.49 billion barrels)[158] and estimated "unproved technically recoverable resources" (18.5 billion barrels)[159] claimed for these two plays. Unproved technically recoverable resources have no price constraints applied and are loosely constrained by geological parameters; to assume the recovery of 77% of proved reserves plus unproved resources by 2040 is extremely optimistic.

Moreover, the EIA's Bakken and Eagle Ford forecast amounts to the recovery of 40% more oil than this report's analysis suggests those plays can produce by 2040 (assuming capital will even be available to drill more than 51,000 additional wells in these plays at a cost of some $410 billion). The EIA's assumption that production from the Bakken and Eagle Ford will still be at more than one million barrels per day in 2040, after producing over 19.6 billion barrels since 2000, strains credibility to the limit.

The large difference between this report's projections and the EIA's forecasts for the Bakken and Eagle Ford, coupled with the high to very high optimism bias in the EIA's forecast for most of the other plays analyzed, suggests that the EIA's total U.S. tight oil forecast is likely to be seriously overstated, and hence very difficult or impossible to achieve. Figure 2-96 illustrates the production that would be required from all other tight oil plays to meet the EIA's reference case tight oil forecast from 2012 through 2040 (43.6 billion barrels), after accounting for this report's "Most Likely" scenario forecasts for the Bakken and Eagle Ford (which are 5.3 billion barrels less than the EIA's through 2040). The result is 29.7 billion barrels that must be made up from other tight oil plays, or two times the projected recovery from the Bakken and Eagle Ford by 2040 (13.9 billion barrels), over this period.

---

[158] EIA, "U.S. Crude Oil and Natural Gas Proved Reserves," April 2014, http://www.eia.gov/naturalgas/crudeoilreserves/index.cfm.
[159] EIA, *Assumptions to the Annual Energy Outlook 2014*, http://www.eia.gov/forecasts/aeo/assumptions/pdf/oilgas.pdf.

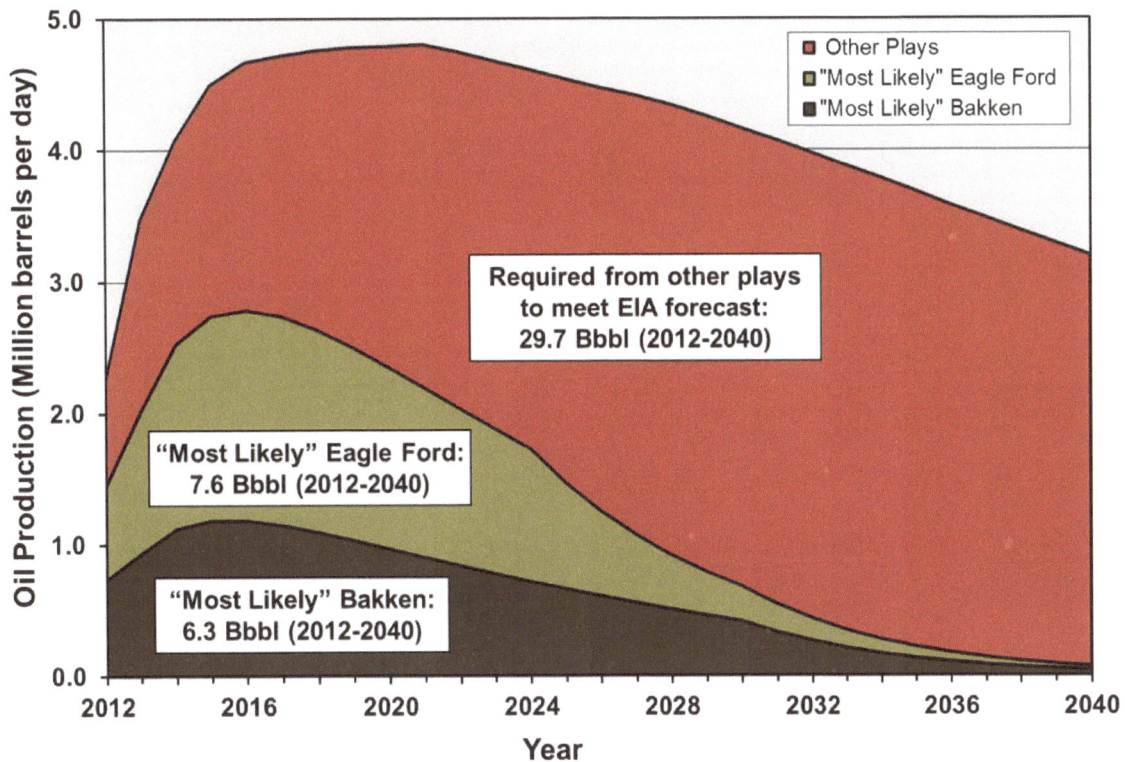

Figure 2-96. "Most Likely" scenario projections of oil production for the Bakken and Eagle Ford plays[160] with the remaining amount of production that would be required from other plays to meet the EIA's total reference case forecast.[161]

The EIA forecasts 43.6 billion barrels of U.S. tight oil will be recovered from 2012 to 2040. After subtracting the 13.9 billion barrels projected by this report for the Bakken and Eagle Ford, 29.7 billion barrels would remain to be produced from all other tight oil plays—5.3 billion barrels more than the EIA's already optimistic forecast for these plays.

---

[160] Data from Drillinginfo retrieved May 2014.
[161] EIA, *Annual Energy Outlook 2014*, Unpublished tables from AEO 2014 provided by the EIA.

2.7 SUMMARY AND IMPLICATIONS

The growth of U.S. tight oil production is one of the few bright spots contributing to global oil production growth. Geopolitical turmoil in parts of the Middle East and northern and western Africa, coupled with production declines in other major producers such as Russia[162], has kept oil prices persistently near historic highs. Investments by oil majors in upstream oil and gas production have increased three-fold since 2000 yet production is up just 14%.[163] Economist Mark Lewis points out that "the damage has been masked so far as big oil companies draw down on their cheap legacy reserves", but that "they are having to look for oil in the deepwater fields off Africa and Brazil, or in the Arctic, where it is much more difficult. The marginal cost for many shale plays is now $85 to $90 a barrel."[164]

Given these factors it is important to understand the long term supply limitations of U.S. tight oil. The analysis presented herein, which is based on one of the best commercial databases of well production information available[165], finds that the longevity of U.S. tight oil production at meaningful rates is highly questionable. Certainly production will rise in the short term, but with the very likely peaking of the Bakken and Eagle Ford plays (which provide 62% of current U.S. tight oil output) in the 2016-2017 timeframe, maintaining production or even stemming the decline will require ever greater amounts of drilling, along with the capital input to sustain it. This will require higher prices, for the nature of shale plays is that the sweet spots get drilled first and progressively lower quality rock gets drilled last.

Furthermore, much of the purported "tight oil" production outside of the Bakken and Eagle Ford comes from long-established plays benefiting from the application of new technology, not new discoveries. Tens of thousands of wells have been drilled in these plays over the past 40 or more years and they have produced much oil and gas, yet the EIA forecast expects them to produce 4-5 times their historical production in the next 26 years. These plays have well qualities as defined by initial productivity and EUR of less than half of the Bakken and Eagle Ford on average. The concept that high quality tight oil plays like the Bakken and Eagle Ford are widespread is false.

The EIA, which is viewed as perhaps the most authoritative source of U.S. energy production forecasts, has consistently overestimated future production.[166] The analysis presented herein suggests that this is the case with respect to tight oil. A play-by-play analysis of the data with respect to the EIA forecasts reveals a high to very high "optimism bias". The EIA assumes that 65% to 85% of its "proved reserves and unproved technically recoverable resources as of January 1, 2012" will be recovered by 2040 for most plays. Unproved resources have no price constraints applied and are loosely constrained compared to "reserves" which are proven to be recoverable with existing technology and economic conditions. Not only do the EIA's projections demonstrate a high or very high optimism bias, they also assume that the U.S. will exit 2040 with tight oil production comparable to today, at 3.2 MMbbl/d. This is highly unlikely given a thorough analysis of the data.

The Bakken and the Eagle Ford have produced just under 2 billion barrels of oil to date and will continue to produce much more oil, assuming drilling rates and the capital input to sustain them will be maintained. This report projects that they will produce 13.9 billion barrels from 2012 to 2040, with marginal production under

---

[162] Reuters, "UPDATE 1-Russian oil output down for fourth month in a row," May 2, 2014, http://uk.reuters.com/article/2014/05/02/russia-energy-production-idUKL6N0NO0UL20140502.

[163] Mark Lewis of Kepler Cheuvreux cited in Ambrose Evans-Pritchard, "Fossil industry is the subprime danger of this cycle", Telegraph, July 9, 2014, http://www.telegraph.co.uk/finance/comment/ambroseevans_pritchard/10957292/Fossil-industry-is-the-subprime-danger-of-this-cycle.html.

[164] Ambrose Evans-Pritchard, "Fossil industry is the subprime danger of this cycle", Telegraph, July 9, 2014.

[165] DI Desktop (formerly HDPI), produced by Drillinginfo.

[166] See Figure 25 in J. David Hughes, *Drill Baby Drill: Can Unconventional Fuels Usher in a New Era of Energy Abundance?*, Post Carbon Institute, 2013, http://www.postcarbon.org/publications/drill-baby-drill.

0.08 MMbbl/d in 2040, given unconstrained capital input. In contrast, the EIA forecasts 19.2 billion barrels of cumulative production from these plays over the same period, with production of just over 1 MMbbl/d in 2040. Figure 2-97 illustrates the stark difference between the EIA's projections and this report's projections of Bakken and Eagle Ford tight oil production.

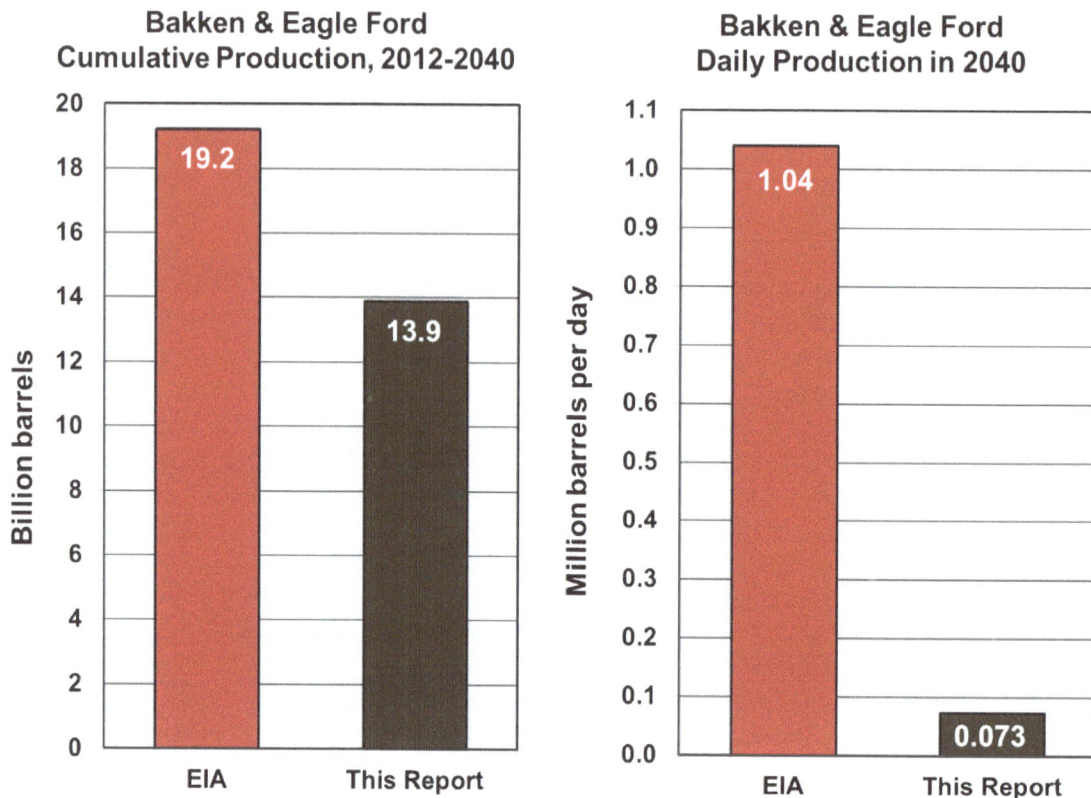

Figure 2-97. Bakken and Eagle Ford plays projected cumulative oil production from 2012 to 2040 and daily oil production in 2040, EIA projection[167] versus this report's projection.

The EIA's forecast strains credibility, given the known decline rates, well quality by area, available drilling locations, and the number of wells that would need to be drilled to make that happen. Given this report's "Most Likely" scenario estimate for the Bakken and Eagle Ford based on the analysis in this report, the remaining significant U.S. tight oil plays would need to produce 29.7 billion barrels of oil between 2012 and 2040 to meet the EIA's forecast—more than twice as much as the Bakken and Eagle Ford combined (see Figure 2-96). However, the EIA projects that these plays will produce just 23.5 billion barrels between 2014 and 2040. A more realistic best-case estimate, assuming capital inputs are not a constraint, is for these plays to produce about ten billion barrels over this period, which, coupled with 12.7 billion barrels from the Eagle Ford and Bakken, is just over half of the EIA's forecast by 2040—if everything goes right. Producing this much oil from these plays will require much higher oil prices than today's in the latter part of the 2014-2040 period. Most troubling from an energy security point of view is that much of the tight oil production will occur in the early years of this period, making supply ever more problematic later on.

---

[167] EIA, *Annual Energy Outlook 2014*, http://www.eia.gov/forecasts/aeo.

The consequences of getting it wrong on future tight oil production are immense. The EIA projects that the U.S. will be a significant oil importer in 2040 (Figure 2-2). Although the flush of tight oil production is likely to peak before 2020 and decline thereafter at much more rapid rates than projected by the EIA, there is increasing pressure by industry to allow crude oil exports.[168] The longer term geopolitical complications certain to arise given increased competition for available oil exports in a shrinking export market should be obvious. Rather than viewing tight oil as an unlimited bounty, it should be viewed for what it is—a short term reprieve from the inexorable decline in U.S. oil production. A sensible energy policy would be based on this prospect.

---

[168] IHS, "U.S. Crude Oil Export Decision," *Crude Oil Export Report*, 2014, http://www.ihs.com/info/0514/crude-oil.aspx.

PART 3: SHALE GAS

PART 3: SHALE GAS - CONTENTS

PART 3: SHALE GAS -FIGURES

PART 3: SHALE GAS -TABLES

3.1 INTRODUCTION

3.1.1 Overview

The widespread adoption of hydraulic fracturing ("fracking") and horizontal drilling in the United States to extract oil and natural gas from previously inaccessible shale formations has been termed the "shale revolution." U.S. natural gas production, thought to be in terminal decline as recently as 2005, has exceeded its all-time 1973 peak. The U.S. Energy Information Administration (EIA) now projects domestic gas production to reach nearly 38 trillion cubic feet per year by 2040, which is 55% above 2013 levels.

Although the U.S. is still a net importer of gas from Canada, there is now a rush to export natural gas overseas. Four liquefied natural gas (LNG) export terminals have been approved—one of which is under construction at Sabine Pass in Louisiana—with a further 13 "proposed" and an additional 13 under consideration as "potential".[1] The enthusiasm for LNG exports is based on the assumption that the North American gas supply will continue to grow for the foreseeable future and prices will remain low, resulting in an attractive differential with much higher gas prices in Europe and Asia.

The environmental, health, and quality of life impacts of shale development have stoked controversy across the country. In contrast, the expectation of long-term domestic natural gas abundance—driven by optimistic forecasts from industry and government—has been widely reported and little questioned, despite the myriad economic and policy consequences. There is no question that the development of shale gas has created a surge in production. However, a look at the fundamentals of shale plays reveals that they come with serious drawbacks, both in terms of environmental impact and the sustainability of long term production.

This report investigates whether the EIA's expectation of long-term domestic gas abundance is founded. It aims to gauge the likely future production of U.S. shale gas, based on an in-depth assessment of actual well production data from the major shale plays. It determines future production profiles given assumed rates of drilling, average well quality by area, well- and field-decline rates, and the estimated number of available drilling locations. This analysis is based on all drilling and production data available through early- to mid-2014.

The analysis shows that maintaining U.S. shale gas production, let alone increasing production at rates forecast by the EIA through 2040, will be problematic. Four of the top seven shale gas plays are already in decline. Of the major plays, only the Marcellus, along with associated gas from the Eagle Ford and Bakken tight oil plays, are increasing—and yet, the EIA reference gas forecast calls for plays currently in decline to grow to new production highs, at moderate future prices. Lesser plays like the Utica and others are also counted on for strong growth. Although significantly higher gas prices needed to justify higher drilling rates could temporarily reverse decline in some of these plays, the EIA forecast is unlikely to be realized.

The analysis also underscores the amount of drilling, the amount of capital investment, and the associated scale of environmental and community impacts that will be required to meet these projections. These findings call into question plans for LNG exports and highlight the real risks to long-term U.S. energy security.

---

[1] FERC, July 18, 2014, "LNG," http://www.ferc.gov/industries/gas/indus-act/lng.asp.

3.1.2   Methodology

This report analyzes the top five U.S. shale gas plays—the Barnett, Haynesville, Fayetteville, Woodford and Marcellus—as well as associated gas production from the top two tight oil plays, the Bakken and Eagle Ford. Together these plays make up 88% of shale gas production through 2040 in the EIA's 2014 Annual Energy Outlook (AEO 2014).

The primary source of data for this analysis is Drillinginfo, a commercial database of well production data widely used by industry and government, including the EIA.[2] Drillinginfo also provides a variety of analytical tools which proved essential for the analysis.

A detailed analysis of well production data for the major shale gas plays reveals several fundamental characteristics that will determine future production levels:

1. **Rate of well production decline:** Shale gas plays have high well production decline rates, typically in the range of 75-85% in the first three years.

2. **Rate of field production decline:** Shale gas plays have high field production declines, typically in the range of 30-45% per year, which must be replaced with more drilling to maintain production levels.

3. **Average well quality:** All shale gas plays invariably have "core" areas or "sweet spots", where individual well production is highest and hence the economics are best. Sweet spots are targeted and drilled off early in a play's lifecycle, leaving lesser quality rock to be drilled as the play matures (requiring higher gas prices to be economic); thus the number of wells required to offset field decline inevitably increases with time. Although technological innovations including longer horizontal laterals, more fracturing stages, more effective additives and higher-volume frack treatments have increased well productivity in the early stages of the development of all plays, they have provided diminishing returns over time, and cannot compensate for poor quality reservoir rock.

4. **Number of potential wells:** Plays are limited in area and therefore have a finite number of locations to be drilled. Once the locations run out, production goes into terminal decline.

5. **Rate of drilling:** The rate of production is directly correlated with the rate of drilling, which is determined by the level of capital investment.

The basic methodology used is as follows:

- Historical production, number of currently producing wells and total wells drilled, the split between horizontal and vertical/directional wells, and the overall play area were determined for all plays. Average well decline for wells, both horizontal and vertical/directional, and the average estimated ultimate recovery (EUR), were also assessed for all plays. These parameters were assessed at both the play level and at the county level (the top counties in terms of the number of producing wells were analyzed individually, whereas counties with few wells were aggregated).

- Field decline rates and the number of available drilling locations were determined at the county- and play-level for all plays.

---

[2] See http://info.drillinginfo.com.

- First-year average production was established from type decline curves (i.e., average well decline profiles) constructed for all wells drilled in the year in question; 2013 was the year used as representative of future average first-year production levels per well. Average first-year production is used to determine the number of wells needed to offset field decline each year, and to determine the production trajectory over time given various drilling rates. In determining future production rates, the current trends in well productivity over time were considered; for example if recent well quality trends were increasing, it was assumed for plays in early stages of development that well quality would increase somewhat in the future before declining as drilling moves into lower quality outlying portions of plays.

- Projections of future production profiles were made for all plays based on various drilling rate scenarios. These projections assume a gradation over time from the well quality observed in the current top counties of a play to the well quality observed in the outlying counties as available drilling locations are used up. The different drilling rate scenarios were prepared so that the effect of a high drilling rate, presumably due to favorable economic conditions, compared to a low or a "Most Likely" drilling rate, could be assessed, both in terms of production over time and cumulative gas recovery from the play by 2040.

- Production projections and the production history and cumulative production for all plays were then compared to the EIA forecasts to assess the likelihood that these forecasts could be met.

- All plays were then compared to each other in terms of well quality and other parameters and an overall assessment of the likely long-term sustainability of shale gas production was determined.

Although public pushback against hydraulic fracturing ("fracking") due to health and environmental concerns has limited access to drilling locations in states like New York and Maryland and several municipalities, as well as triggered lawsuits, this report assumes there will be no restrictions to access due to environmental concerns. It also assumes there will be no restrictions on access to the capital required to meet the various drilling rate scenarios. In these respects, it presents a "best case," as any restrictions on access to drilling locations or to the capital needed to drill wells would reduce forecast production levels.

3.2 THE CONTEXT OF U.S. GAS PRODUCTION

3.2.1 U.S. Gas Production Forecasts

The EIA's Annual Energy Outlook 2014 provides various scenarios of future U.S. gas production, as well as price projections and stated assumptions in terms of available technically recoverable reserves and resources, play areas, well productivity, and so forth.

Figure 3-1 illustrates the range of the EIA's gas production forecasts through 2040 compared to historical production. These scenarios project U.S. gas production to rise anywhere from 37% to 71% above 2013 levels by 2040 and recover between 856 and 971 trillion cubic feet of gas over the 2013-2040 period. This amounts to 2.5-2.9 times the proved reserves that existed as of 2012[3] (proved reserves are generally considered to be economically recoverable with current technology). Adding in unproved resources, which are uncertain estimates without price constraints, between 37% and 42% of remaining potentially recoverable gas in the U.S. will be consumed over the next 26 years according to the EIA projections. This amounts to the equivalent of 85% to 99% of all the gas produced over the 54 years between 1960 and 2013.

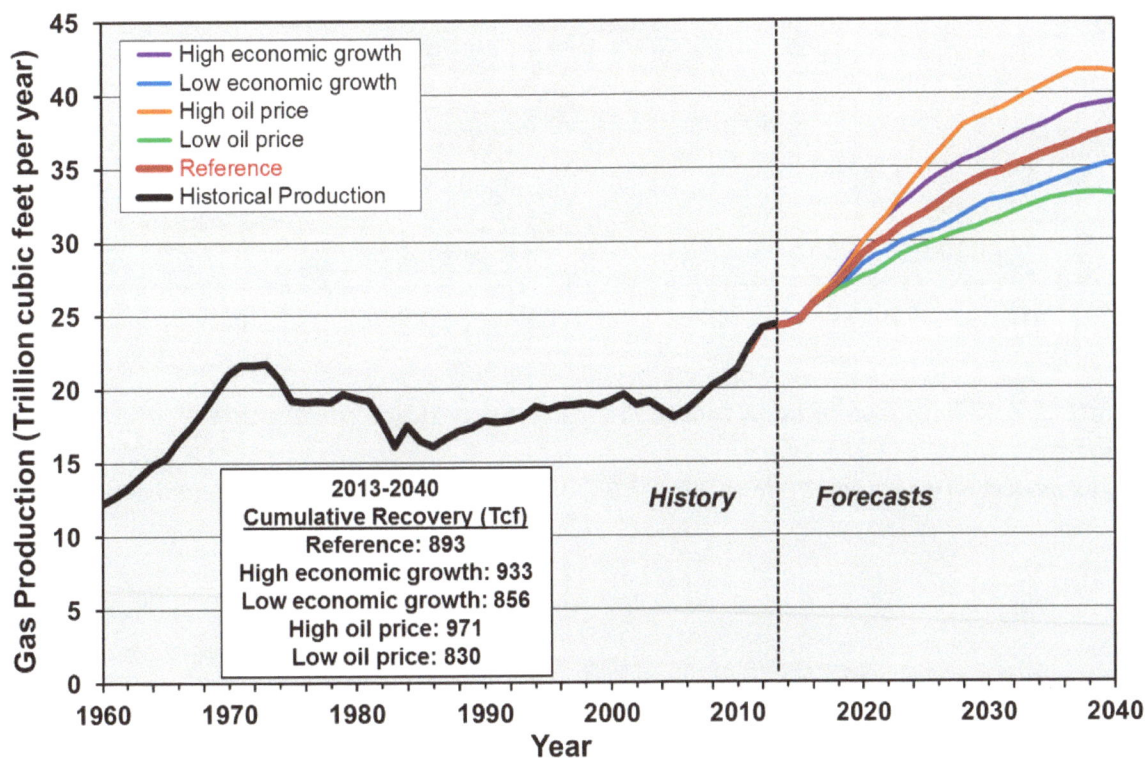

Figure 3-1. Scenarios of U.S. gas production through 2040 from the EIA's Annual Energy Outlook 2014[4] compared to historical production from 1960.

---

[3] EIA, *Assumptions to the Annual Energy Outlook 2014*, http://www.eia.gov/forecasts/aeo/assumptions/pdf/oilgas.pdf.
[4] EIA, *Annual Energy Outlook 2014*, http://www.eia.gov/forecasts/aeo.

The source of this optimism in future gas production is the application of high-volume, multi-stage, hydraulic fracturing technology ("fracking") in horizontal wells, which has unlocked previously inaccessible gas trapped in highly impermeable shales. Figure 3-2 illustrates the EIA's reference case gas production projection by source through 2040. Although conventional production is forecast to be flat or grow only slightly over the period, shale gas is forecast to more than double from 2013 levels and be 53% of a much expanded supply by 2040. Gas prices in this reference case are forecast to remain below $5 per million Btu (MMBtu) (2012 dollars) through 2024 and $6/MMBtu through 2030. Some 15% of production is forecast to be available for LNG and other exports in 2040, and net imports from Canada will cease by 2018.

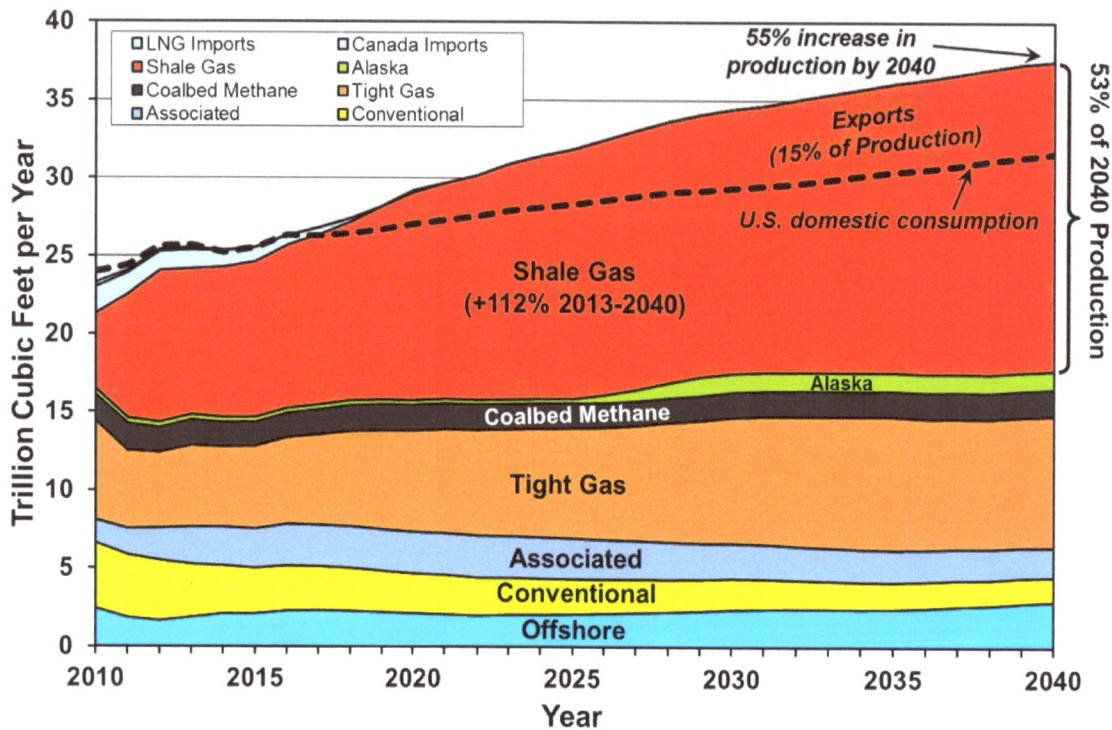

Figure 3-2. EIA reference case forecast of U.S. natural gas production by source through 2040.[5]

Overall production increases 55% from 2013 to 2040, whereas shale gas increases 112% over the same period.

---

[5] EIA, *Annual Energy Outlook 2014*, http://www.eia.gov/forecasts/aeo.

Figure 3-3 illustrates EIA forecasts for shale gas production in several cases. These assume the extraction of between 66% and 79% of the EIA's estimated 611 trillion cubic feet of proved shale gas reserves and unproved resources by 2040[6] (unproved resources have no implied price required for extraction and are highly uncertain compared to proved reserves which are recoverable with current technology under current economic conditions).

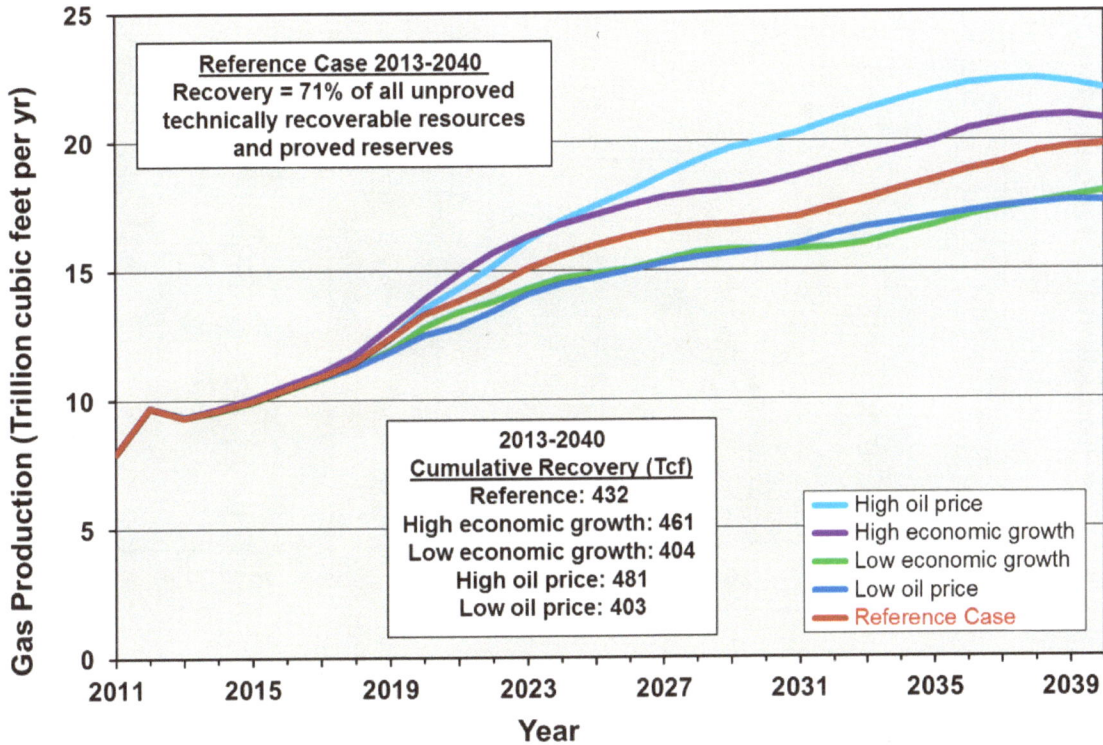

Figure 3-3. EIA scenarios of U.S. shale gas production through 2040.[7]

Unproved technically recoverable resources are estimated by the EIA at 489 trillion cubic feet and proved reserves at 122 trillion cubic feet[8], so these scenarios amount to the recovery of 66% to 79% of all proved reserves and unproved resources by 2040.

---

[6] EIA, *Annual Energy Outlook 2014*, http://www.eia.gov/forecasts/aeo.

[7] EIA, *Annual Energy Outlook 2014*, http://www.eia.gov/forecasts/aeo.

[8] EIA, *Assumptions to the Annual Energy Outlook 2014*, http://www.eia.gov/forecasts/aeo/assumptions/pdf/oilgas.pdf.

Figure 3-4 illustrates how the EIA reference case projections for shale gas production are divided between plays.

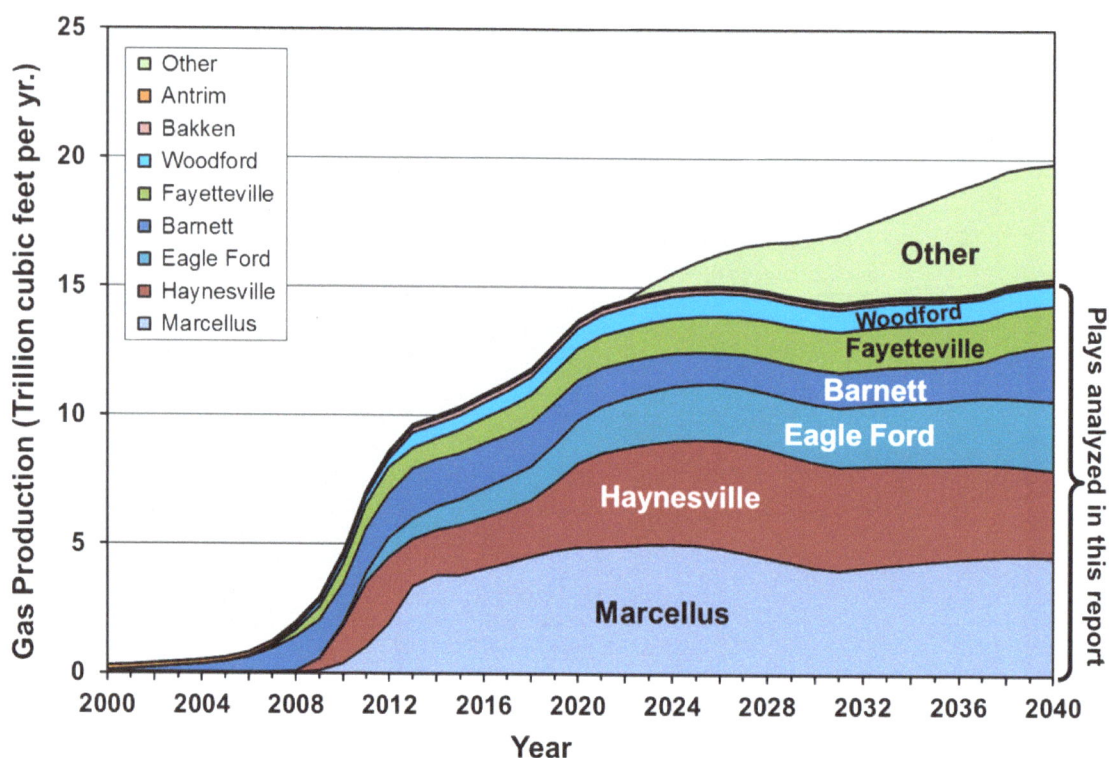

Figure 3-4. EIA reference case forecast of shale gas production divided by play through 2040.[9]

This report analyzed the seven most productive plays, which account for 88% of EIA's reference case shale gas production forecast to 2040.

The EIA reference case clearly expects the seven shale gas plays analyzed in this report to provide the bulk of production through 2040, with "other" plays increasing significantly after 2020. Shale gas production in all these plays has risen quickly due to rapid increases in drilling rates and sustained high levels of capital input; however, four of them are now in decline. High well- and field-decline rates, coupled with a finite number of drilling locations, suggest that production will be problematic to sustain, let alone grow at these forecast rates. Section 3 of this report explores the realistic production potential for these plays in depth.

[9] EIA, *Annual Energy Outlook 2014*, unpublished tables from AEO 2014 provided by the EIA.

3.2.2   Current U.S. Shale Gas Production

Production of shale gas began in the Barnett play of eastern Texas in the late 1990s and early 2000s. With the widespread application of horizontal drilling and hydraulic fracturing ("fracking") beginning in 2003, production grew rapidly. The Haynesville play of Louisiana and east Texas was unknown as recently at 2007, and became the largest shale play in the U.S. at its peak in late 2011—although production has subsequently declined by 46%. The distribution of shale plays in the U.S. lower 48 states is illustrated in Figure 3-5.

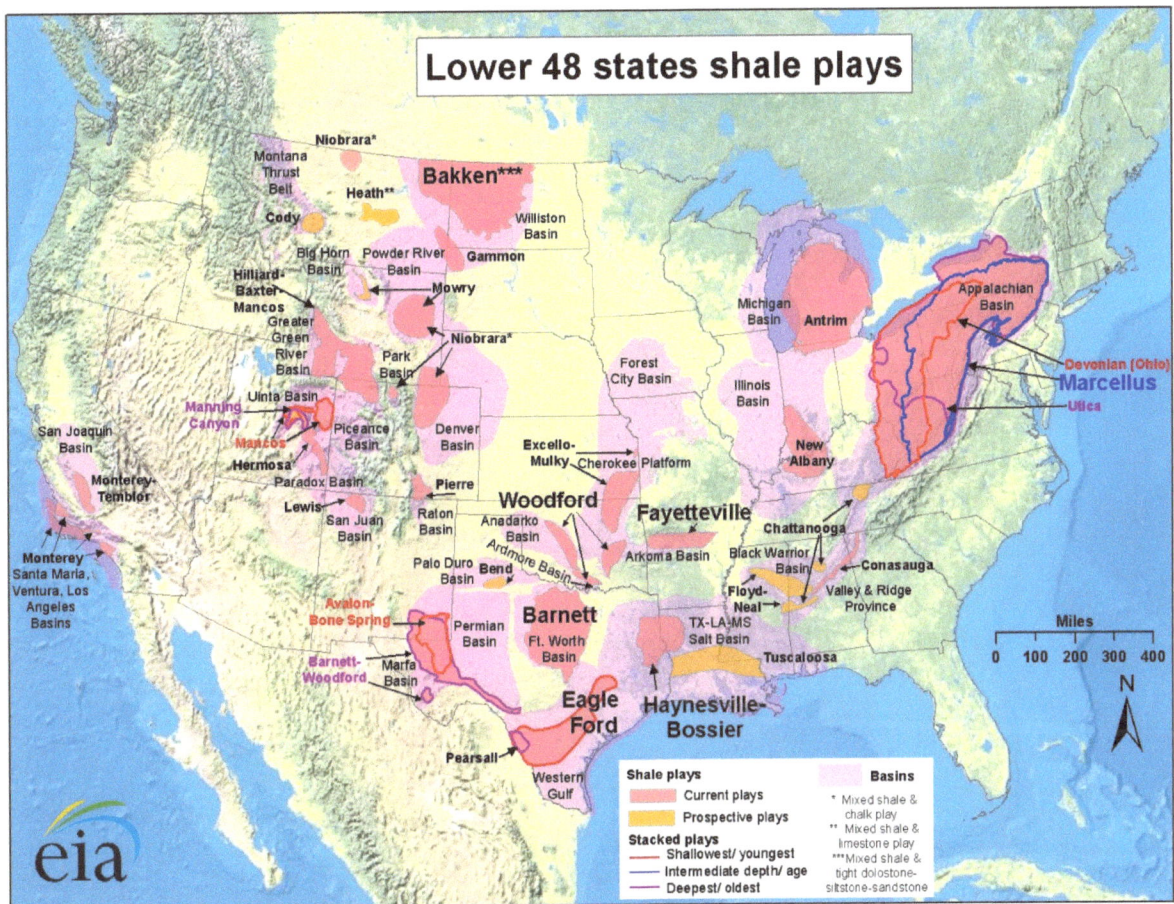

Figure 3-5. Distribution of lower 48 states shale gas and oil plays.[10]

---

[10] EIA, "Shale Gas and Oil Plays, Lower 48 States," http://www.eia.gov/pub/oil_gas/natural_gas/analysis_publications/maps/maps.htm.

Current production from U.S. shale gas plays is estimated by the EIA at 37 billion cubic feet per day. Despite the apparent widespread nature of shale plays in Figure 3-5, nearly half of this production comes from just two plays—the Barnett and the Marcellus—and 78% comes from just five plays. Figure 3-6 illustrates shale gas production by play from 2000 through August 2014 according to the EIA.

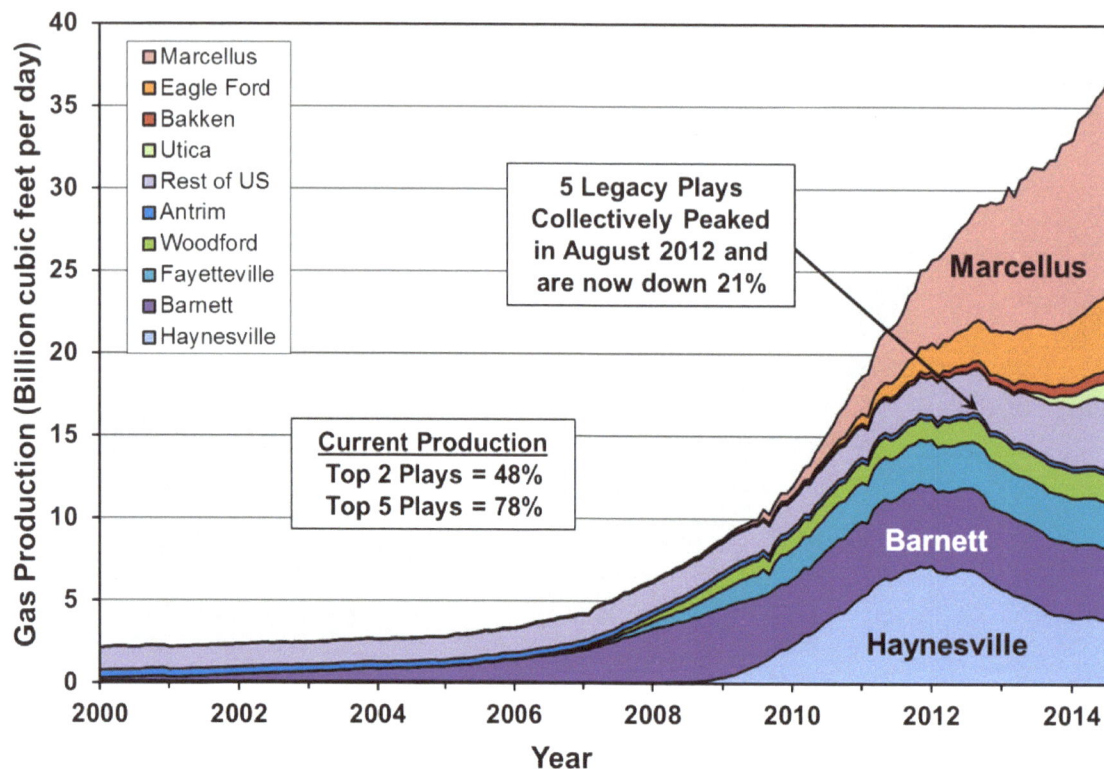

Figure 3-6. U.S. shale gas production by play from 2000 through July 2014, according to the EIA.[11]

---

[11] EIA estimates obtained in October 2014 from http://www.eia.gov/naturalgas/weekly.

3.3 MAJOR U.S. SHALE GAS PLAYS

3.3.1 Barnett Play

The EIA forecasts recovery of 53 Tcf of gas from the Barnett play by 2040. The analysis of actual production data presented below suggests that this forecast is unlikely to be realized.

The Barnett play is where shale gas production got its start in the late 1990s and the combination of horizontal drilling with multi-stage hydraulic fracturing ("fracking") was first applied at scale. Shale fracking was commercialized here by Mitchell Energy, a company headed by the late George Mitchell, "the father of fracking."[12] Figure 3-7 illustrates the distribution of wells as of early 2014. Over 19,600 wells have been drilled to date of which 15,906 were producing at the time of writing. The play covers parts of 24 counties although most of the drilling is concentrated in five counties in east Texas surrounding the city of Dallas/Fort Worth.

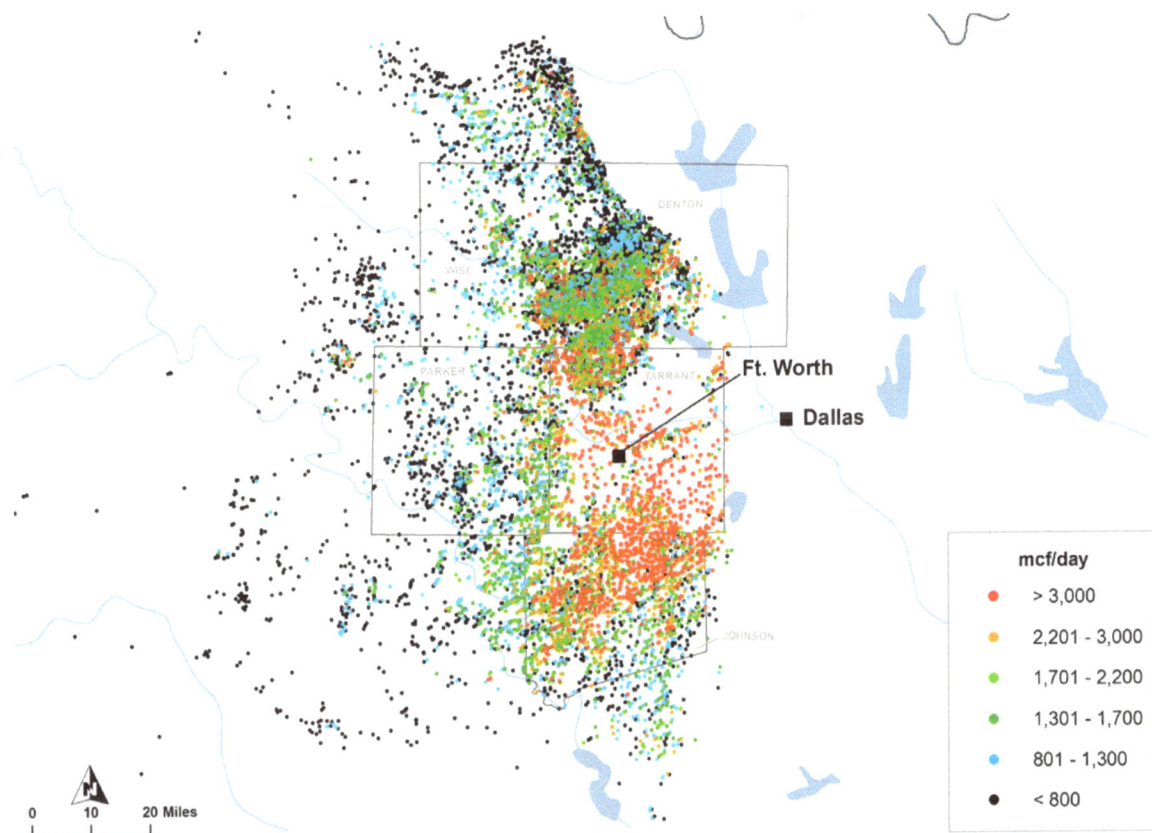

Figure 3-7. Distribution of wells in the Barnett play as of early 2014, illustrating highest one-month gas production (initial productivity, IP).[13]

Well IPs are categorized approximately by percentile; see Appendix.

---

[12] *The Economist*, August 3, 2013, "The father of fracking," http://www.economist.com/news/business/21582482-few-businesspeople-have-done-much-change-world-george-mitchell-father.
[13] Data from Drillinginfo retrieved August 2014.

Production in the Barnett peaked at nearly six billion cubic feet per day in December 2011 as illustrated in Figure 3-8. Ninety-four percent of current production is from horizontal fracked wells. The rate of drilling grew from about 500 (mainly vertical) wells per year in 2002 to a peak of over 2,800 (mainly horizontal) wells per year in 2008. It has since fallen to about 400 wells per year which is insufficient to offset field decline. Drilling rates required to keep production flat at current production levels are about 1,161 wells per year.

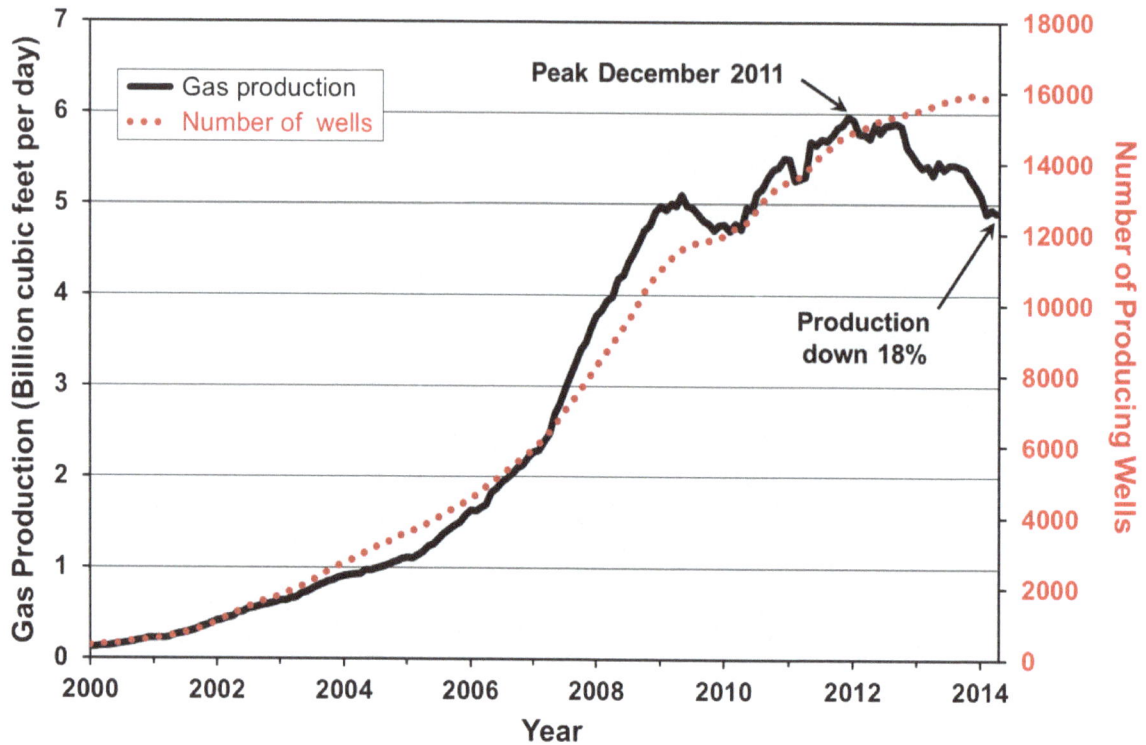

Figure 3-8. Barnett play shale gas production and number of producing wells, 2000 to 2014.[14]

Gas production data are provided on a "raw gas" basis.

[14] Data from Drillinginfo retrieved August 2014. Three-month trailing moving average.

Vertical wells played a significant role in the early development of the Barnett play and still produce some oil and gas, although new wells are predominantly horizontal. The evolution of the Barnett began in Denton and adjacent counties with vertical and directional wells before moving to horizontal wells as the limits of the play were defined, as illustrated in Figure 3-9.

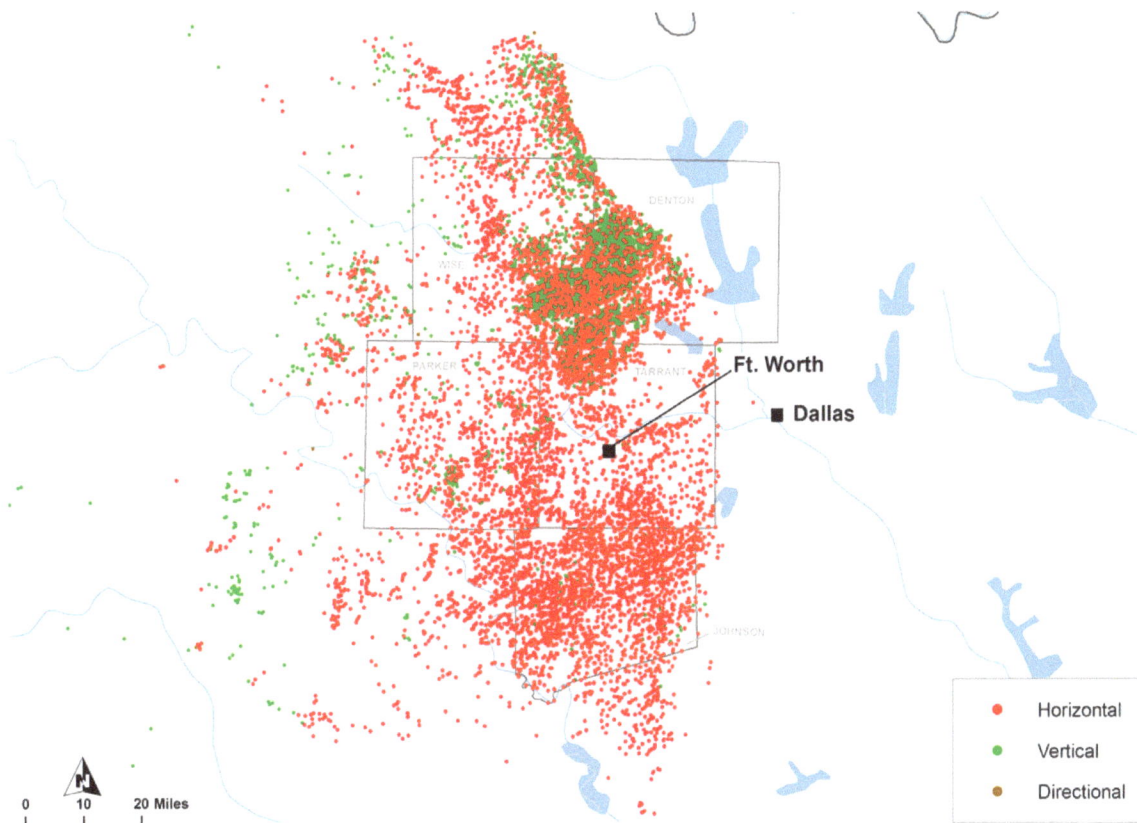

Figure 3-9. Distribution of gas wells in Barnett play categorized by drilling type, as of early 2014.[15]

Development began with vertical and directional wells in Denton County before expanding to largely horizontal drilling as the play's limits were defined.

---

[15] Data from Drillinginfo retrieved August 2014.

Production by well type is illustrated in Figure 3-10. There were still 3,366 producing vertical or directional wells, or 21% of the 15,906 producing wells in the play at the time of writing—yet these now produce less than 6% of total gas output. Very few vertical/directional wells are being drilled today; the future of the play lies in horizontal fracked wells. The dramatic growth in production from horizontal wells is noted in Figure 3-10.

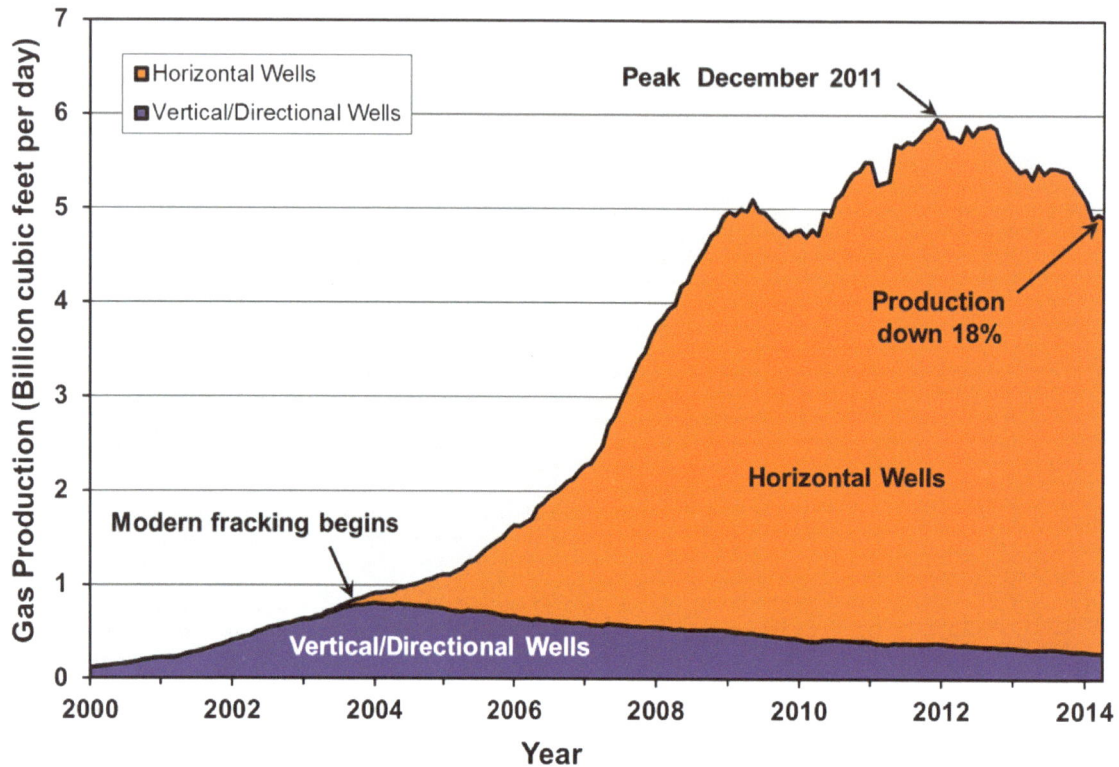

Figure 3-10. Gas production from the Barnett play by well type, 2000 to 2014.[16]
Fracking of horizontal wells at scale got underway in the Barnett in 2003.

---

[16] Data from Drillinginfo retrieved August 2014. Three-month trailing moving average.

3.3.1.1 Well Decline

The first key fundamental in determining the life cycle of Barnett production is the *well decline rate*. Barnett wells exhibit high decline rates in common with all shale plays. Figure 3-11 illustrates the average decline rate of Barnett horizontal and vertical/directional wells. Decline rates are steepest in the first year and are progressively less in the second and subsequent years. The decline rate over the first three years of average well life is 75%, which is considerably higher than most conventional wells. As can be seen, vertical/directional wells have much lower productivity than horizontal wells and hence are being phased out.

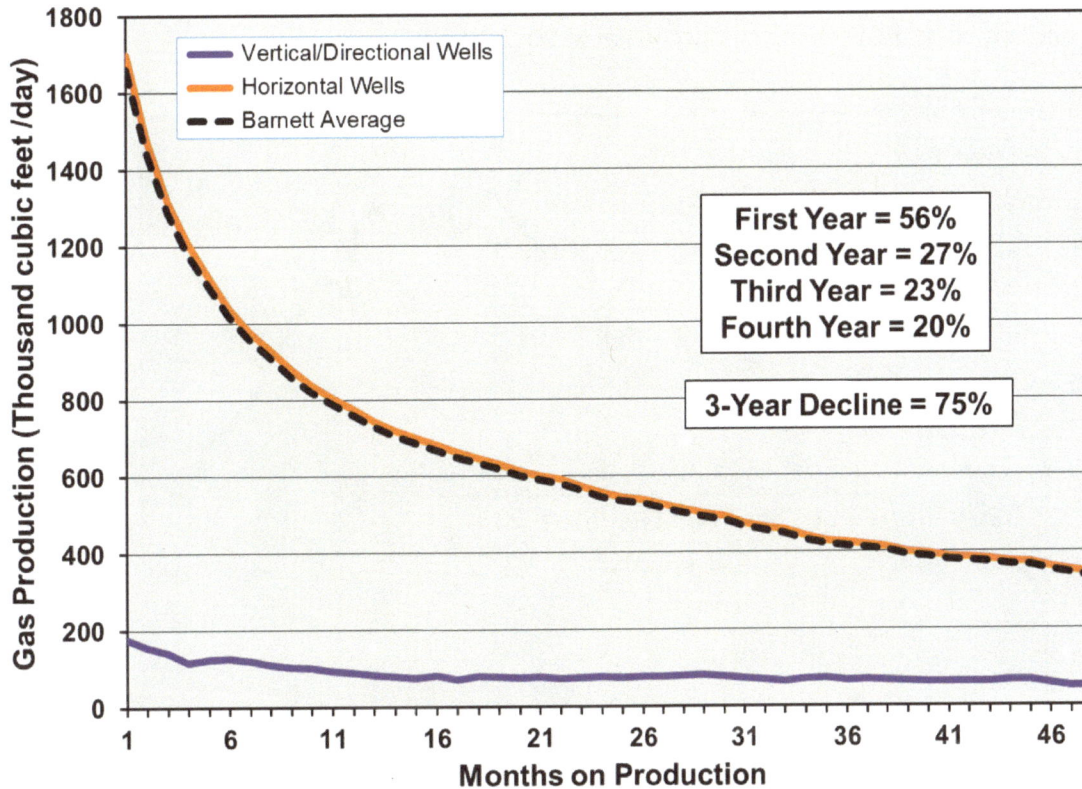

Figure 3-11. Average decline profile for gas wells in the Barnett play.[17]

Decline profile is based on all shale gas wells drilled since 2009.

---

[17] Data from Drillinginfo retrieved August 2014.

3.3.1.2 Field Decline

A second key fundamental is the overall *field decline rate*, which is the amount of production that would be lost for the entire play in a year without more drilling. Figure 3-12 illustrates production from the 12,000 horizontal wells drilled prior to 2013. The first-year decline rate is 23%. This is lower than the well decline rate as the field decline is made up of both new wells declining at high rates and older wells declining at lesser rates. It is also one of the lowest field decline rates observed in any shale field. Assuming new wells will produce in their first year at the average first-year rates observed for wells drilled in 2013, 1,161 new wells each year would be required to offset field decline at current production levels. At an average cost of $3.5 million per well,[18] this would represent a capital input of about $4 billion per year, exclusive of leasing and other ancillary costs, just to keep production flat at 2013 levels.

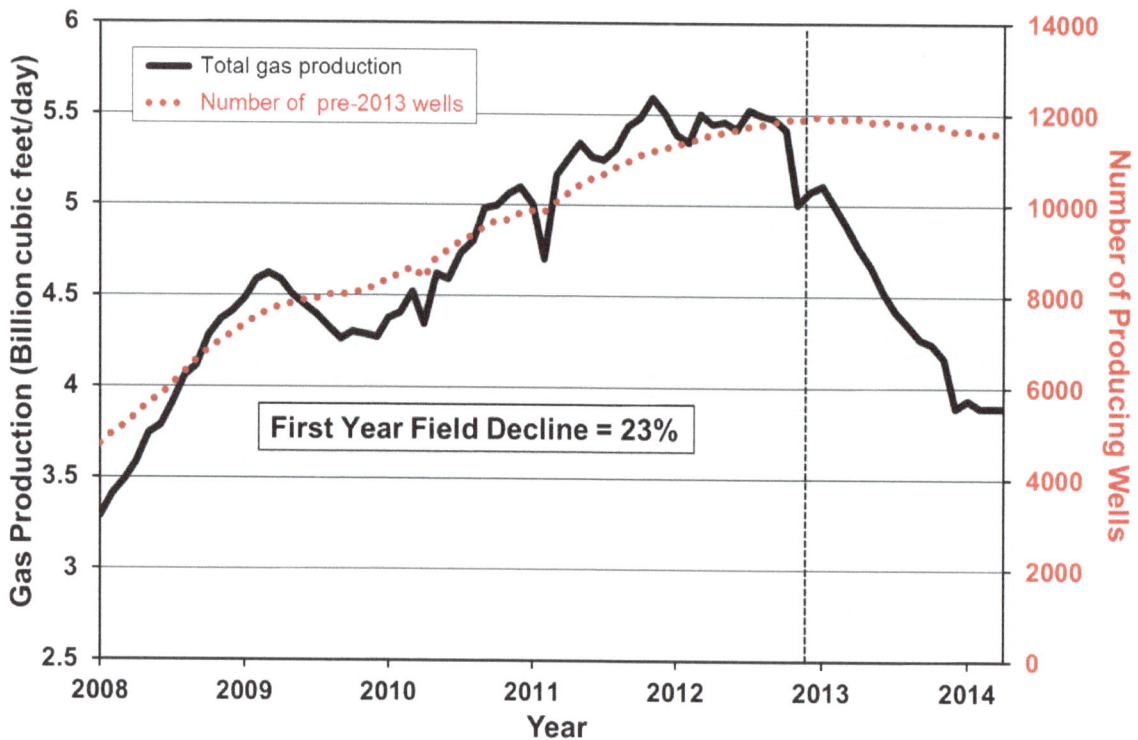

Figure 3-12. Production rate and number of horizontal shale gas wells drilled in the Barnett play prior to 2013, 2008 to 2014.[19]

This defines the field decline for the Barnett play, which is 23% per year (only production from horizontal wells is analyzed as few vertical/directional wells are likely to be drilled in the future).

---

[18] Browning et al., 2013, Barnett Gas Production Outlook,
http://www.searchanddiscovery.com/pdfz/documents/2013/10541browning/ndx_browning.pdf.html.
[19] Data from Drillinginfo retrieved August 2014.

3.3.1.3   Well Quality

The third key fundamental is the *average well quality* by area and its trend over time. Petroleum engineers tell us that technology is constantly improving, with longer horizontal laterals, more frack stages per well, more sophisticated mixtures of proppants and other additives in the frack fluid injected into the wells, and higher-volume frack treatments. This has certainly been true over the past few years, along with multi-well pad drilling which has reduced well costs. It is, however, approaching the limits of diminishing returns, and improvements in average well quality are non-existent in the Barnett. The average first-year production rate of Barnett wells is down 17%from what it was in 2011, as illustrated in Figure 3-13. This is clear evidence that geology is winning out over technology, as drilling moves into lower-quality locations as investigated further below.

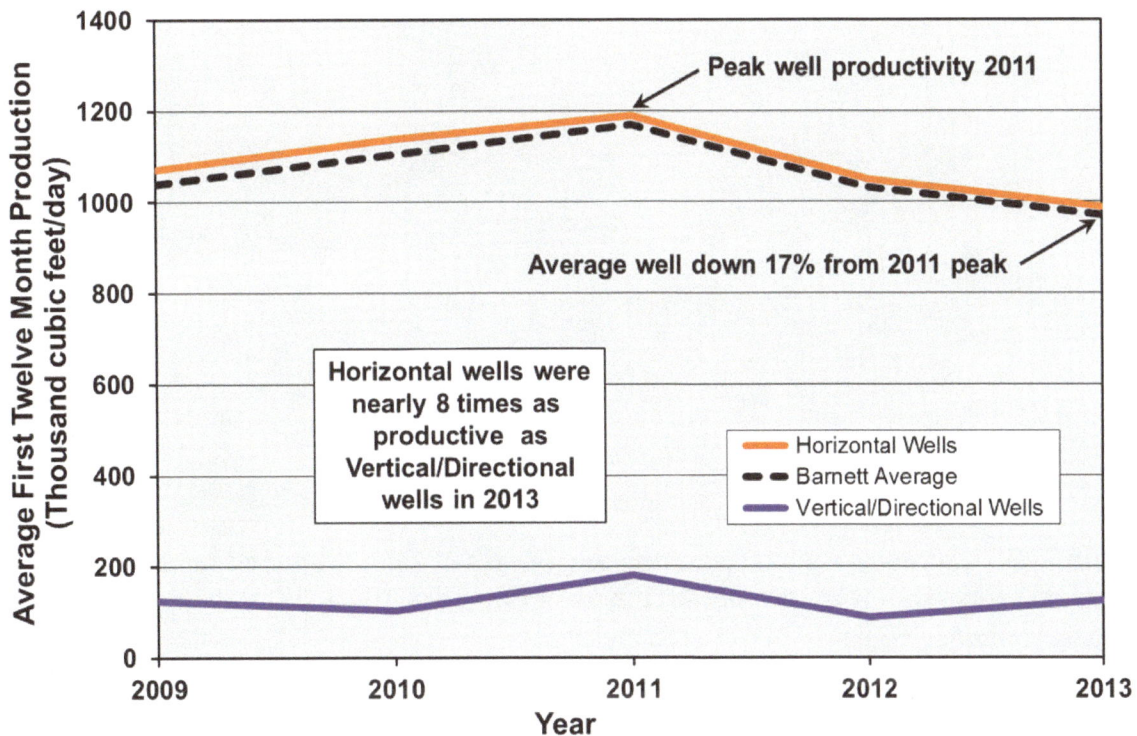

Figure 3-13. Average first-year production rates for Barnett horizontal and vertical/directional gas wells, 2009 to 2013.[20]

Average well quality has fallen by 17% from 2011, a clear indication that geology is trumping technology in this mature shale play.

---

[20] Data from Drillinginfo retrieved August 2014.

Another measure of well quality is cumulative production and well life. More than 14% of the horizontal wells that have been drilled in the Barnett are no longer productive. Figure 3-14 illustrates the cumulative production of these shut-down wells over their lifetime. At a mean lifetime of 37 months and a mean cumulative production of 0.38 billion cubic feet, these wells would in large part be economic losers.

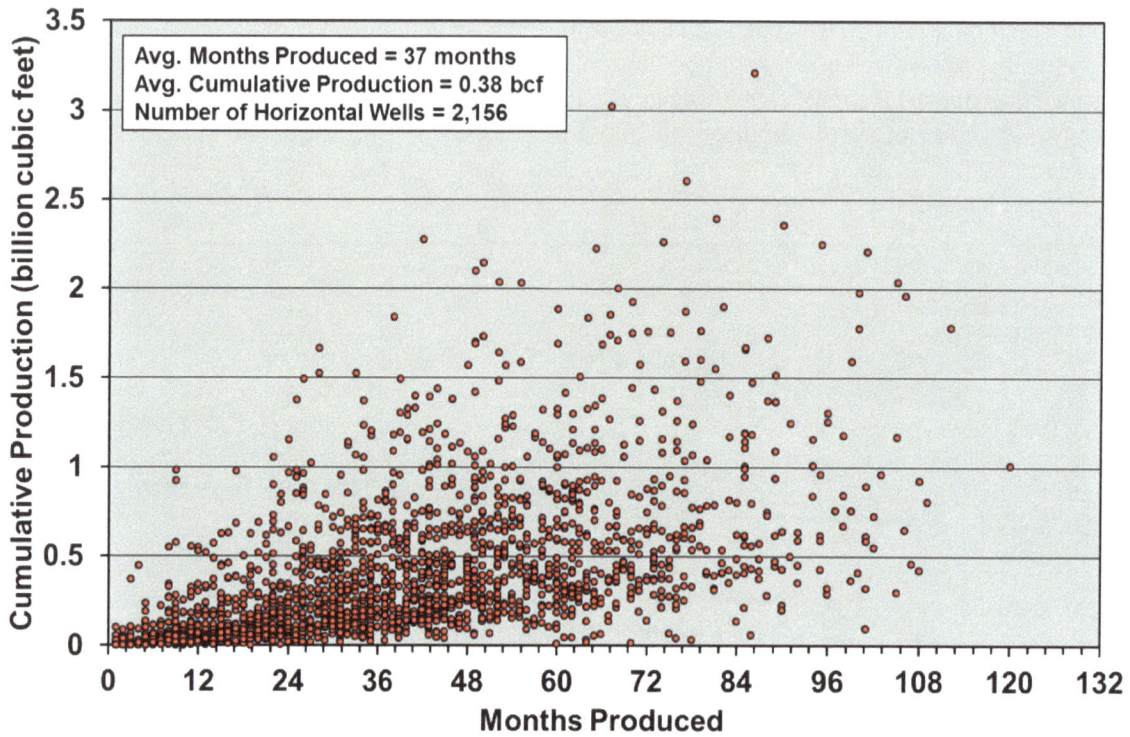

Figure 3-14. Cumulative gas production and length of time produced for Barnett horizontal wells that were not producing as of February 2014.[21]

These well constitute more than 14% of all horizontal wells drilled; most would be economic failures, given the mean life of 37 months and average cumulative production of 0.38 billion cubic feet when production ended.

---

[21] Data from Drillinginfo retrieved August 2014.

Figure 3-15 illustrates the cumulative production of all horizontal wells that were producing in the Barnett as of March 2014. Although it can be seen that there are a few very good wells that recovered large amounts of gas in the first few years, and undoubtedly were great economic successes, the average well had produced just 0.95 billion cubic feet over a lifespan averaging 58 months. Just 1% of these wells are more than 10 years old.

The lifespan of wells is another key parameter as many operators assume a minimum well life of 30 years and longer; this is conjectural given the lack of data and the large numbers of wells that have been shut down after less than 10 years.

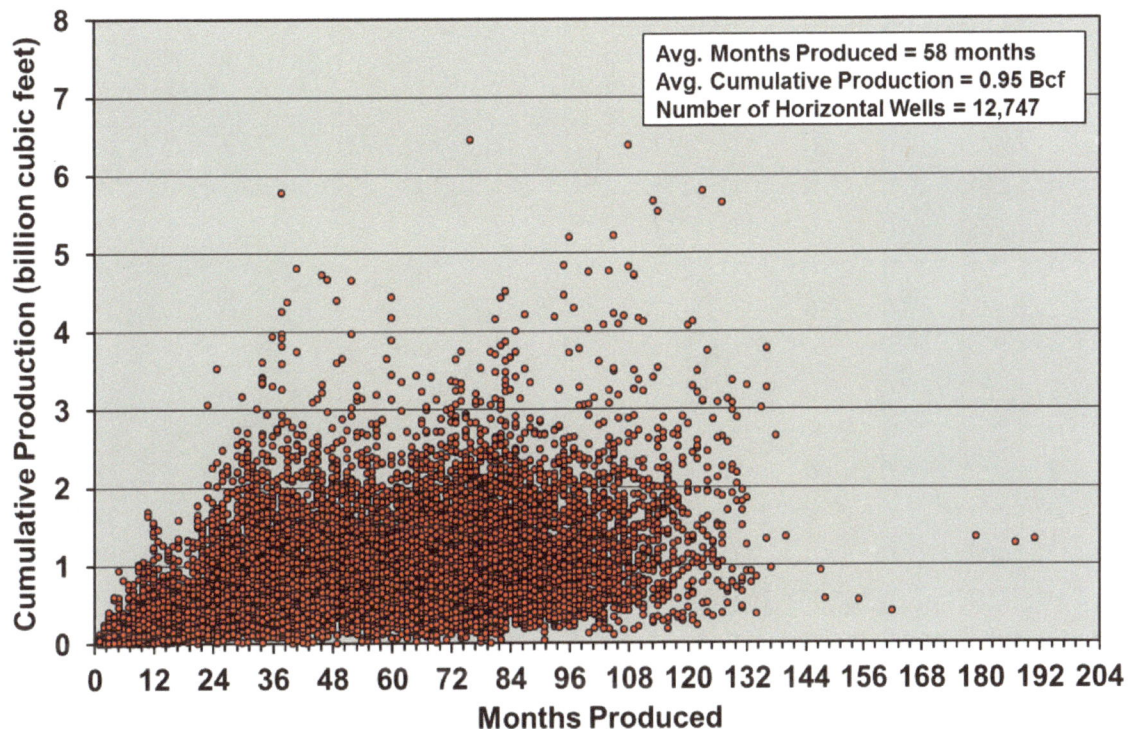

Figure 3-15. Cumulative gas production and length of time produced for Barnett horizontal wells that were producing as of March 2014.

These well constitute 86% of all horizontal wells drilled. Very few wells are greater than ten years old, with a mean age of 58 months and a mean cumulative recovery of 0.95 billion cubic feet.[22]

---

[22] Data from Drillinginfo retrieved August 2014.

Cumulative production of course depends on how long a well has been producing, so looking at young wells in not necessarily a good indication of how much gas these wells will produce over their lifespan (although production is heavily weighted to the early years of well life). A measure of well quality independent of age is initial productivity (IP), which is often focused on by operators. Figure 3-16 illustrates the average daily output over the first six months of production for all wells in the Barnett play (six-month IP). Again, as with cumulative production, there are a few exceptional wells—one percent produced more than 4 million cubic feet per day (MMcf/d)—but the average for all wells drilled since 1995 is just 1.04 MMcf/d. Figure 3-7 illustrates the distribution of IPs in map form.

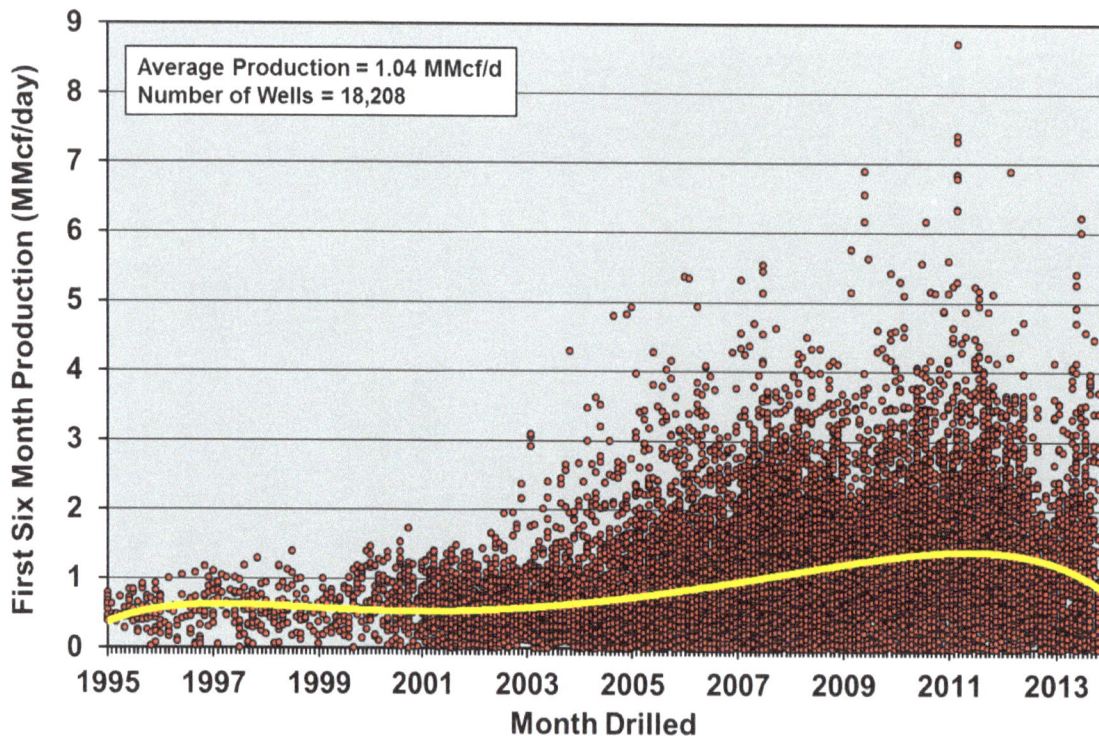

Figure 3-16. Average gas production over the first six months for all wells drilled in the Barnett play, 1995 to 2014.

Although there are a few exceptional wells, the average well produced 1.04 million cubic feet per day over this period.[23] The trend line indicates mean productivity over time.

---

[23] Data from Drillinginfo retrieved August 2014.

Different counties in the Barnett display markedly different well quality characteristics which are critical in determining the most likely production profile in the future. Figure 3-17, which illustrates production over time by county, shows that as of April 2014, the top two counties produced 57% of the total, the top five produced 88%, and the remaining 19 counties produced just 12%.

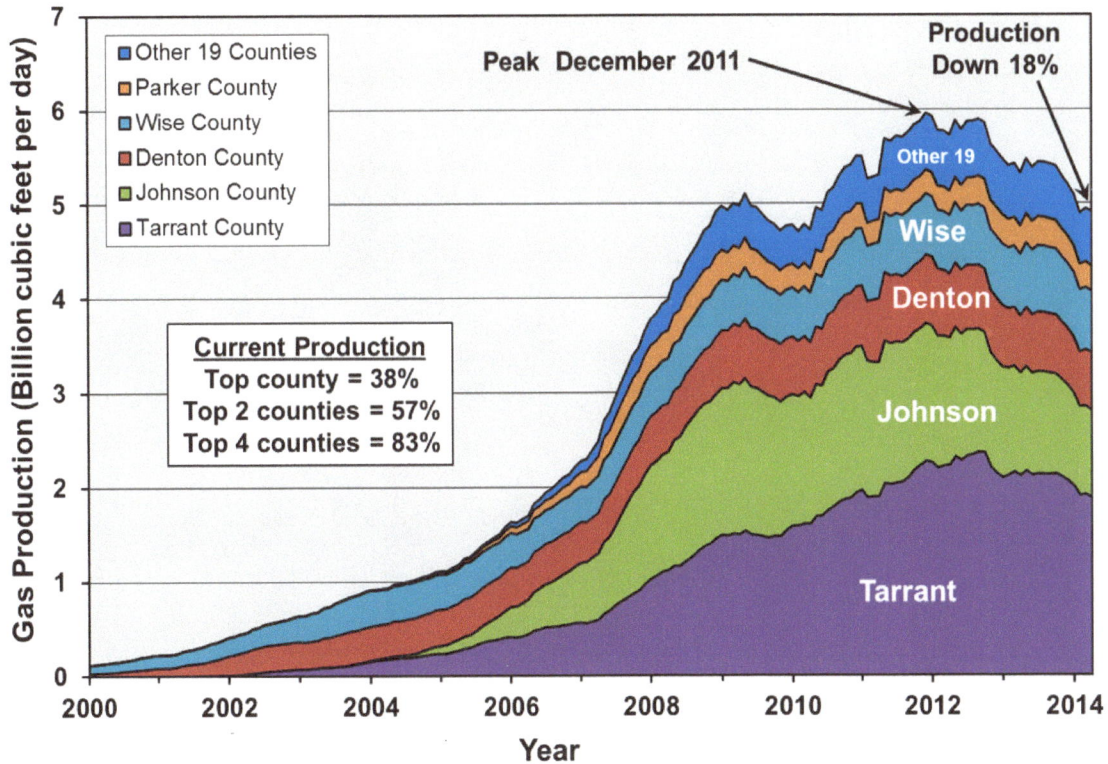

Figure 3-17. Gas production by county in the Barnett play, 2000 through 2014.[24]

The top five counties produced 88% of production in April 2014.

---

[24] Data from Drillinginfo retrieved August 2014. Three-month trailing moving average.

The same trend holds in terms of cumulative production since the field commenced. As illustrated in Figure 3-18, the top two counties have produced 56% of the gas and the top five have produced 92%. All of the counties have peaked, although with increased drilling rates some could conceivably resume production growth.

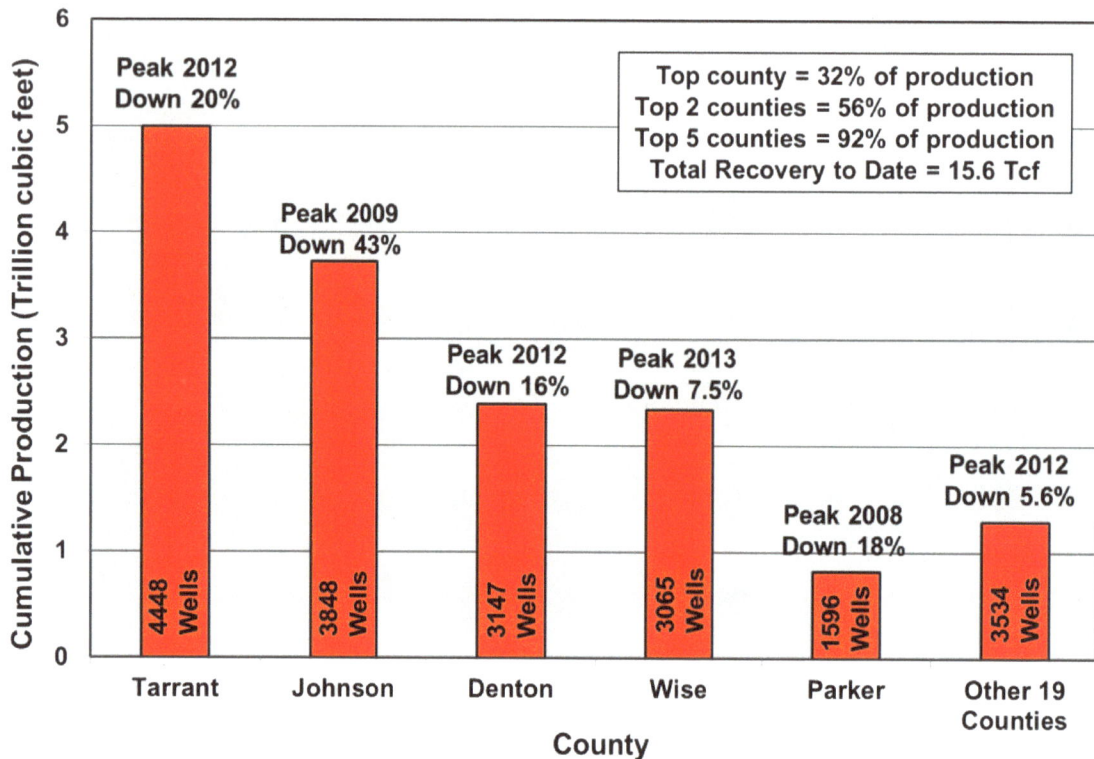

Figure 3-18. Cumulative gas production by county in the Barnett play through 2014.[25]
The top five counties have produced 92% of the 15.6 trillion cubic feet of gas produced to date.

---

[25] Data from Drillinginfo retrieved August 2014.

The Barnett also produces limited amounts of natural gas liquids and oil. Most liquids production is not within the top five counties but is located in the northern and western extremities of the play as illustrated in Figure 3-19. Some 59 million barrels of liquids have been produced since 2000, and although it has somewhat improved economics in marginal counties for gas production, in the big picture liquids production from the Barnett is relatively insignificant (Figure 3-20).

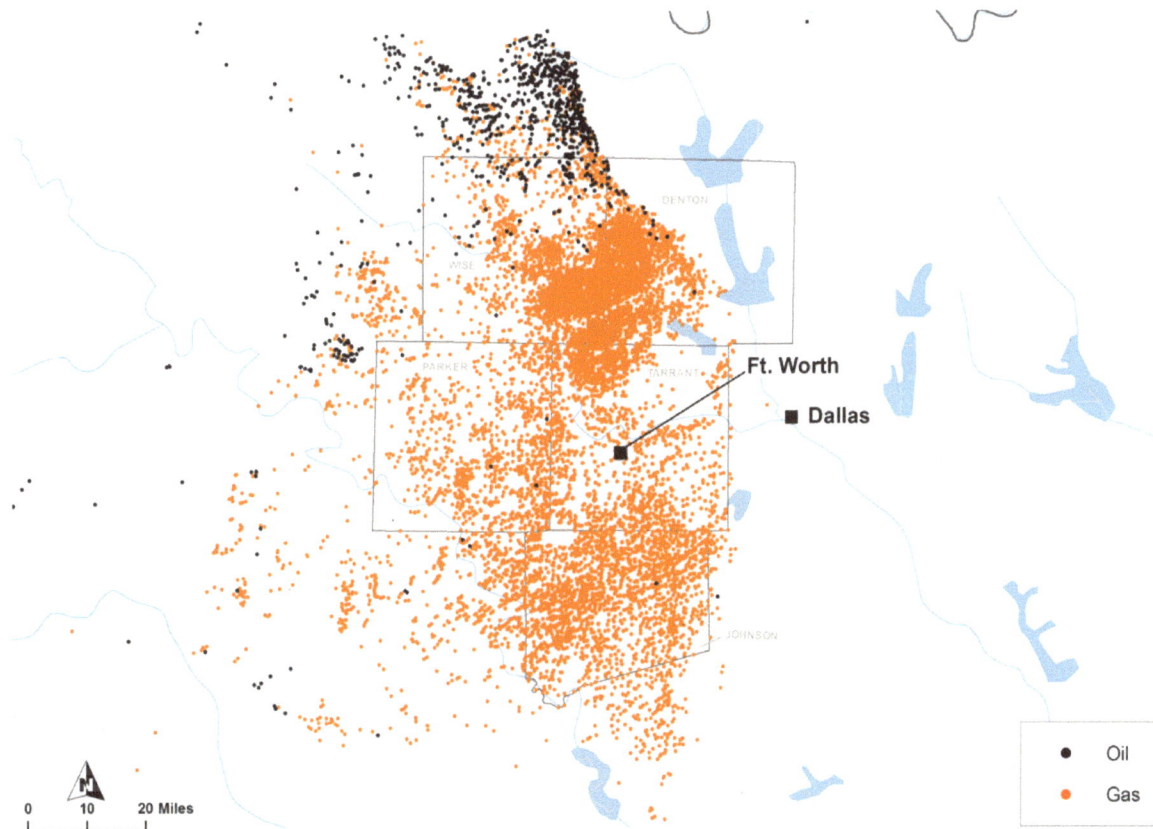

Figure 3-19. Distribution of gas and oil wells in the Barnett play as of early 2014.[26]

Liquids production from wells classified as "oil" occurs mainly in the northern and western extremities of the play.

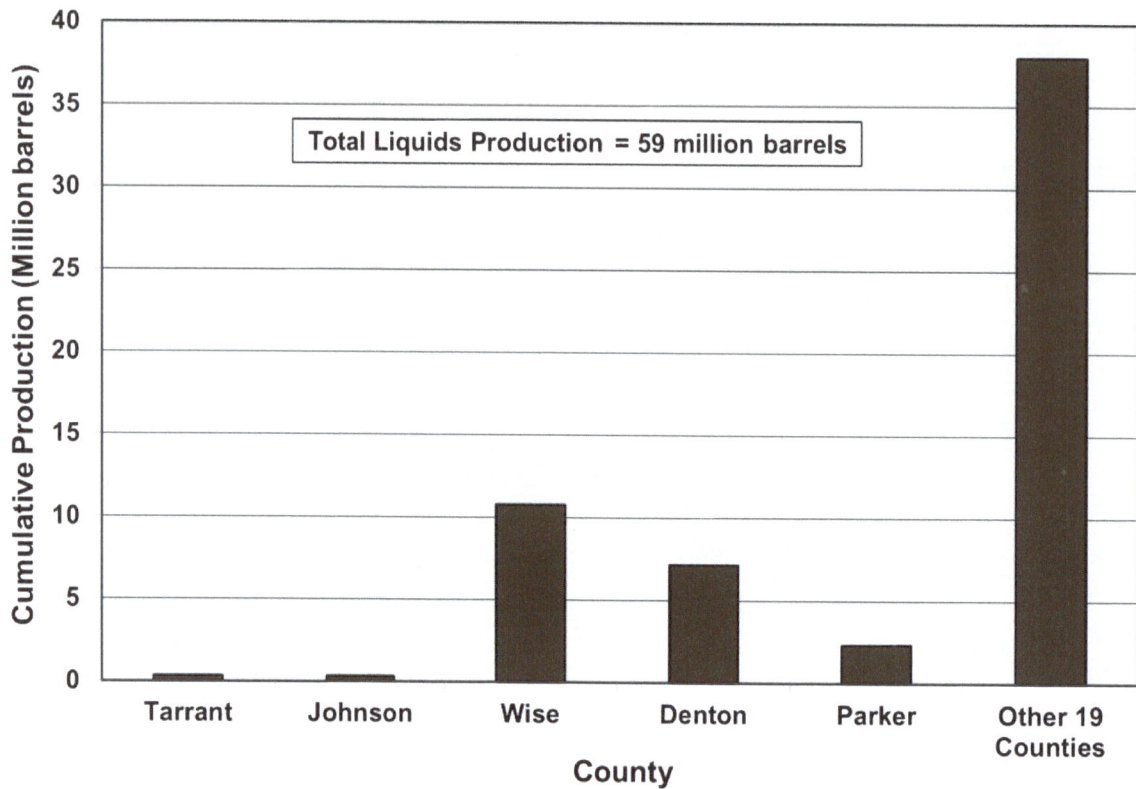

Figure 3-20. Cumulative liquids production by county in the Barnett play through 2014.[27]
The "other 19" counties account for 65% of the 59 million barrels produced to date.

---

[27] Data from Drillinginfo retrieved August 2014.

Operators are highly sensitive to the economic performance of the wells they drill, which typically cost on the order of $3.5 million or more each in the Barnett, not including leasing costs and other expenses. The areas of highest quality—the "core" or "sweet spots"—have now been well defined. Figure 3-21 illustrates average horizontal well decline curves by county, which are a measure of well quality (recognizing that future gas production from the Barnett will be from horizontal, not vertical, wells). Initial well productivities (IPs) from Tarrant and Johnson counties are double those of Wise and Parker counties and quadruple those of the outlying 19 counties. The decline curves from the top three counties are all above the Barnett average, hence these counties are attracting the bulk of the drilling and investment—but they are nearly saturated with wells. Future drilling will have to focus more and more on lesser-quality counties.

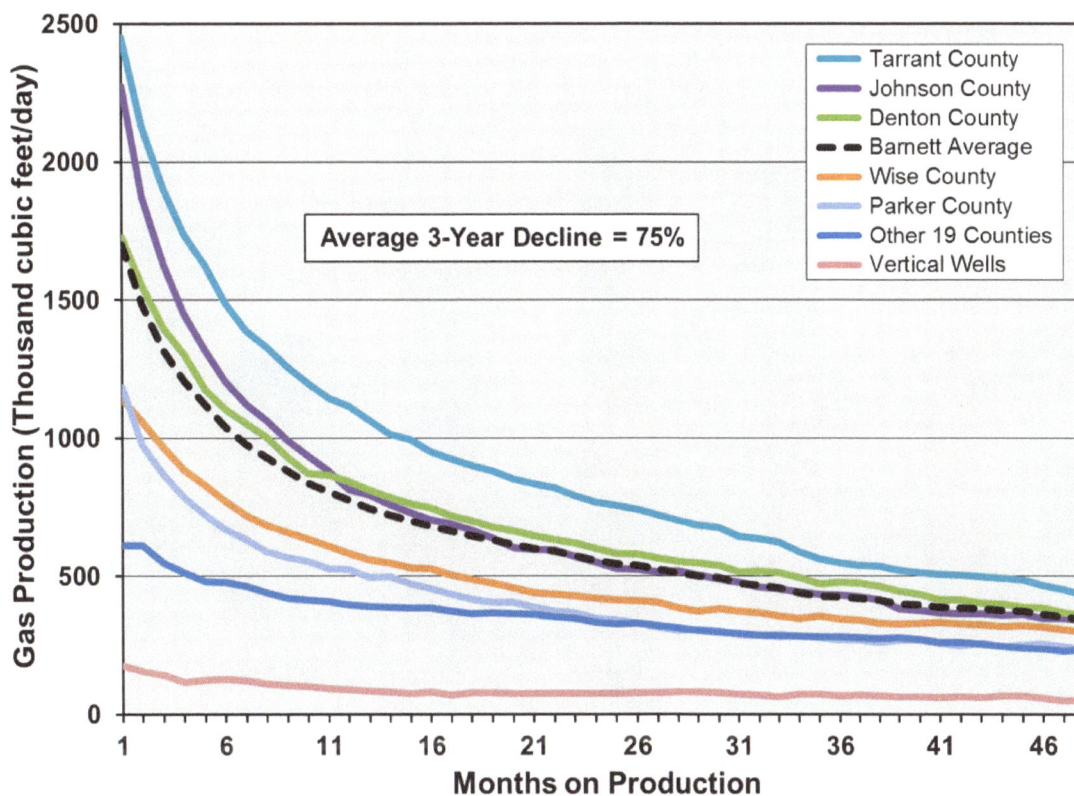

Figure 3-21. Average horizontal gas well decline profiles by county for the Barnett play.[28]

The top three counties, which have produced much of the gas in the Barnett, are clearly superior.

Another measure of well quality is "estimated ultimate recovery" or EUR—the amount of gas a well will recover over its lifetime. To be clear, no one knows what the lifespan of an average Barnett well is, given that few of them are more than ten years old (see Figure 3-14 and Figure 3-15), and some 14% of horizontal wells drilled have ceased production at an average age of just over three years. Operators fit hyperbolic and/or exponential curves to data such as presented in Figure 3-21, assuming well life spans of 30-50 years (as is typical for conventional wells), but so far this is speculation, given the nature of the extremely low permeability reservoirs and the completion technologies used in the Barnett. Nonetheless, for comparative well quality purposes only, one can use the data in Figure 3-21, which exhibits steep initial decline with

---

[28] Data from Drillinginfo retrieved August 2014.

progressively more gradual decline rates, and assume a constant terminal decline rate thereafter to develop a theoretical EUR.

Figure 3-22 illustrates theoretical EURs by county for the Barnett for comparative purposes of well quality. These range from 1.01 to 2.34 billion cubic feet per well, which are somewhat higher than the 0.19 to 1.62 billion cubic feet assumed by the EIA.[29] The steep initial well production declines mean that well payout, if it is achieved, comes in the first few years of production, as between 51% and 58% of an average well's lifetime production occurs in the first four years.

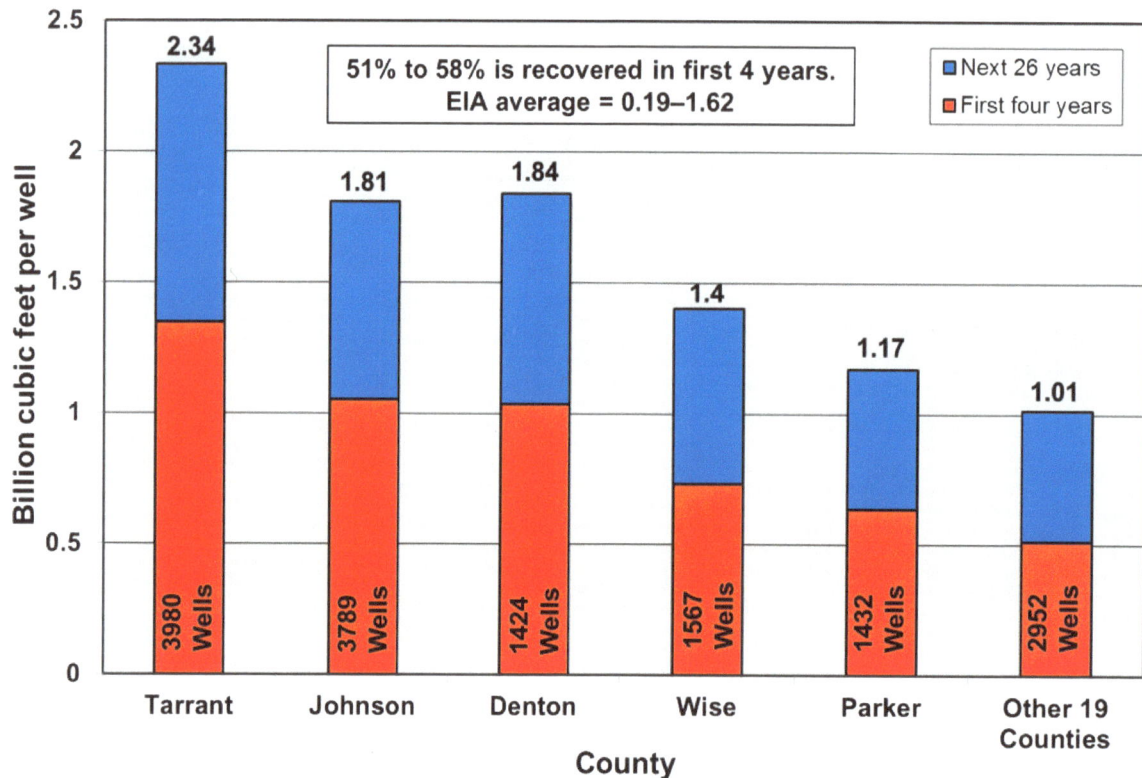

Figure 3-22. Estimated ultimate recovery of gas per well by county for the Barnett play.[30]

EURs are based on average well decline profiles (Figure 3-21) and a terminal decline rate of 15%. These are for comparative purposes only as it is highly uncertain if wells will last for 30 years. The steep decline rates mean that most production occurs early in well life.

---

[29] EIA, *Assumptions to the Annual Energy Outlook 2014*, http://www.eia.gov/forecasts/aeo/assumptions/pdf/oilgas.pdf.
[30] Data from Drillinginfo retrieved August 2014.

Well quality can also be expressed as the average rate of production over the first year of well life. If we know both the rate of production in the first year of the average well and the field decline rate, we can calculate the number of wells that need to be drilled each year to offset field decline in order to maintain production. Given that drilling is currently focused on the highest quality counties, the average first-year production rate per well will fall as drilling moves into lower-quality counties as the best locations are drilled off. As average well quality falls, the number of wells that must be drilled to offset field decline must rise, until the drilling rate can no longer offset decline and the field peaks.

Figure 3-23 illustrates the average first year production rate of wells by county. Notwithstanding modest recent gains in the top two counties—which are also those that are most densely drilled—the average well quality is flat or falling, as progressively more wells are drilled in lower quality parts of individual counties and in the play overall.

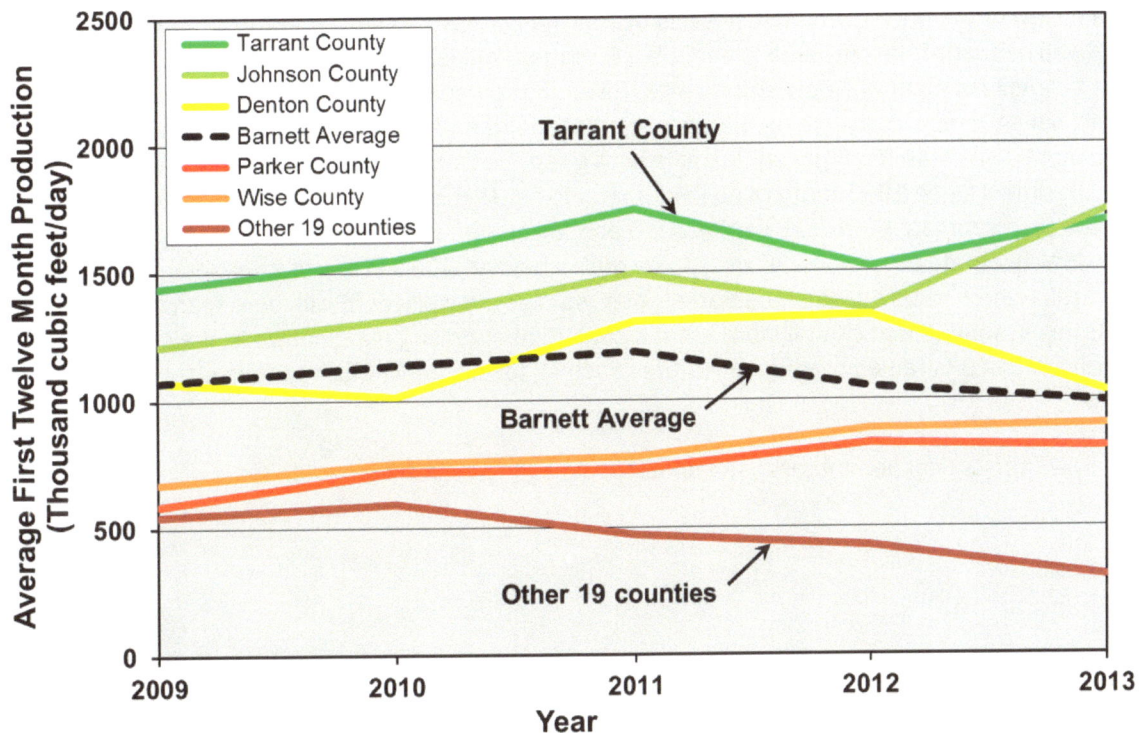

Figure 3-23. Average first-year gas production rates of wells by county for the Barnett play, 2009 to 2013.[31]

Well quality is rising modestly in Tarrant and Johnson counties and falling or flat in other counties. First year production rate in the lowest 19 counties, where the bulk of remaining drilling locations are, is less than a quarter of the top two counties, and is falling.

---

[31] Data from Drillinginfo retrieved August 2014.

3.3.1.4   Number of Wells

The fourth key fundamental is the number of wells that can ultimately be drilled. The Bureau of Economic Geology at the University of Texas at Austin has done a detailed analysis of the Barnett in which they suggest a total of 29,217 wells will be drilled by 2030 in its base case (including 15,144 wells drilled through 2010 and 14,073 new wells to be drilled through 2030).[32] The range of total estimated wells in the University of Texas study was from 20,636 for its low case to 40,267 for its high case. The EIA, on the other hand, suggests that there are 6,725 square miles that can be drilled at a density of 8 wells per square mile for a total of 53,797 wells.[33] However, more than two-thirds of the EIA's estimated wells occur in counties with very low production potential (EUR estimated by the EIA of just 0.19 Bcf per well)—hence it is questionable if many of these wells would ever be drilled. It is also not clear if the EIA's drillable area excludes areas already drilled, which, if so, would increase the total area of the play and the number of wells that ultimately would be drilled.

A careful review of the drilling production levels by well in Figure 3-7 reveals that the limits of the Barnett play are quite well defined. Total play area is about 5,140 square miles, which translates to 41,121 locations if drilled at a density of eight wells per square mile. Given that prospective parts of Denton County now exceed eight wells per square mile (averaging 8.86 per square mile) the ultimate total well count would be 41,426 (i.e., 305 more wells than the 8 per square mile limit given the Denton County overshoot), which includes 3,732 wells drilled since 1995 that are no longer producing. This is considerably higher than the University of Texas base case estimate of wells drilled by 2030 and lower than the EIA estimate (although the Browning et al. study does not state the number of wells to be drilled beyond 2030 in any of its cases). It assumes that 21,788 wells remain to be drilled in the Barnett play, so that the well count will more than double from current levels assuming that capital input is not a constraint in drilling marginal wells. It also assumes that drilling will not be constrained by surface features such as towns, parks etc. and thus is a best case estimate.

Table 3-1 lists the critical parameters used for determining the future production rates of the Barnett play.

---

[32] Browning et al., 2014, *Oil and Gas Journal*, "BARNETT SHALE MODEL-2 (Conclusion): Barnett study determines full-field reserves, production forecast," http://www.ogj.com/articles/print/volume-111/issue-9/drilling-production/barnett-study-determines-full-field-reserves.html.
[33] EIA, *Assumptions to the Annual Energy Outlook 2014*, http://www.eia.gov/forecasts/aeo/assumptions/pdf/oilgas.pdf.

Parameter	County						Total
	Denton	Johnson	Parker	Tarrant	Wise	Other 19	
Production April 2014 (Bcf/d)	0.61	0.92	0.27	1.88	0.65	0.57	4.91
% of Field Production	13	19	6	38	13	12	100
Cumulative Gas (Tcf)	2.39	3.73	0.82	5	2.34	1.29	15.57
Cumulative Liquids (MMBBL)	7.14	0.32	2.31	0.31	10.76	37.99	58.85
Number of Wells	3147	3848	1596	4448	3065	3534	19638
Number of Producing Wells	2678	3028	1135	3735	2608	2722	15906
Average EUR per well (Bcf)	1.84	1.81	1.17	2.34	1.4	1.01	1.70
Field Decline (%)	19.05	23.81	25.75	24.86	22.56	20.58	23.37
3-Year Well Decline (%)	72	81	77	78	70	55	75
Peak Year	Jan-12	May-09	Dec-08	Sep-12	Oct-13	May-12	Dec-11
% Below Peak	16	43	18	20	7.5	5.6	18
Average First Year Production in 2013 (Mcf/d)	1032	1740	812	1701	900	308	988
New Wells Needed to Offset Field Decline	113	126	86	275	163	382	1161
Area in square miles	888	729	904	864	905	19000	23290
% Prospective	40	90	90	80	80	10	22
Net Square Miles	355.2	656.1	813.6	691.2	724	1900	5140
Well Density per square mile	8.86	5.86	1.96	6.44	4.23	1.86	3.82
Additional locations to 8/sq. Mile	0	1401	4913	1082	2727	11666	21788
Population	584238	126811	88495	1446219	48793	N/A	N/A
Total Wells 8/sq. Mile	3147	5249	6509	5530	5792	15200	41426
Total Producing Wells 8/sq. Mile	2678	4429	6048	4817	5335	14388	37694

Table 3-1. Parameters for projecting Barnett production, by county.

Area in square miles under "Other" is estimated.

3.3.1.5   Rate of Drilling

Given known well- and field-decline rates, well quality by area, and the number of available drilling locations, the most important parameter in determining future production levels is the rate of drilling—the fifth key fundamental. Figure 3-24 illustrates the historical drilling rates in the Barnett. Horizontal drilling rates peaked in 2008 at 2,707 wells per year and have fallen to current levels of less than 400 wells per year. Current drilling rates are far less than the 1,161 wells per year required to maintain production at current levels, hence each new well drilled now serves only to slow the overall production decline of the play.

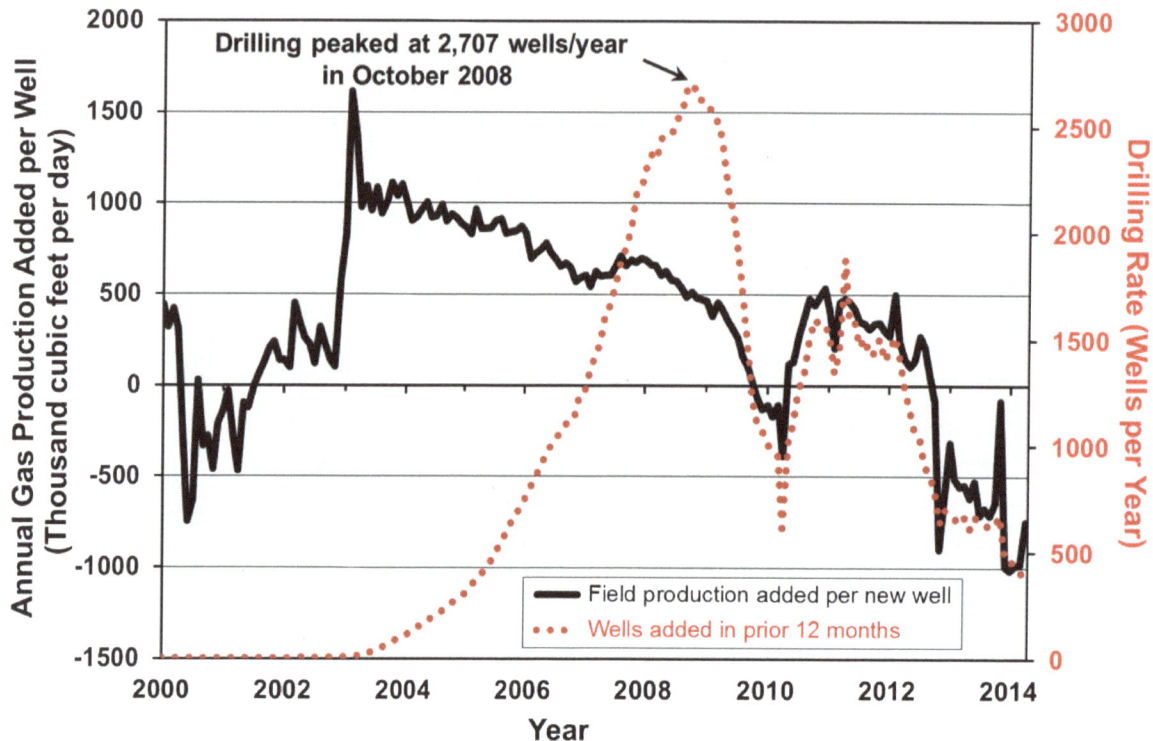

Figure 3-24. Annual gas production added per new horizontal well and annual drilling rate in the Barnett play, 2000 through 2014.[34]

Drilling rate peaked in 2008 and is now far below the level needed to keep production flat, hence each new well now only serves to slow the overall production decline of the play.

---

[34] Data from Drillinginfo retrieved August 2014. Three-month trailing moving average.

3.3.1.6 Future Production Scenarios

Based on the five key fundamentals outlined above, several production projections for the Barnett play were developed to illustrate the effects of changing the rate of drilling. Figure 3-25 illustrates the production profiles of four drilling rate scenarios if 100% of the prospective play area is drillable at eight wells per square mile. These scenarios are:

1. MOST LIKELY RATE scenario: Drilling increases from the current rate to 600 wells per year, then gradually declines to 500 wells per year

2. LOW RATE scenario: Drilling continues at current level of 400 wells per year, holding constant.

3. TRIPLE RATE scenario: Drilling increases to 1,200 wells per year, then gradually declines to 600 wells per year.

4. QUINTUPLE RATE scenario: Drilling increases to 2,000 per year, then gradually declines to 1,000 wells per year.

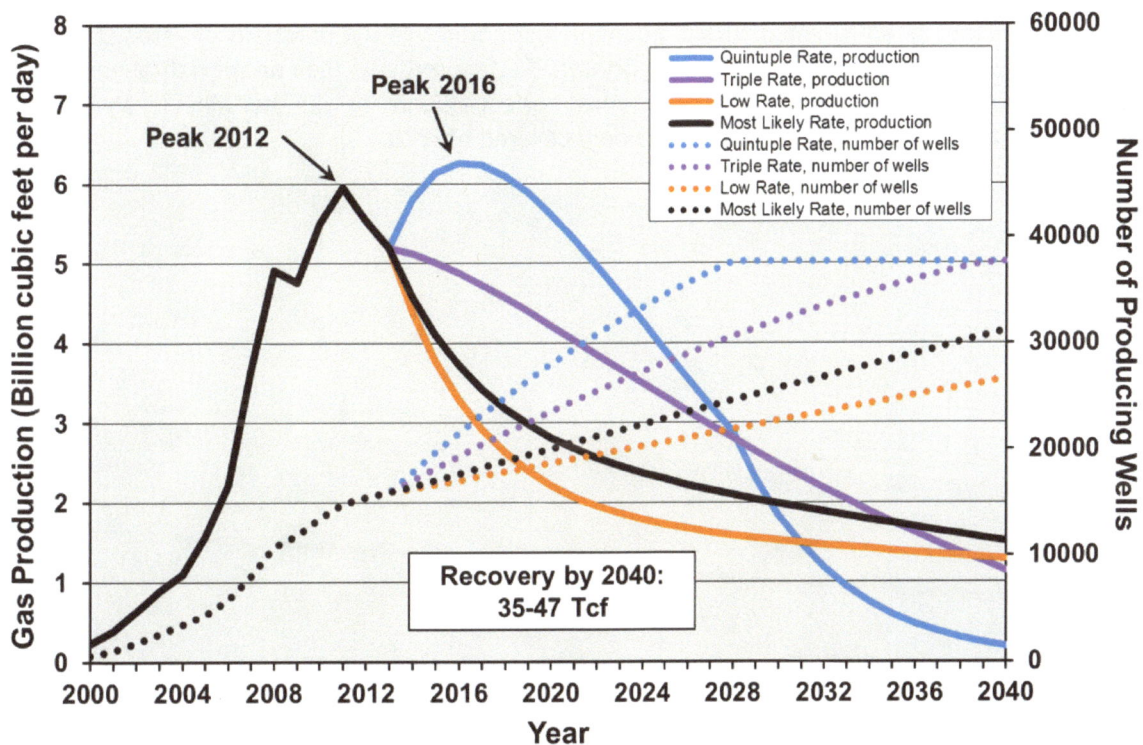

Figure 3-25. Four drilling rate scenarios of Barnett gas production (assuming 100% of the area is drillable at eight wells per square mile).[35]

"Most Likely Rate" scenario: drilling increases to 600 wells/year, declining to 500 wells/year.
"Low Rate" scenario: drilling continues at 400 wells/year, holding constant.
"Triple Rate" scenario: drilling increases to 1,200 wells/year, declining to 600 wells/year.
"Quintuple Rate" scenario: drilling increases to 2,000 wells/year, declining to 1,000 wells/year.
Although the peak month was December 2011, on a total year production basis the peak year is 2012.

[35] Data from Drillinginfo retrieved August 2014.

The drilling rate scenarios have the following results:

1. MOST LIKELY RATE scenario: The drilling rate declines after its initial increase as drilling moves into poorer quality locations. Total gas recovery by 2040 would be 39.2 trillion cubic feet, and drilling would continue beyond 2040.

2. LOW RATE scenario: Total gas recovery by 2040 would be 34.8 trillion cubic feet, and drilling would continue beyond 2040.

3. TRIPLE RATE scenario: Total gas recovery by 2040 would be 45.6 trillion cubic feet, and drilling would end by 2039.

4. QUINTUPLE RATE scenario: The current production decline would be reversed and grow to a new peak in 2016; however, drilling locations would run out by 2028 followed by a steep production decline, making the supply situation much worse in later years than in the "Most Likely Rate" scenario. Total gas recovery by 2040 would be 46.7 trillion cubic feet.

Both the recovery of 39.2 trillion cubic feet by 2040 in the "Most Likely Rate" scenario and the recovery of 46.7 trillion cubic feet in the "Quintuple Rate" scenario agree well with the University of Texas study, which calculates an ultimate recovery of 45 Tcf for the Barnett.[36] (They continue their analysis through 2050 for their ultimate recovery estimate, hence there is almost perfect agreement with the "Most Likely Rate" scenario given that considerably more gas would be recovered after 2040).

---

[36] Browning et al., 2014, *Oil and Gas Journal*, "BARNETT SHALE MODEL-2 (Conclusion): Barnett study determines full-field reserves, production forecast," http://www.ogj.com/articles/print/volume-111/issue-9/drilling-production/barnett-study-determines-full-field-reserves.html.

3.3.1.7   Comparison to EIA Forecast

Figure 3-26 illustrates the EIA's projection for Barnett production through 2040 compared to the "Most Likely Rate" scenario. The EIA projects a recovery by 2040 of 53.3 Tcf to meet its reference case forecast (44.4 Tcf between 2012 and 2040). Not only is this far higher than the projections of this report and the University of Texas study, it *projects a new high in production in 2040*, which implies very considerable future production after 2040. Furthermore, this amounts to *the complete recovery* of all of the EIA's estimated 20.3 Tcf of proved reserves by 2040[37] plus 23.7 Tcf of unproved resources (44 Tcf in total).[38] This strains credibility to the limit; how can all the proved and unproved resources and reserves be extracted and still have production at all-time highs in 2040?

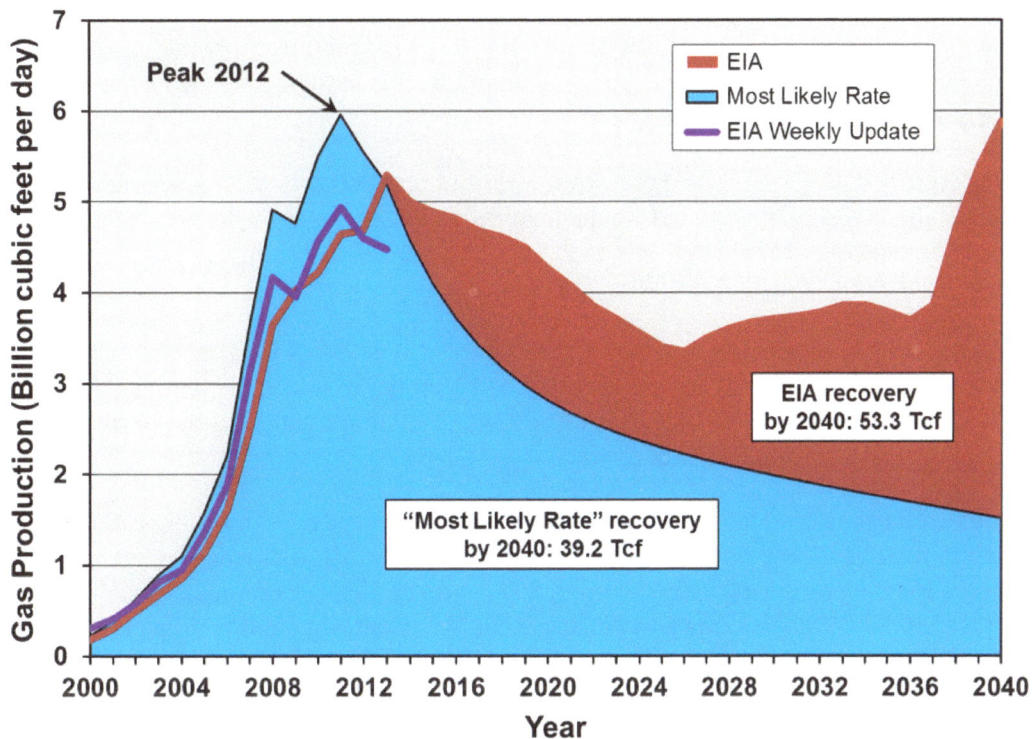

Figure 3-26. "Most Likely Rate" scenario of Barnett gas production compared to the EIA reference case, 2000 to 2040.[39]

The EIA assumes the Barnett will reach a new all-time high by 2040 after producing all proved reserves and unproved resources, and presumably produce a great deal more gas in the post-2040 period. Note that although the peak month was December 2011, on a total year production basis the peak year is 2012. The EIA forecast is made on a "dry gas" basis, whereas the "Most Likely Rate" scenario forecast is made on a "raw gas" basis. The EIA production data are also shown on a dry basis; the difference between the EIA's data and the Drillinginfo data used in this report may be due to the shrinkage factor between "raw" and "dry" gas.[40]

---

[37] EIA, 2014, Principal shale gas plays: natural gas production and proved reserves, 2011-12, http://www.eia.gov/naturalgas/crudeoilreserves/excel/table_4.xls.

[38] EIA, *Assumptions to the Annual Energy Outlook 2014*, http://www.eia.gov/forecasts/aeo/assumptions/pdf/oilgas.pdf.

[39] EIA, *Annual Energy Outlook 2014*, unpublished tables from AEO 2014 provided by the EIA.

[40] EIA, *Natural Gas Weekly Update*, retrieved October 2014, http://www.eia.gov/naturalgas/weekly

3.3.1.8 Barnett Play Analysis Summary

Several things are clear from this analysis:

1. Drilling rates have fallen markedly in the Barnett due to gas prices and to saturation of sweet spots with wells.

2. High well- and field-decline rates mean a continued high rate of drilling is required to maintain, let alone increase, production. Current drilling rates of 384 wells per year are far below the level of 1,161 wells per year required to maintain production, which would require the investment of $4 billion per year for drilling (assuming $3.5 million per well). Future production profiles are most dependent on drilling rate and, to a lesser extent, on the number of drilling locations (i.e., greatly increasing the number of drilling locations would not change the production profile nearly as much as changing the drilling rate). Maintaining or growing production in the Barnett would require much higher gas prices to justify higher drilling rates.

3. Quintupling current drilling rates could reverse the current production decline and raise production to a new peak in the 2016 timeframe, but would increase cumulative recovery only by 19% by 2040 and wouldn't change the ultimate recovery of the play. Increasing drilling rates effectively recovers the gas sooner, making the supply situation worse later.

4. The projected recovery of 39.2 Tcf by 2040 in this report's "Most Likely Rate" scenario is comparable to the University of Texas study's ultimate recovery of 45 Tcf (given that considerable gas would be recovered in the "Most Likely Rate" scenario after 2040).[41] Both are significantly less than the EIA's reference case projection of 53.3 Tcf by 2040.

5. This report's projections are optimistic in that they assume the capital will be available for the drilling treadmill that must be maintained. They also assume that 100% of the prospective area is drillable. This is not a sure thing as drilling in the poorer quality parts of the play will require much higher gas prices to be economic. Failure to maintain drilling rates will result in a steeper drop off in production.

6. More than double the current number of producing wells will need to be drilled to meet the production projection of the "Most Likely Rate" scenario over the next several decades.

7. The EIA projection for future Barnett gas production included in its reference case forecast for AEO 2014[42] strains credibility to the limit. It is highly unlikely to be realized, especially at the gas prices the EIA forecasts.[43]

---

[41] Browning et al., 2014, *Oil and Gas Journal*, "BARNETT SHALE MODEL-2 (Conclusion): Barnett study determines full-field reserves, production forecast," http://www.ogj.com/articles/print/volume-111/issue-9/drilling-production/barnett-study-determines-full-field-reserves.html.

[42] EIA, *Annual Energy Outlook 2014*, unpublished tables from AEO 2014 provided by the EIA.

[43] EIA, *Annual Energy Outlook 2014*, http://www.eia.gov/forecasts/aeo/.

3.3.2   Haynesville Play

The EIA forecasts recovery of 102 Tcf of gas from the Haynesville play by 2040. The analysis of actual production data presented below suggests that this forecast is highly unlikely to be realized.

The Haynesville play was discovered in 2007 and production rapidly increased until it became the largest shale gas play in the U.S. at its peak in early 2012. Figure 3-27 illustrates the distribution of wells as of early 2014. Over 3,500 wells have been drilled to date, of which 3,274 were producing at the time of writing. The play covers parts of 16 counties although most of the drilling is concentrated in Caddo, DeSoto, and Red River parishes in Louisiana and Panola County in east Texas.

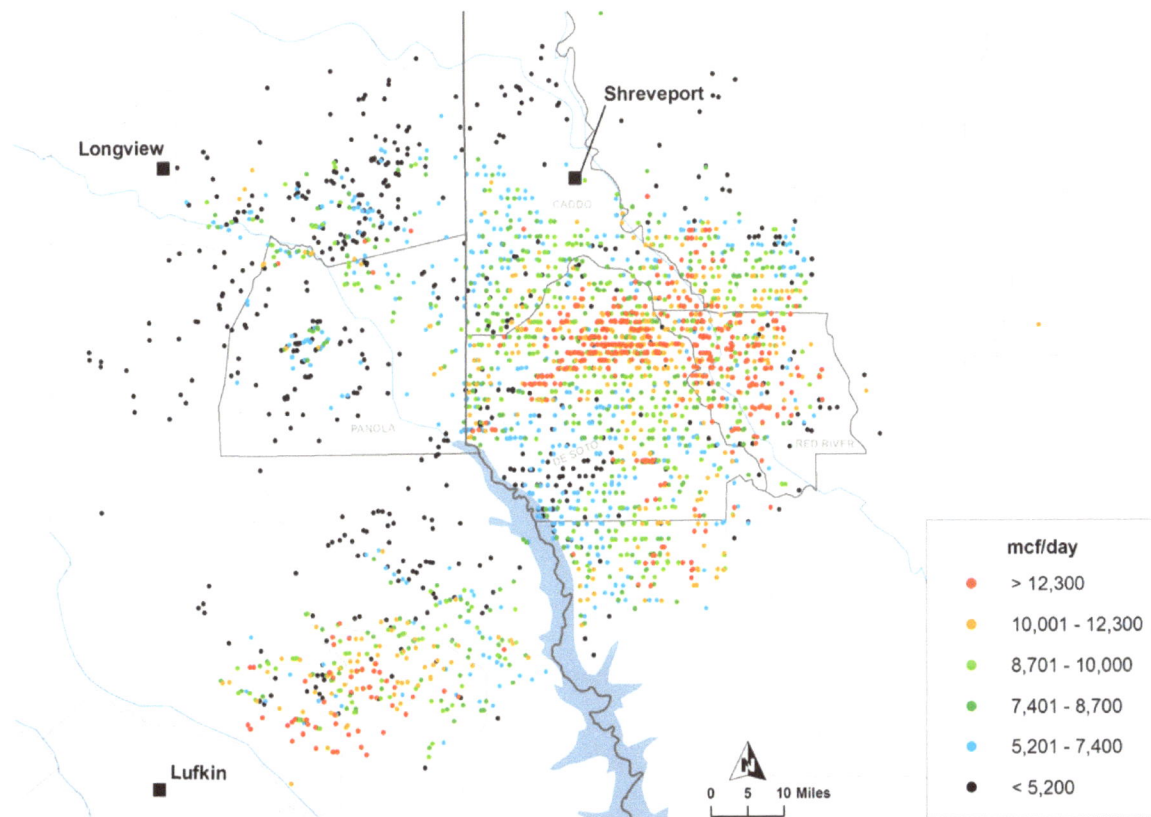

Figure 3-27. Distribution of wells in the Haynesville play as of early 2014, illustrating highest one-month gas production (initial productivity, IP).[44]

Well IPs are categorized approximately by percentile; see Appendix.

---

[44] Data from Drillinginfo retrieved April 2014.

Production in the Haynesville peaked at more than 7 billion cubic feet per day in January 2012 as illustrated in Figure 3-28. Ninety-five percent of current production is from horizontal fracked wells. Horizontal drilling grew from virtually nothing in 2008 to a peak rate of 1,050 wells per year in mid-2011. It has since fallen to 215 wells per year, which is insufficient to offset field decline. Drilling rates required to keep production flat at current production levels are about 400 wells per year.

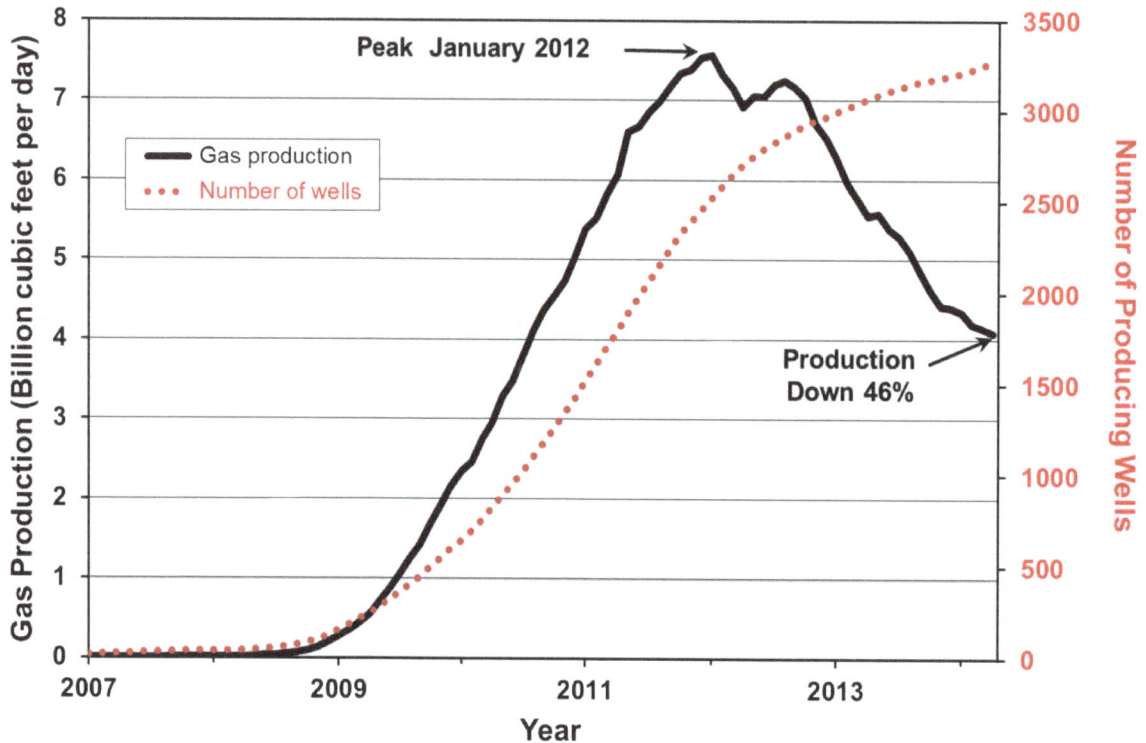

Figure 3-28. Haynesville play shale gas production and number of producing wells, 2007 to 2014.[45]

Gas production data are provided on a "raw gas" basis.

---

[45] Data from Drillinginfo retrieved August 2014. Three-month trailing moving average.

Although vertical and directional wells played a role in the early development of the Haynesville play and still produce some oil and gas, new wells are predominantly horizontal. There are still 417 producing vertical and directional wells at the time of writing, or 14% of the 3,274 producing wells in the play, yet they produce less than 5% of gas output. Production by well type is illustrated in Figure 3-29. Very few vertical/directional wells are being drilled today—the future of the play lies in horizontal fracked wells.

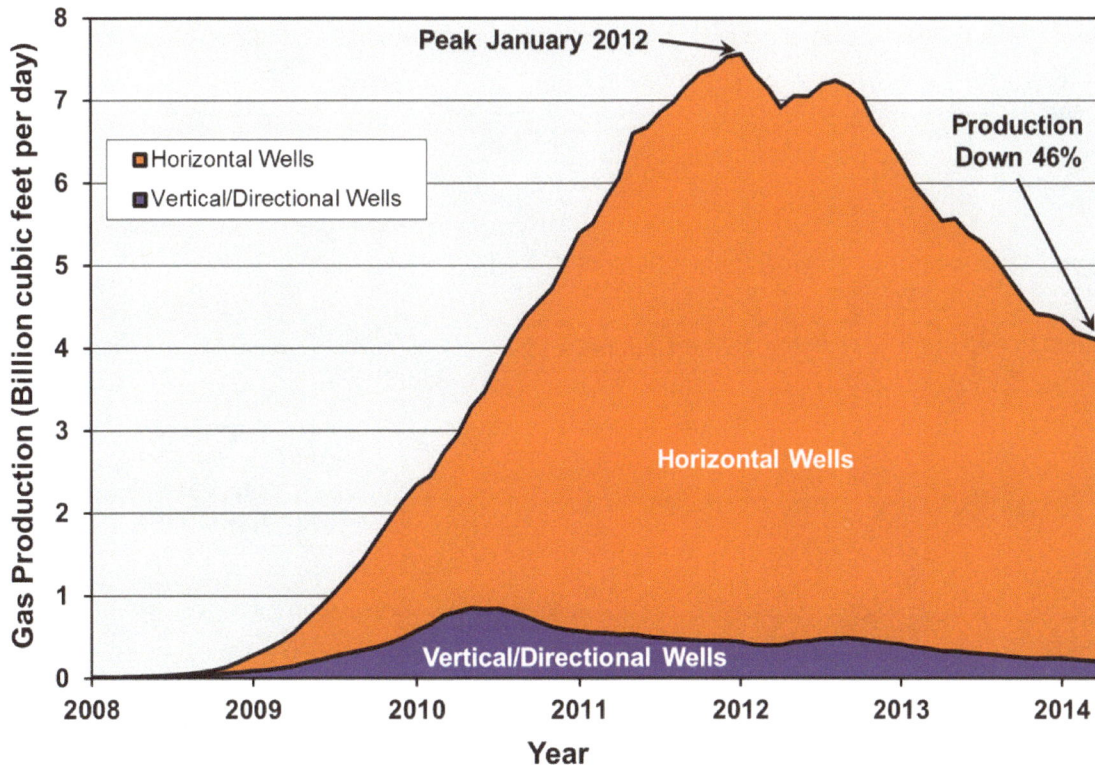

Figure 3-29. Gas production from the Haynesville play by well type, 2008 to 2014.[46]

---

[46] Data from Drillinginfo retrieved August 2014. Three-month trailing moving average.

3.3.2.1   Well Decline

The first key fundamental in determining the life cycle of Haynesville production is the *well decline rate*. Haynesville wells exhibit high decline rates in common with all shale plays. Figure 3-30 illustrates the average decline rate of the most recent Haynesville horizontal and vertical/directional wells. Decline rates are steepest in the first year and are progressively less in the second and subsequent years. The average decline rate over the first three years of well life is 88%, one of the highest of the plays analyzed. As can be seen, vertical/directional wells have lower productivity than horizontal wells and hence are being phased out.

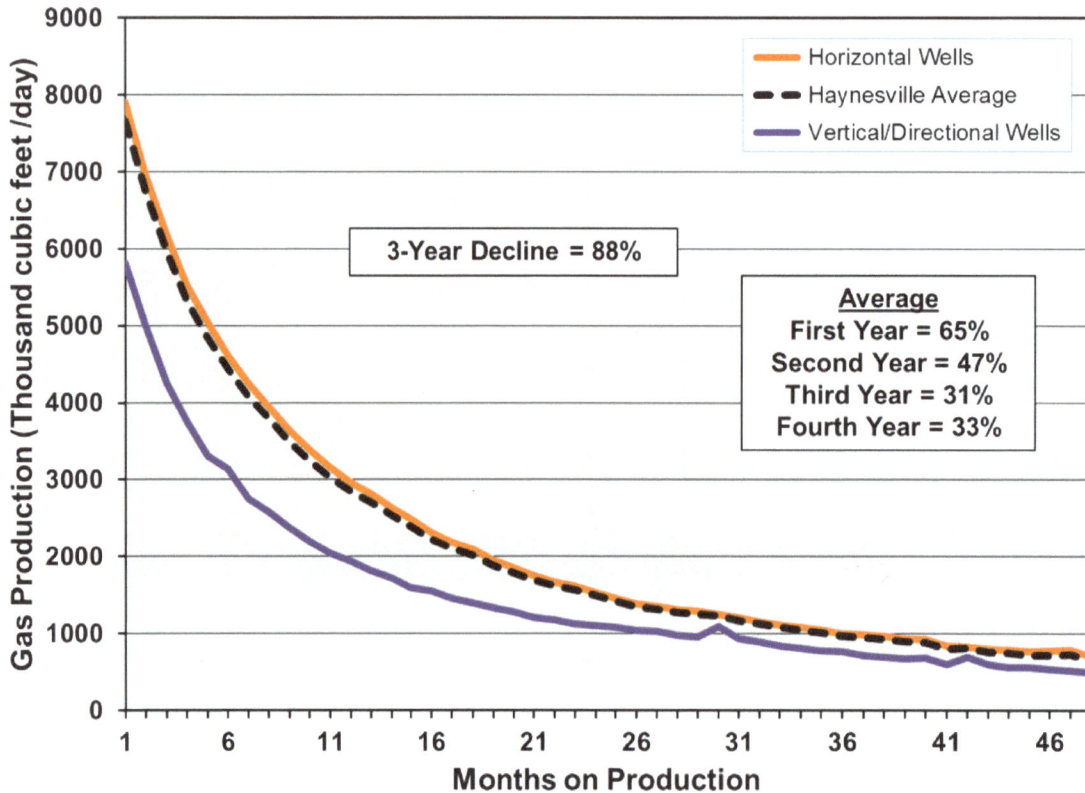

Figure 3-30. Average decline profile for gas wells in the Haynesville play.[47]

Decline profile is based on all shale gas wells drilled since 2009.

[47] Data from Drillinginfo retrieved August 2014.

3.3.2.2   Field Decline

A second key fundamental is the overall *field decline rate*, which is the amount of production that would be lost in a year in the Haynesville without more drilling. Figure 3-31 illustrates production from the 2,600 horizontal wells drilled prior to 2013. The first-year decline is 49%. This is lower than the well decline rate as the field decline is made up of both new wells declining at high rates and older wells declining at lesser rates. It is also one of the highest field decline rates observed in any shale field. Assuming new wells will produce in their first year at the average first-year rates observed for wells drilled in 2013, approximately 400 new wells each year would be required to offset field decline at current production levels. At an average cost of $9 million per well[48], this would represent a capital input of about $3.6 billion per year, exclusive of leasing and other ancillary costs, just to keep production flat at 2014 levels.

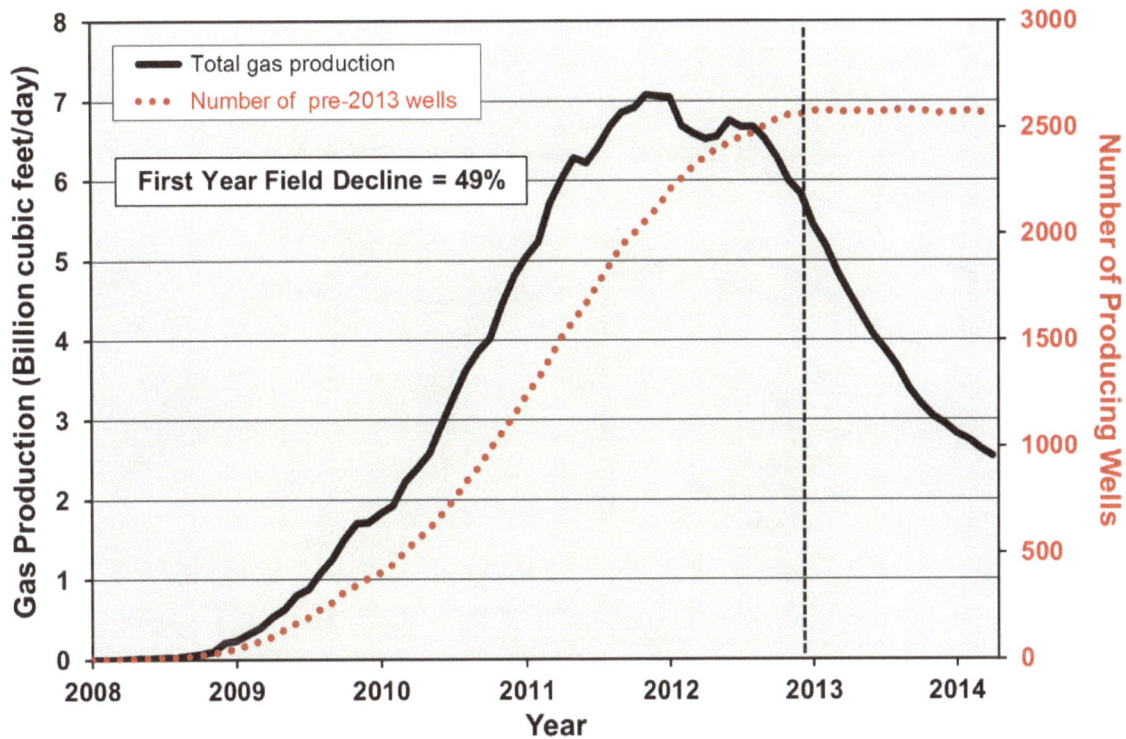

Figure 3-31. Production rate and number of horizontal shale gas wells drilled in the Haynesville play prior to 2013, 2008 to 2014.[49]

This defines the field decline for the Haynesville play, which is 49% per year (only production from horizontal wells is analyzed as few vertical/directional wells are likely to be drilled in the future).

---

[48] Mark J. Kaiser, June 2014, *Oil and Gas Journal*, "HAYNESVILLE UPDATE—2: North Louisiana drilling costs vary slightly 2007-12," http://www.ogj.com/articles/print/volume-112/issue-1/exploration-development/north-louisiana-drilling-costs-vary-slightly-2007-12.html.
[49] Data from Drillinginfo retrieved August 2014.

3.3.2.3   Well Quality

The third key fundamental is the *average well quality* by area and its trend over time. Petroleum engineers tell us that technology is constantly improving, with longer horizontal laterals, more frack stages per well, more sophisticated mixtures of proppants and other additives in the frack fluid injected into the wells, and higher-volume frack treatments. This has certainly been true over the past few years, along with multi-well pad drilling which has reduced well costs. It is, however, approaching the limits of diminishing returns, and improvements in average well quality appear to have ended in the Haynesville. The average first-year production rate of Haynesville wells has been flat over the past year after rising significantly in the early years of the play, as illustrated in Figure 3-32. This is clear evidence that geology is winning out over technology, as drilling moves into lower-quality locations (as investigated further below), given that operators tend to apply more sophisticated technology over time.

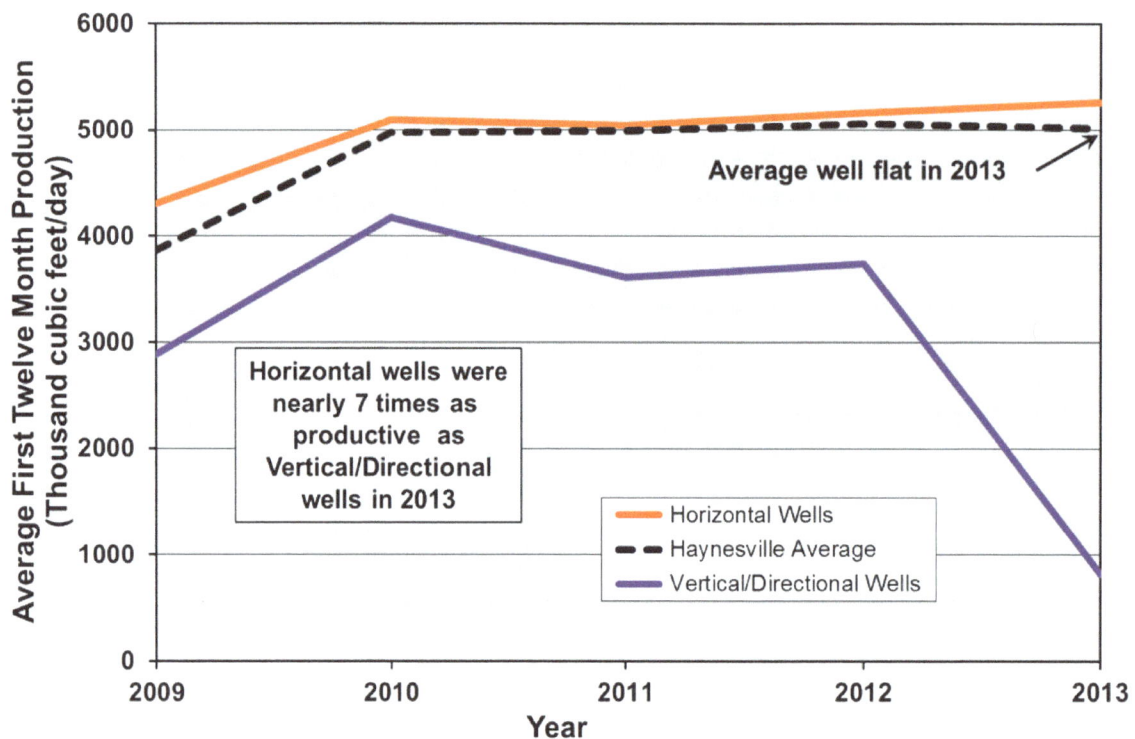

Figure 3-32. Average first-year production rates for Haynesville horizontal and vertical/directional gas wells, 2009 to 2013.[50]

Average well quality is flat in the most recent year after rising significantly in the early years of the play.

[50] Data from Drillinginfo retrieved August 2014.

Another measure of well quality is cumulative production and well life. Nearly 5% of the wells that have been drilled in the Haynesville are no longer productive. Figure 3-33 illustrates the cumulative production of these shut-down wells over their lifetime. At a mean lifetime of 21 months and a mean cumulative production of 1.1 billion cubic feet, many of these wells would be economic losers, although wells that produced more than three billion cubic feet were likely economic despite their short lifespan.

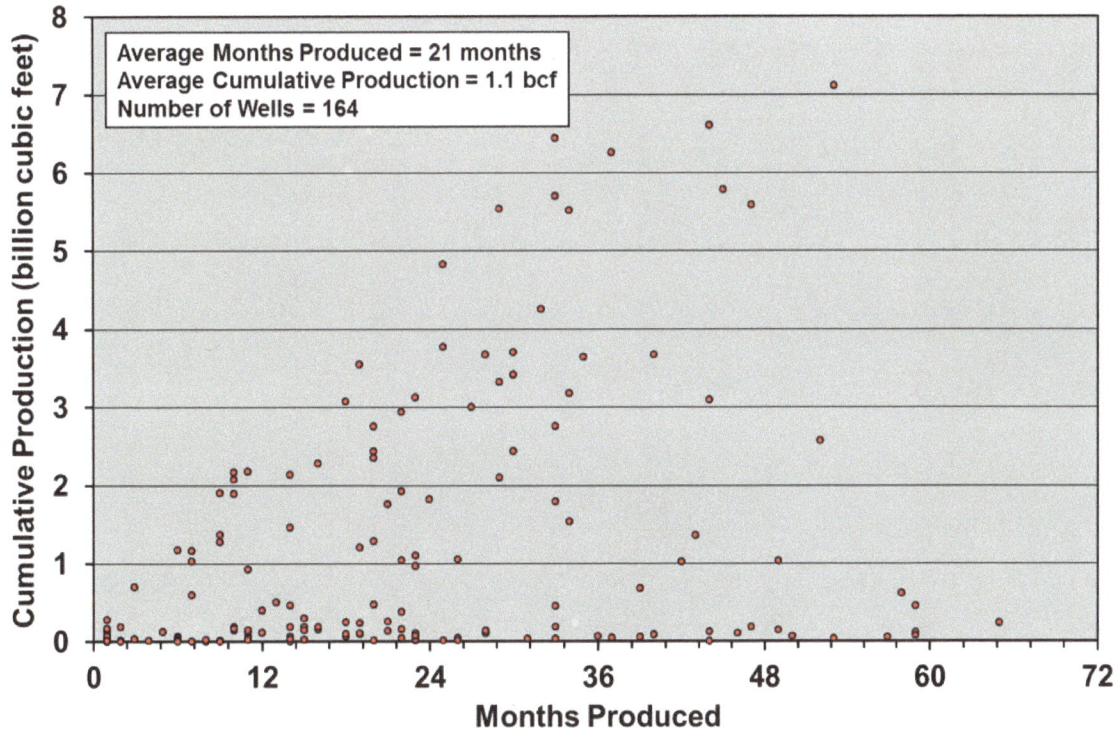

Figure 3-33. Cumulative gas production and length of time produced for Haynesville wells that were not producing as of February 2014.[51]

These well constitute nearly 5% of all horizontal wells drilled; many would be economic failures, given the mean life of 21 months and average cumulative production of 1.1 billion cubic feet when production ended.

---

[51] Data from Drillinginfo retrieved August 2014.

Figure 3-34 illustrates the cumulative production of all wells that were producing in the Haynesville in March 2014. Roughly 18% of the wells have produced more than 4 billion cubic feet over a relatively short lifespan and are clearly economic; however, 33% have yet to produce 2 billion cubic feet. The average well has produced 2.8 billion cubic feet over a lifespan averaging 38 months. Just 8% of these wells are more than 5 years old.

The lifespan of wells is another key parameter as many operators assume a minimum life of 30 years and longer; this is conjectural at this point given the lack of long-term production data.

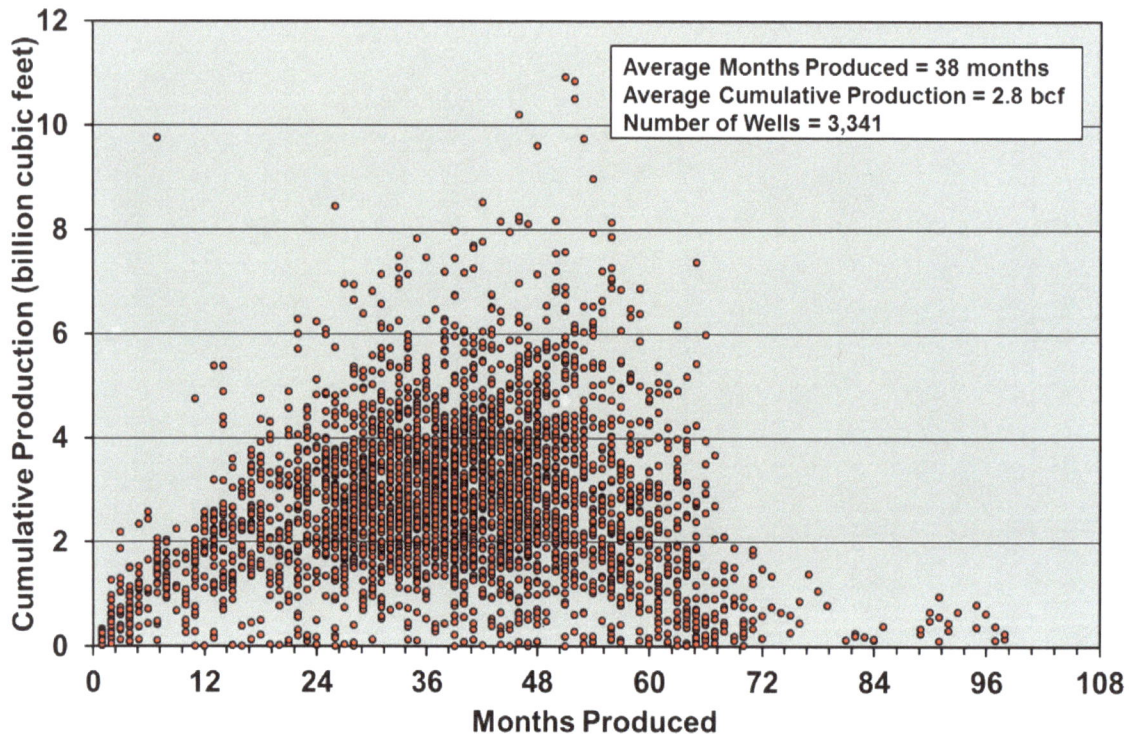

Figure 3-34. Cumulative gas production and length of time produced for Haynesville wells that were producing as of March 2014.[52]

These well constitute 95% of all wells drilled. Very few wells are greater than five years old, with a mean age of 38 months and a mean cumulative recovery of 2.8 billion cubic feet.

[52] Data from Drillinginfo retrieved August 2014.

Cumulative production of course depends on how long a well has been producing, so looking at young wells is not necessarily a good indication of how much gas these wells will produce over their lifespan (although production is heavily weighted to the early years of well life). A measure of well quality, independent of age, is initial productivity (IP), which is often focused on by operators. Figure 3-35 illustrates the average daily output over the first six months of production for all wells in the Haynesville play (six month IP). Again, as with cumulative production, there are a few exceptional wells—3% produced more than 12 million cubic feet per day (MMcf/d)—but the average for all wells drilled since 2009 is 5.72 MMcf/d. Figure 3-27 illustrates the distribution of IPs in map form.

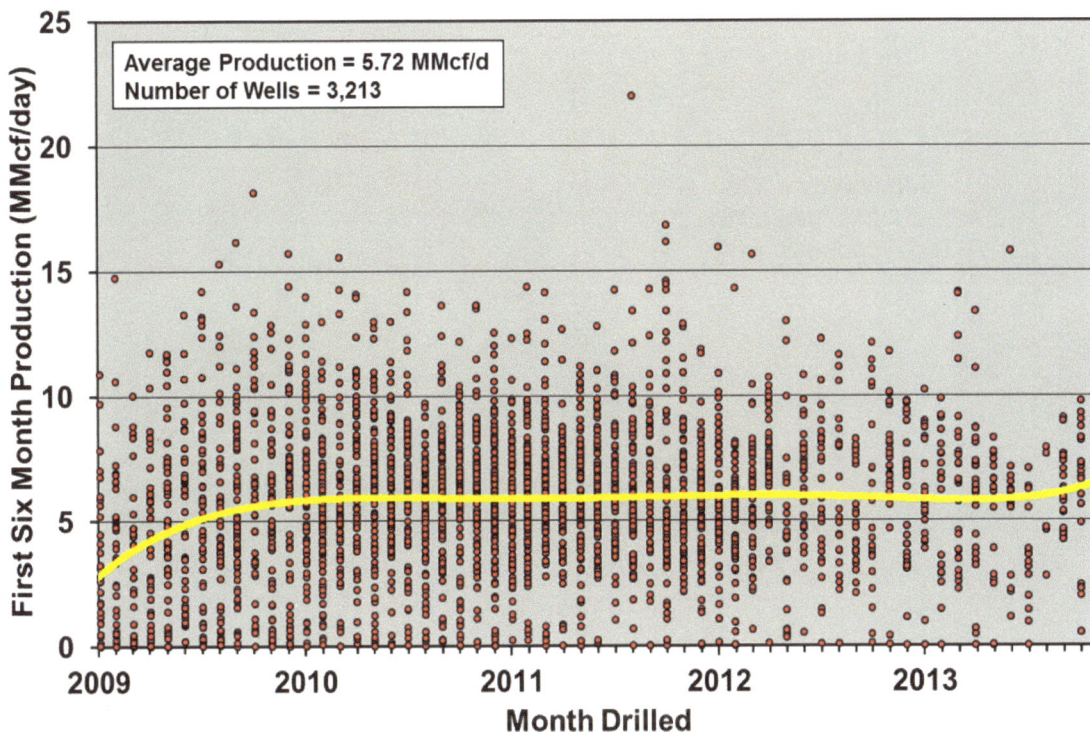

Figure 3-35. Average gas production over the first six months for all wells drilled in the Haynesville play, 2009 to 2014.[53]

Although there are a few exceptional wells, the average well produced 5.48 million cubic feet per day over this period. The trend line indicates mean productivity over time

---

[53] Data from Drillinginfo retrieved August 2014.

Different counties in the Haynesville display different well quality characteristics which are critical in determining the most likely production profile in the future. Figure 3-36, which illustrates production over time by county, shows that, as of April 2014, the top two counties produced 56% of the total, the top four produced 74%, and the remaining 12 counties produced just 26%.

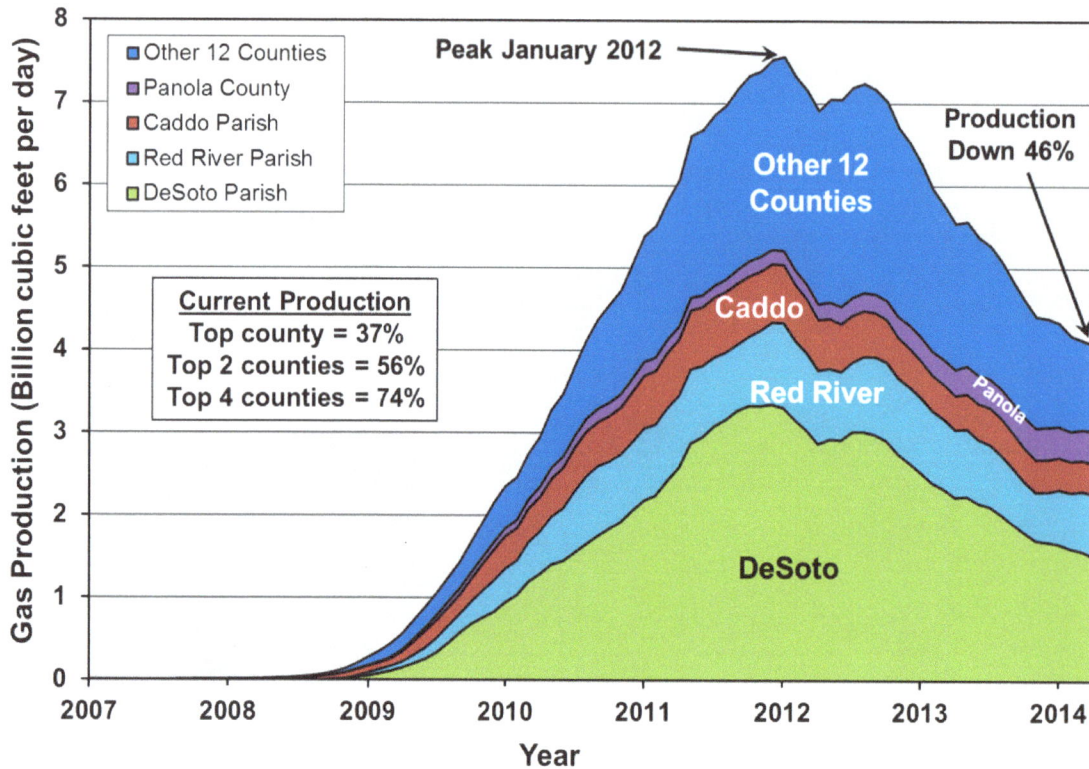

Figure 3-36. Gas production by county in the Haynesville play, 2007 through 2014.[54]
The top four counties produced 74% of production in April 2014.

[54] Data from Drillinginfo retrieved August 2014. Three-month trailing moving average.

The same trend holds in terms of cumulative production since the field commenced. As illustrated in Figure 3-37, the top two counties have produced 56% of the gas and the top four have produced 70%. All of the counties except Panola in Texas have peaked although with increased drilling rates some could conceivably resume production growth. Production in the top county—DeSoto—is down 55% from peak and production in the other counties is down from 26% to 59%.

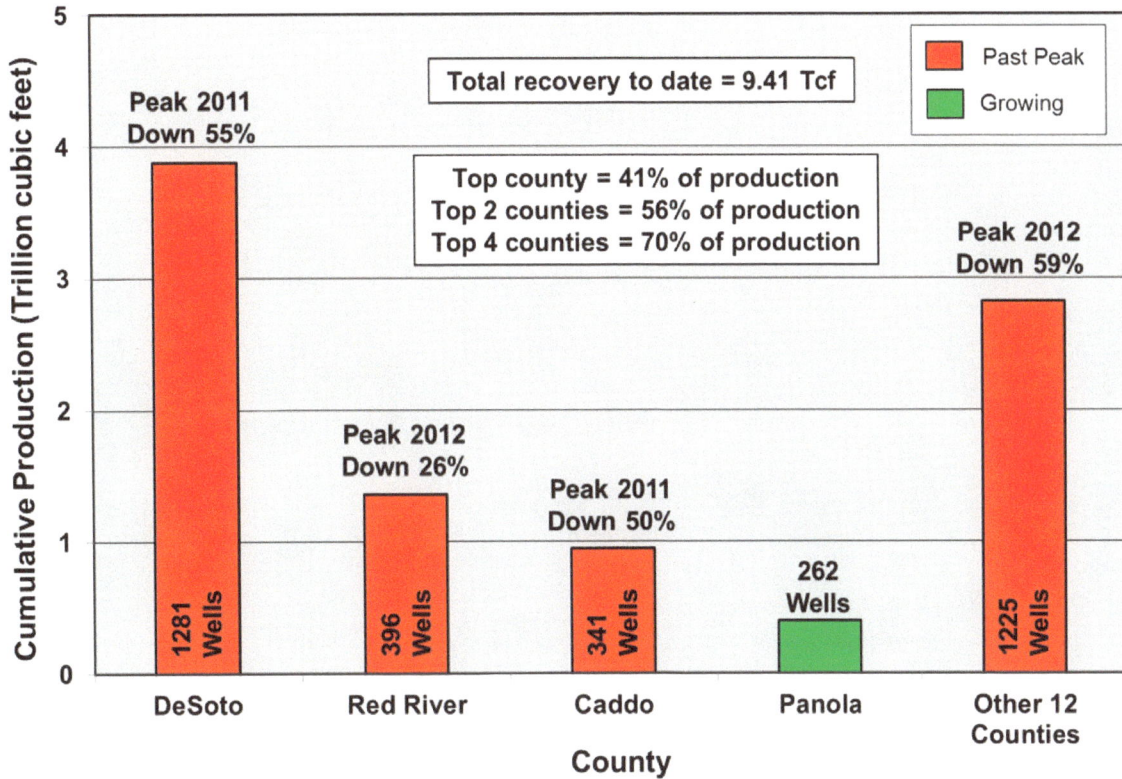

Figure 3-37. Cumulative gas production by county in the Haynesville play through 2014.
The top four counties have produced 70% of the 9.4 trillion cubic feet of gas produced to date.[55]

---

[55] Data from Drillinginfo retrieved August 2014.

The Haynesville also produces very limited amounts of natural gas liquids and oil. Most liquids production is not within the top four counties as illustrated in Figure 3-38. Some 1.5 million barrels of liquids have been produced since 2006, and although it has somewhat improved economics in marginal counties for gas production, in the big picture liquids production from the Haynesville is insignificant.

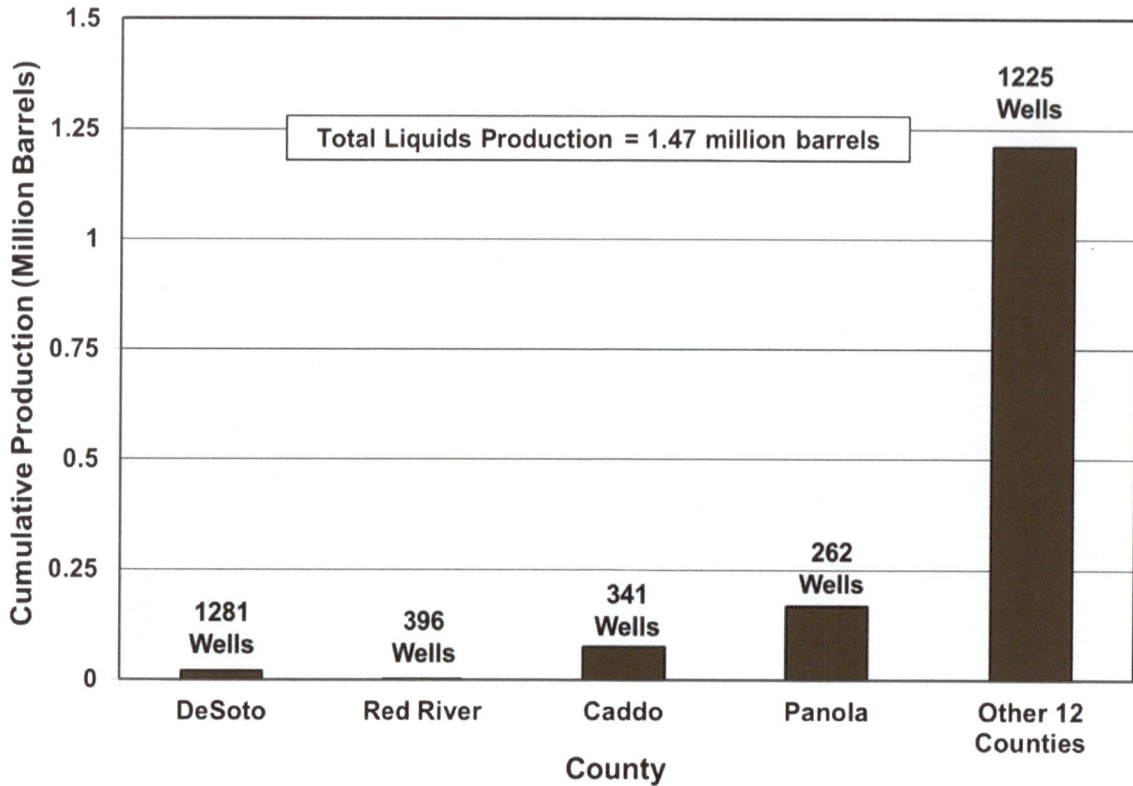

Figure 3-38. Cumulative liquids production by county in the Haynesville play through 2014.

The "other 12" counties account for 82% of the 1.5 million barrels produced to date.[56]

---

[56] Data from Drillinginfo retrieved August 2014.

Operators are highly sensitive to the economic performance of the wells they drill, which typically cost on the order of $9 million or more each[57], not including leasing costs and other expenses. The areas of highest quality—the "core" or "sweet spots"—have now been well defined. Figure 3-39 illustrates average well decline curves by county, which are a measure of well quality. Initial well productivities (IPs) are more closely grouped than in the Barnett, however the top producing counties—DeSoto and Red River in Louisiana, —which are in steep decline—are significantly better than Panola County in Texas, which is the only county growing in production. There are still a significant number of locations in which to drill wells in the top producing counties, although the overall play area of the Haynesville is smaller than plays like the Barnett and is dwarfed by the Marcellus.[58]

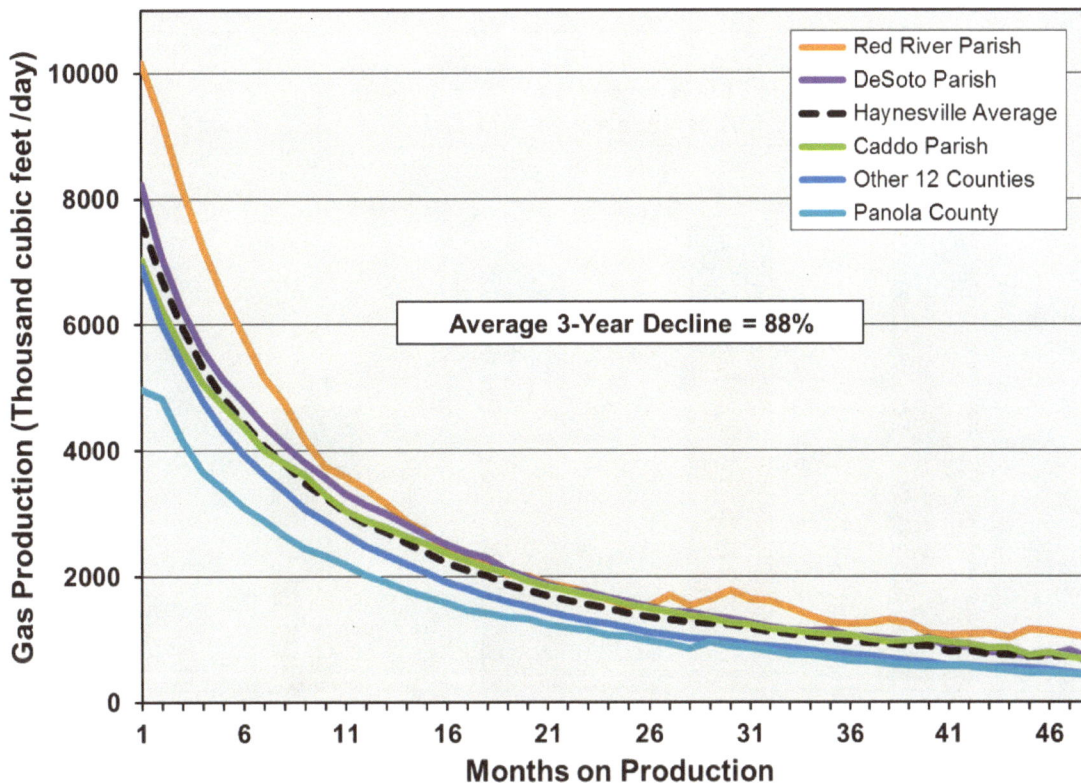

Figure 3-39. Average horizontal gas well decline profiles by county for the Haynesville play.

The top two counties, which have produced much of the gas in the Haynesville, are clearly superior.[59]

Another measure of well quality is "estimated ultimate recovery" or EUR—the amount of gas a well will recover over its lifetime. To be clear, no one knows what the lifespan of an average Haynesville well is, given that few of them are more than five years old (see Figure 3-33 and Figure 3-34), and some 5% of wells drilled have ceased production at an average age of under two years. Operators fit hyperbolic and/or exponential curves to data such as presented in Figure 3-39, assuming well life spans of 30-50 years (as is typical for conventional wells), but so far this is speculation given the nature of the extremely low permeability

[57]Mark J. Kaiser, June 2014, *Oil and Gas Journal*, "HAYNESVILLE UPDATE–2: North Louisiana drilling costs vary slightly 2007-12," http://www.ogj.com/articles/print/volume-112/issue-1/exploration-development/north-louisiana-drilling-costs-vary-slightly-2007-12.html.
[58] EIA, *Assumptions to the Annual Energy Outlook 2014*, http://www.eia.gov/forecasts/aeo/assumptions/pdf/oilgas.pdf.
[59] Data from Drillinginfo retrieved August 2014.

reservoirs and the completion technologies used in the Haynesville. Nonetheless, for comparative well quality purposes only, one can use the data in Figure 3-39, which exhibits steep initial decline with progressively more gradual decline rates, and assume a constant terminal decline rate thereafter to develop a theoretical EUR.

Figure 3-40 illustrates theoretical EURs by county for the Haynesville for comparative purposes of well quality. These range from 3.0 to 5.9 billion cubic feet per well, which agrees fairly well with the 3.14 to 3.71 billion cubic feet assumed by the EIA.[60] The steep initial well production declines mean that well payout, if it is achieved, comes in the first few years of production, as between 70% and 78% of an average well's lifetime production occurs in the first four years.

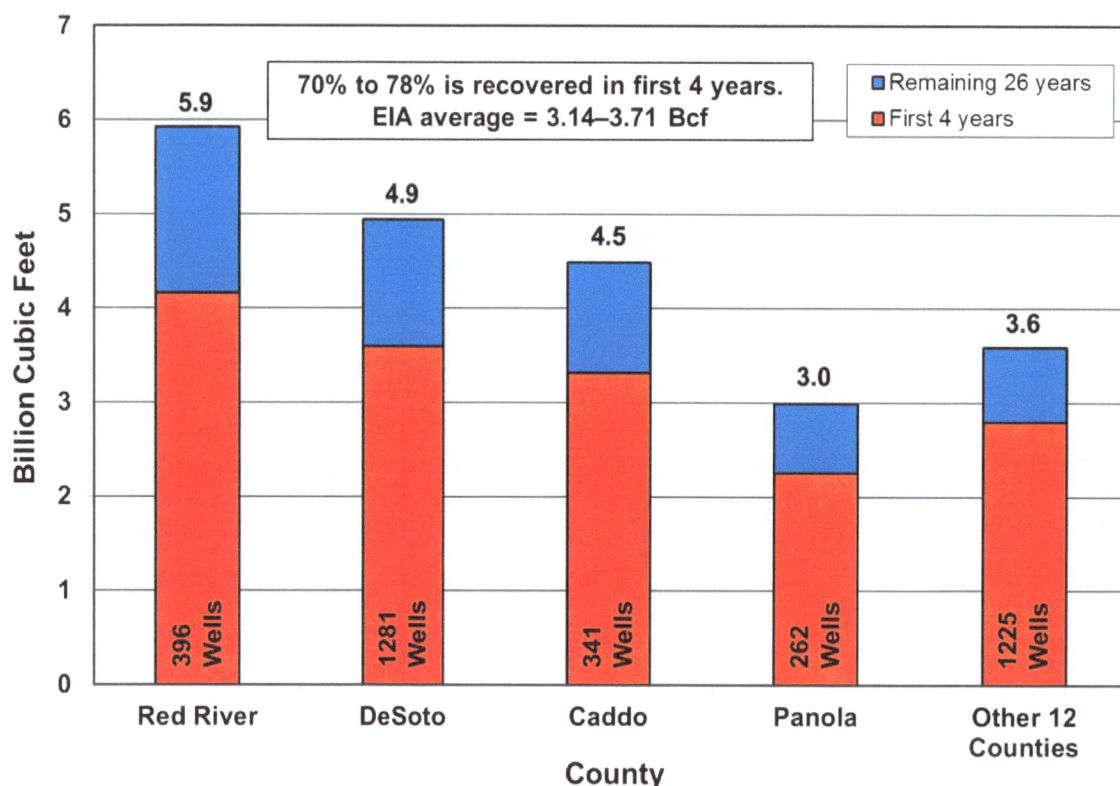

Figure 3-40. Estimated ultimate recovery of gas per well by county for the Haynesville play.[61]

EURs are based on average well decline profiles (Figure 3-39) and a terminal decline rate of 20%. These are for comparative purposes only as it is highly uncertain if wells will last for 30 years. The steep decline rates mean that most production occurs early in well life.

[60] EIA, *Assumptions to the Annual Energy Outlook 2014*, http://www.eia.gov/forecasts/aeo/assumptions/pdf/oilgas.pdf.
[61] Data from Drillinginfo retrieved August 2014.

Well quality can also be expressed as the average rate of production over the first year of well life. If we know both the rate of production in the first year of the average well and the field decline rate, we can calculate the number of wells that need to be drilled each year to offset field decline in order to maintain production. Figure 3-41 illustrates the average first-year production rate of wells by county. Notwithstanding significant gains in Red River Parish (which has the smallest prospective area of the top four counties), the average well quality is flat on average and is declining in Caddo Parish.

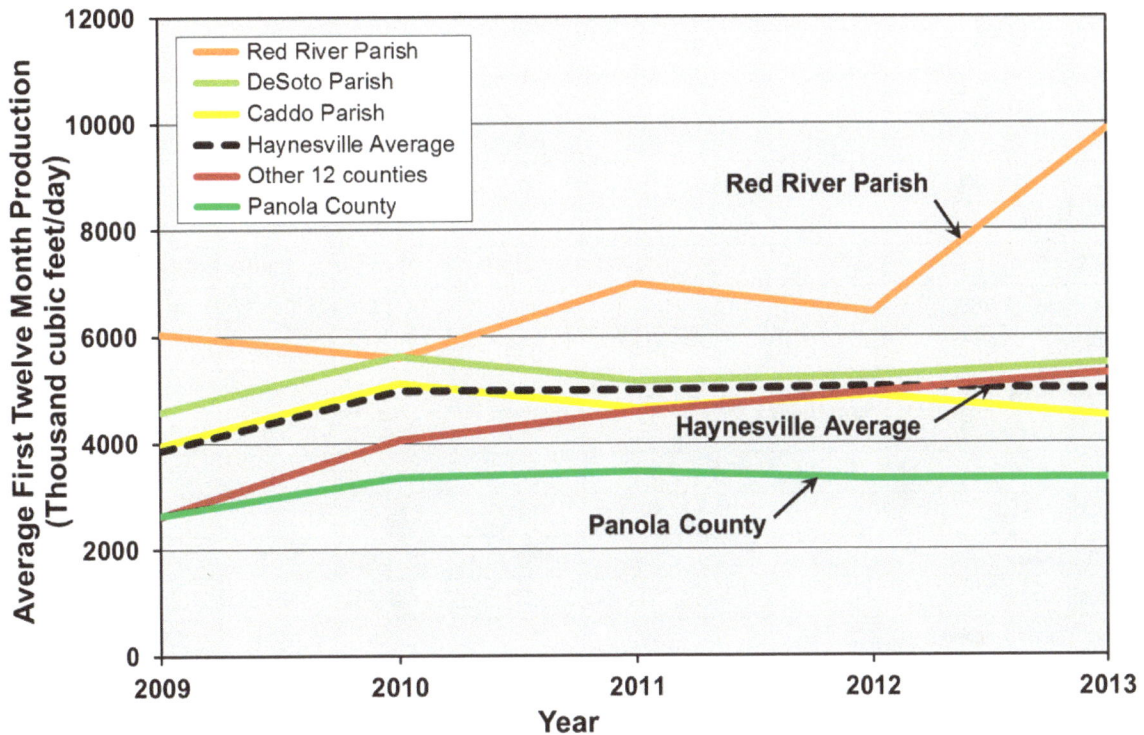

Figure 3-41. Average first year gas production rates of wells by county for the Haynesville play, 2009 to 2013.[62]

Well quality is rising significantly in Red River Parish but is flat on average for the play as a whole. Panola County, which is the only county in which production is rising, had first-year average well production of less than half that of Red River Parish in 2013.

---

[62] Data from Drillinginfo retrieved August 2014.

3.3.2.4 Number of Wells

The fourth key fundamental is the number of wells that can ultimately be drilled in the Haynesville play. The EIA has estimated the total play area in Louisiana and Texas at 3,419 square miles and suggests this can be drilled at a well density of six per square mile, for a total of 20,511 wells. As 3,505 wells have already been drilled this leaves 17,006 yet-to-drill wells.

Table 3-2 breaks down the number of yet-to-drill wells by county along with other critical parameters used for determining the future production rates of the Haynesville play.

Parameter	County					Total
	Caddo	DeSoto	Panola	Red River	Other 12	
Production April 2014 (Bcf/d)	0.37	1.51	0.39	0.76	1.04	4.08
% of Field Production	9.1	37.1	9.6	18.7	25.6	100.0
Cumulative Gas (Tcf)	0.95	3.88	0.40	1.36	2.82	9.41
Cumulative Liquids (MMBBL)	0.07	0.02	0.17	0.00	1.21	1.47
Number of Wells	341	1281	262	396	1225	3505
Number of Producing Wells	326	1216	243	369	1120	3274
Average EUR per well (Bcf)	4.5	4.9	3	5.9	3.6	4.9
Field Decline (%)	34	50	52	49	50	49
3-Year Well Decline (%)	86	87	87	88	89	88
Peak Month	Sep-11	Dec-11	Rising	Jan-12	Jul-12	Jan-12
% Below Peak	50	55	Rising	29	59	46
Average First Year Production in 2013 (Mcf/d)	4492	5493	3330	9881	5286	5011
New Wells Needed to Offset Field Decline	28	138	61	38	99	399
Area in square miles	937	895	801	402	8000	11035
% Prospective	35	90	90	80	15	31
Net square miles	328	806	721	322	1243	3419
Well Density per square mile	1.04	1.59	0.36	1.23	0.99	1.03
Additional locations to 6/sq. Mile	1627	3552	4063	1534	6230	17006
Population	254969	26656	22756	9091	N/A	N/A
Total Wells 6/sq. Mile	1968	4833	4325	1930	7455	20511
Total Producing Wells 6/sq. Mile	1952	4768	4306	1902	7351	20280

Table 3-2. Parameters for projecting Haynesville production, by county.

Area in square miles under "Other" is estimated.

3.3.2.5   Rate of Drilling

Given known well- and field-decline rates, well quality by area, and the number of available drilling locations, the most important parameter in determining future production levels is the rate of drilling—the fifth key fundamental. Figure 3-42 illustrates the historical drilling rates in the Haynesville. Horizontal drilling rates peaked in 2011 at 1,051 wells per year and have fallen to current levels of about 200 wells per year. Current drilling rates are only half the 400 wells per year required to maintain production at current levels, hence each new well drilled now serves only to slow the overall production decline of the play. It appears that the drilling rate is stabilizing at 200 wells per year so production will keep falling until this number of wells is sufficient to offset field decline.

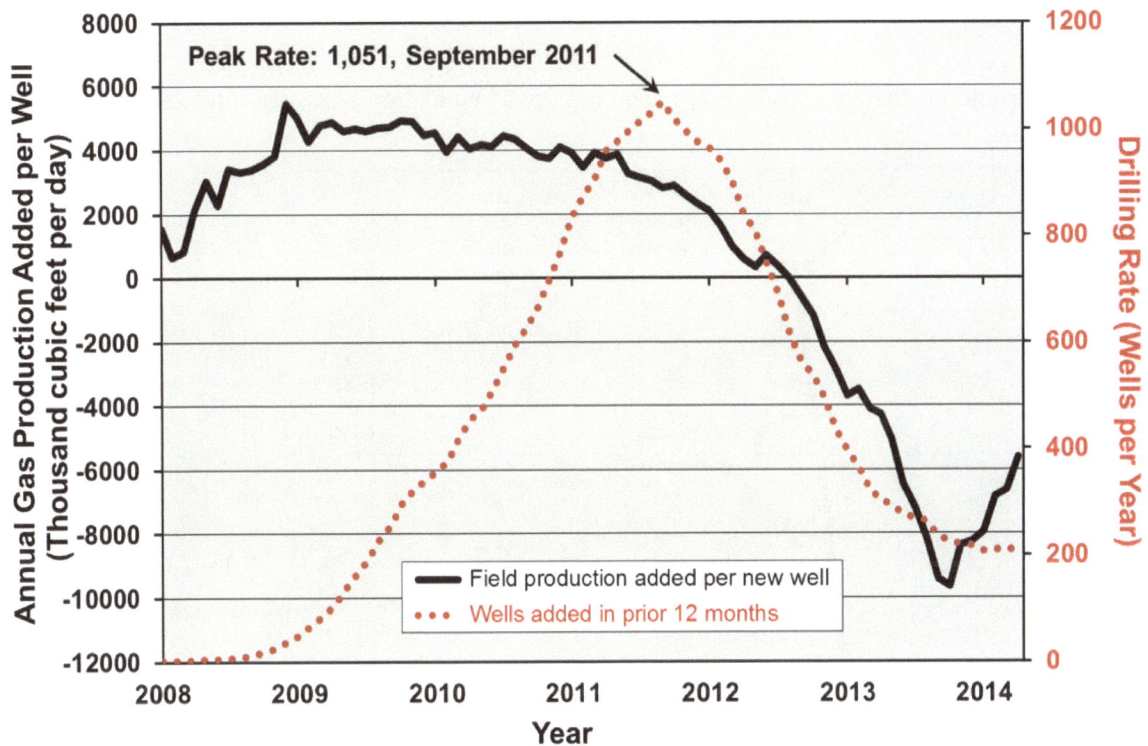

Figure 3-42. Annual gas production added per new horizontal well and annual drilling rate and in the Haynesville play, 2008 through 2014.[63]

Drilling rate peaked in 2011 and is now far below the level needed to keep production flat, hence each new well now only serves to slow the overall production decline of the play.

---

[63] Data from Drillinginfo retrieved August 2014. Three-month trailing moving average.

3.3.2.6 Future Production Scenarios

Based on the five key fundamentals outlined above, several production projections for the Haynesville play were developed to illustrate the effects of changing the rate of drilling. Figure 3-43 illustrates the production profiles of three drilling rate scenarios if 100% of the prospective play area is drillable at six wells per square mile (the EIA estimate of well density as well as drillable area[64]). These scenarios are:

1. MOST LIKELY RATE scenario: Drilling increases by 50% from the current rate to 300 wells per year.

2. LOW RATE scenario: Drilling remains at the current rate of 200 wells per year and holds constant.

3. HIGH RATE scenario: Drilling more than doubles to 500 wells per year, then gradually declines to 300 wells per year.

In all of these scenarios there are sufficient drilling locations to maintain drilling beyond 2040.

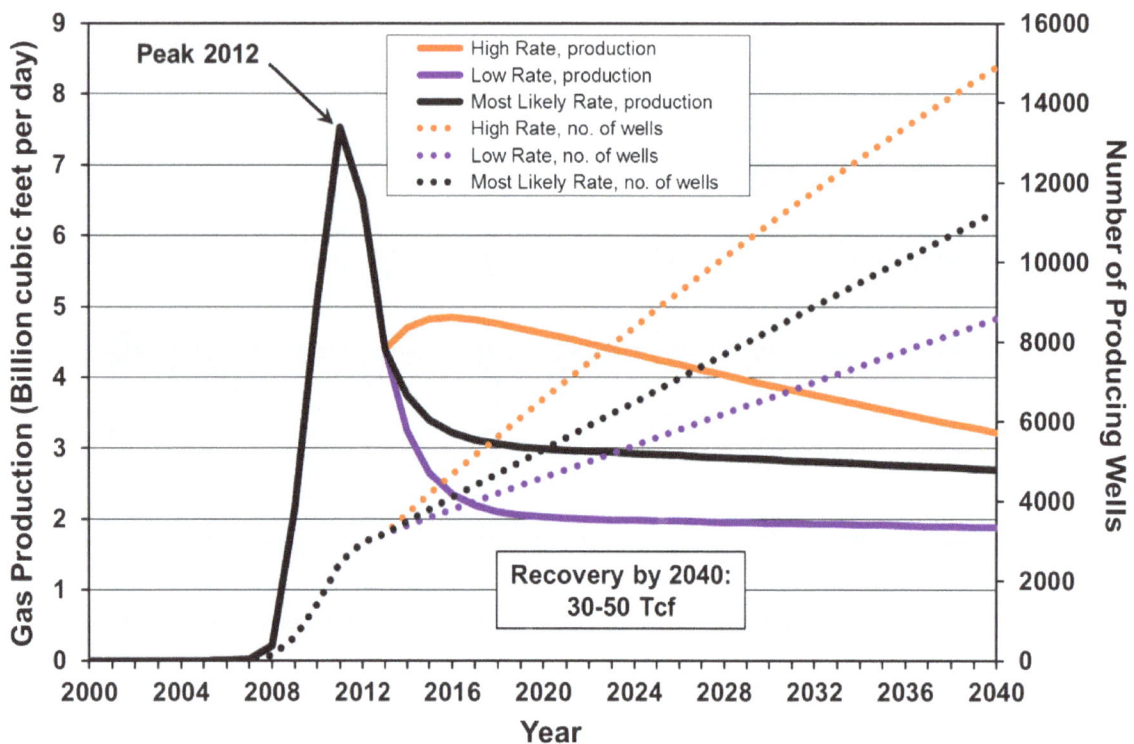

Figure 3-43. Three drilling rate scenarios of Haynesville gas production (assuming 100% of the area is drillable at six wells per square mile).[65]

"Most Likely Rate" scenario: drilling increases to 300 wells/year, holding constant.
"Low Rate" scenario: drilling holds constant at 200 wells/year.
"High Rate" scenario: drilling increases to 500 wells/year, declining to 300 wells/year.

---

[64] EIA, *Assumptions to the Annual Energy Outlook 2014*, http://www.eia.gov/forecasts/aeo/assumptions/pdf/oilgas.pdf.
[65] Data from Drillinginfo retrieved August 2014.

The drilling rate scenarios have the following results:

1. MOST LIKELY RATE scenario: Production will continue to fall until it stabilizes at about 3 billion cubic feet per day—less than half of the Haynesville's peak rate. Total gas recovery by 2040 would be 38.4 trillion cubic feet and drilling would continue beyond 2040.

2. LOW RATE scenario: Production will continue to fall until stabilizing at about 2 billion cubic feet per day—less than a third of peak production rates. Total gas recovery by 2040 would be 29.7 trillion cubic feet and drilling would continue beyond 2040.

3. HIGH RATE scenario: Production decline in the Haynesville could be temporarily reversed and grow somewhat in the short term. Total gas recovery by 2040 would be 49.8 trillion cubic feet and drilling would continue beyond 2040.

Total recovery of 38.4 trillion cubic feet by 2040 in the "Most Likely Rate" scenario is four times what has been recovered so far in the Haynesville, and in the "High Rate" scenario as much as 49.8 trillion cubic feet could be recovered; however, production rates would be far below those projected by the EIA for the Haynesville play.

3.3.2.7   Comparison to EIA Forecast

Figure 3-44 illustrates the EIA's projection for Haynesville production through 2040 compared to the "Most Likely Rate" scenario. The EIA projects a recovery by 2040 of 102 Tcf to meet its reference case forecast, and projects a new peak of the play in 2027 at a level far higher than the early-2012 peak. This represents the recovery of 110% of both proved reserves[66] and unproved resources.[67] Furthermore, the EIA projects that production in 2040 will be higher than the 2012 peak, suggesting that vastly more gas will be recovered beyond 2040. This strains credibility to the limit. How can all the proved and unproved resources and reserves be extracted and still have production above all-time highs in 2040?

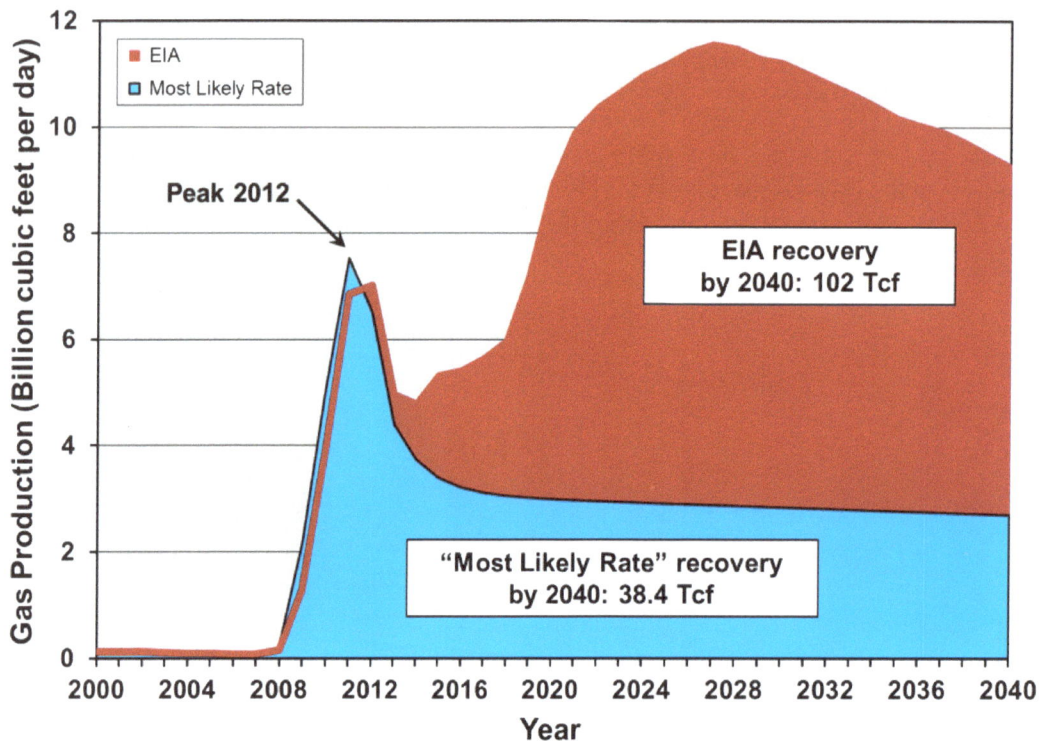

Figure 3-44. "Most Likely Rate" scenario of Haynesville gas production compared to the EIA reference case, 2000 to 2040.[68]

The EIA assumes the Haynesville will reach a new all-time high by 2027, produce 110% of proved reserves and unproved resources by 2040, and presumably produce a great deal more gas in the post-2040 period. . The EIA forecast is made on a "dry gas" basis, whereas the "Most Likely Rate" scenario forecast is made on a "raw gas" basis.

---

[66] EIA, 2014, "Principal shale gas plays: natural gas production and proved reserves, 2011-12,"
http://www.eia.gov/naturalgas/crudeoilreserves/excel/table_4.xls.
[67] EIA, *Assumptions to the Annual Energy Outlook 2014*, http://www.eia.gov/forecasts/aeo/assumptions/pdf/oilgas.pdf.
[68] EIA, *Annual Energy Outlook 2014*, unpublished tables from AEO 2014 provided by the EIA.

3.3.2.8 Haynesville Play Analysis Summary

Several things are clear from this analysis:

1. Drilling rates have fallen markedly in the Haynesville due to gas prices, although there are still locations to drill in the sweet spots.

2. High well- and field-decline rates mean a continued high rate of drilling is required to maintain, let alone increase, production. The Haynesville field decline rate of 49% is the highest observed in any shale gas play. Current drilling rates of 200 wells per year are just half of the level required to maintain production. Maintaining production at current levels would require the investment of $3.6 billion per year for drilling (assuming $9 million per well). Future production profiles are most dependent on drilling rate and, to a lesser extent, on the number of drilling locations (i.e., greatly increasing the number of drilling locations would not change the production profile nearly as much as changing the drilling rate). Maintaining or growing production in the Haynesville would require considerably higher gas prices to justify higher drilling rates.

3. More than doubling current drilling rates could reverse the current production decline temporarily and raise production somewhat, but nowhere near its early 2012 peak. Cumulative recovery by 2040 in this high drilling rate scenario would be increased by 30% over the "Most Likely Rate" scenario but would still be less than half that projected by the EIA in its reference case.

4. The projected recovery of 38.4 Tcf by 2040 in the "Most Likely Rate" scenario represents four times as much gas as has been recovered so far from the Haynesville, yet is only 38% of the 102 Tcf projected by the EIA in its reference case forecast.

5. This report's projections are optimistic in that they assume the capital will be available for the drilling treadmill that must be maintained. They also assume that 100% of the prospective area is drillable. This is not a sure thing as drilling in the poorer quality parts of the play will require higher gas prices to be economic. Failure to increase current drilling rates will result in a steeper drop off in production.

6. Nearly four times the current number of wells will need to be drilled to meet the production projection of the "Most Likely Rate" scenario by 2040.

7. The EIA projection for future Haynesville gas production included in its reference case forecast for AEO 2014,[69] which forecasts recovery of 110% of proved reserves plus unproved resources by 2040, strains credibility to the limit. It is highly unlikely to be realized, especially at the gas prices the EIA forecasts.[70]

---

[69] EIA, *Annual Energy Outlook 2014*, unpublished tables from AEO 2014 provided by the EIA.
[70] EIA, *Annual Energy Outlook 2014*, http://www.eia.gov/forecasts/aeo/.

3.3.3   Fayetteville Play

The EIA forecasts recovery of 41.5 Tcf of gas from the Fayetteville play by 2040. The analysis of actual production data presented below suggests that this forecast is highly unlikely to be realized.

The Fayetteville play was discovered in Arkansas in 2005 and production grew rapidly until its peak in late 2012. Since that time it has been on an undulating production plateau with production down just over 2% since peak. Figure 3-45 illustrates the distribution of wells as of early 2014. Nearly 5,300 wells have been drilled to date of which 4,914 were producing at the time of writing. The play covers parts of 10 counties although most of the drilling is concentrated in Cleburne, Conway, Faulkner, Van Buren and White counties.

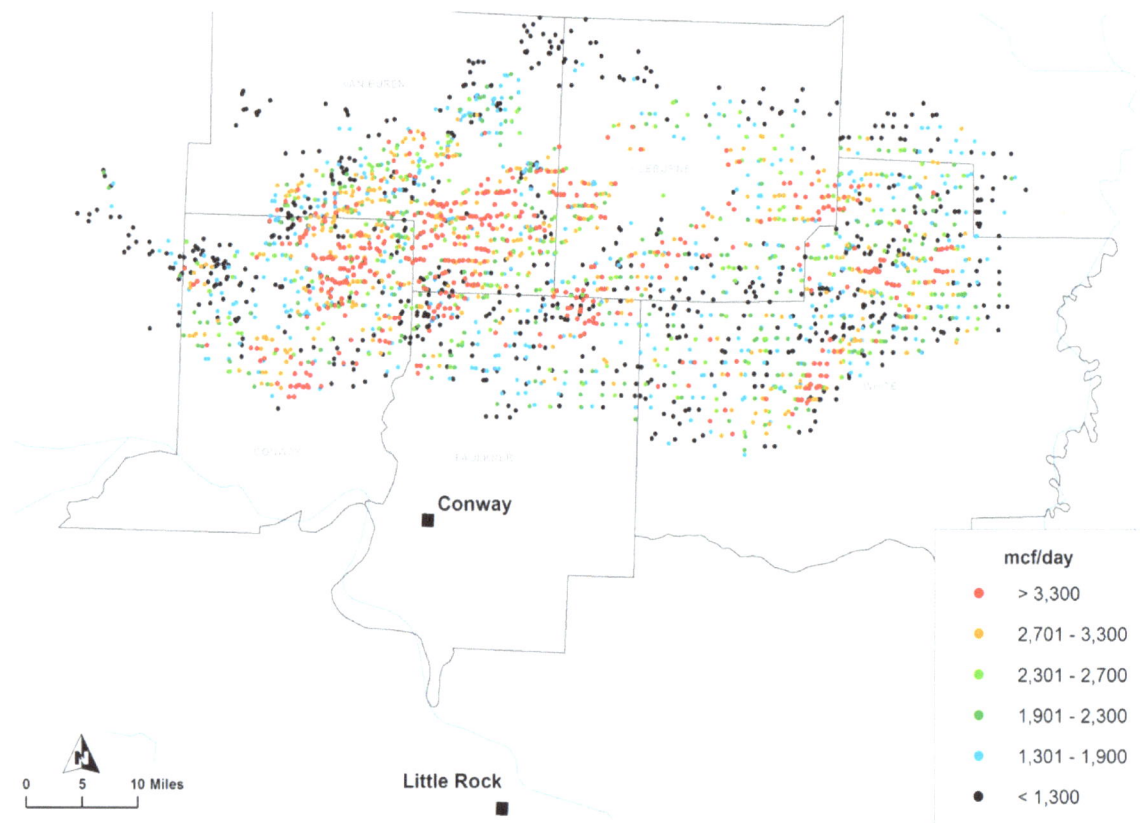

Figure 3-45. Distribution of wells in the Fayetteville play as of early 2014, illustrating highest one-month gas production (initial productivity, IP).[71]

Well IPs are categorized approximately by percentile; see Appendix.

[71] Data from Drillinginfo retrieved April 2014.

Production in the Fayetteville peaked at nearly 3 billion cubic feet per day in December 2012 as illustrated in Figure 3-46. Ninety-nine percent of current production is from horizontal fracked wells. Horizontal drilling grew from virtually nothing in 2006 to a peak rate of nearly 900 wells per year in late-2010. It has since fallen to 500 wells per year, which is insufficient to offset field decline. Drilling rates required to keep production flat at current production levels are about 600 wells per year.

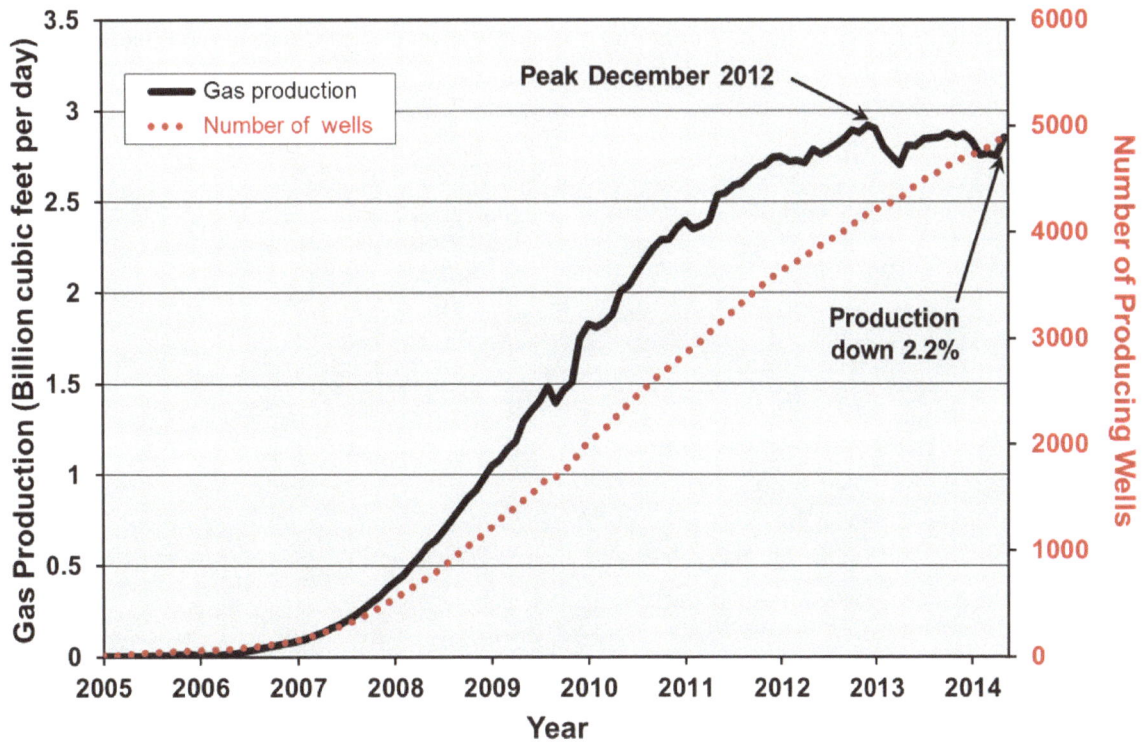

Figure 3-46. Fayetteville play shale gas production and number of producing wells, 2005 to 2014.[72]

Gas production data are provided on a "raw gas" basis.

---

[72] Data from Drillinginfo retrieved August 2014. Three-month trailing moving average.

3.3.3.1   Well Decline

The first key fundamental in determining the life cycle of Fayetteville production is the *well decline rate*. Fayetteville wells exhibit high decline rates in common with all shale plays. Figure 3-47 illustrates the average decline rate of Fayetteville wells. Decline rates are steepest in the first year and are progressively less in the second and subsequent years. The average decline rate over the first three years of well life is 79%, which is well within the typical range of shale plays. Wells are generally more productive than Barnett wells and less so than Haynesville wells. Production is almost exclusively dry gas with no liquids.

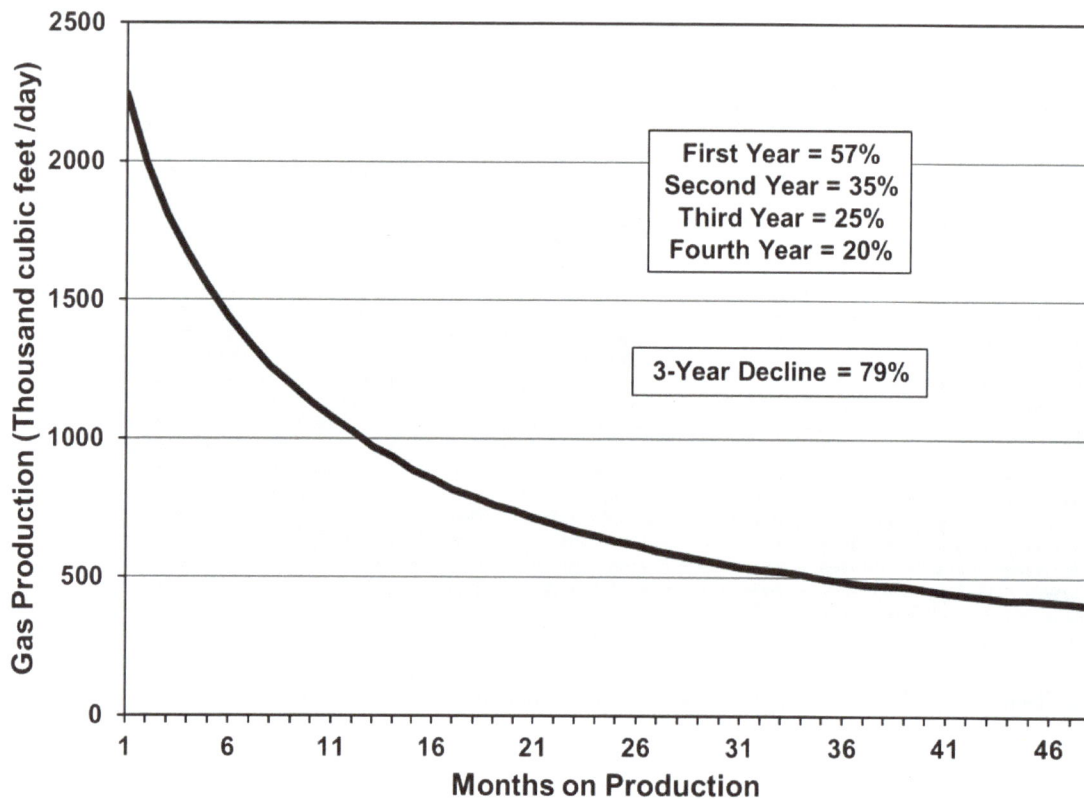

First Year = 57%
Second Year = 35%
Third Year = 25%
Fourth Year = 20%

3-Year Decline = 79%

Figure 3-47. Average decline profile for horizontal gas wells in the Fayetteville play.[73]
Decline profile is based on all shale gas wells drilled since 2009.

[73] Data from Drillinginfo retrieved August 2014.

3.3.3.2 Field Decline

A second key fundamental is the overall *field decline rate*, which is the amount of production in the Fayetteville that would be lost in a year without more drilling. Figure 3-48 illustrates production from the 4,200 wells drilled prior to 2013. The first-year decline rate is 34%. This is lower than the well decline rate as the field decline is made up of both new wells declining at high rates and older wells declining at lesser rates. Assuming new wells will produce in their first year at the average first-year rates observed for wells drilled in 2013, approximately 600 new wells each year would be required to offset field decline at current production levels. At an average cost of $2.4 million per well[74], this would represent a capital input of about $1.4 billion per year, exclusive of leasing and other ancillary costs, to keep production flat at 2013 levels. Fayetteville wells are among the cheapest of any shale play and this is likely what has allowed relatively high rates of drilling to be maintained.

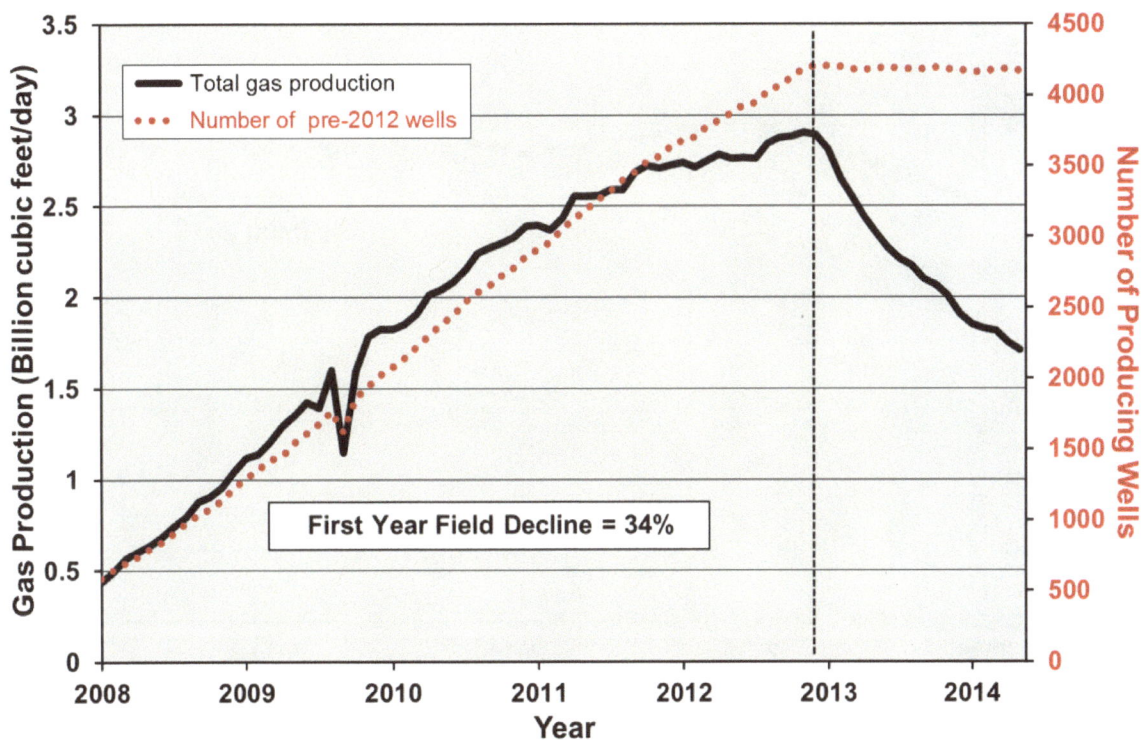

Figure 3-48. Production rate and number of horizontal shale gas wells drilled in the Fayetteville play prior to 2013, 2008 to 2014.[75]

This defines the field decline for the Fayetteville play, which is 34% per year.

[74] Fayetteville Shale, Southwest Energy, 2014, http://www.swn.com/operations/pages/fayettevilleshale.aspx.
[75] Data from Drillinginfo retrieved August 2014.

3.3.3.3 Well Quality

The third key fundamental is the *average well quality* in the Fayetteville by area and its trend over time. Petroleum engineers tell us that technology is constantly improving, with longer horizontal laterals, more frack stages per well, more sophisticated mixtures of proppants and other additives in the frack fluid injected into the wells, and higher-volume frack treatments. This has certainly been true over the past few years, along with multi-well pad drilling which has reduced well costs. It is, however, approaching the limits of diminishing returns, with average well productivity in the Fayetteville up just 2% in 2013, after rising significantly in the early years of the play, as illustrated in Figure 3-49. Given the propensity of operators to drill their best locations first, the slight increase in average quality may have as much to do with concentrating drilling on the highest quality locations as with improvements in technology.

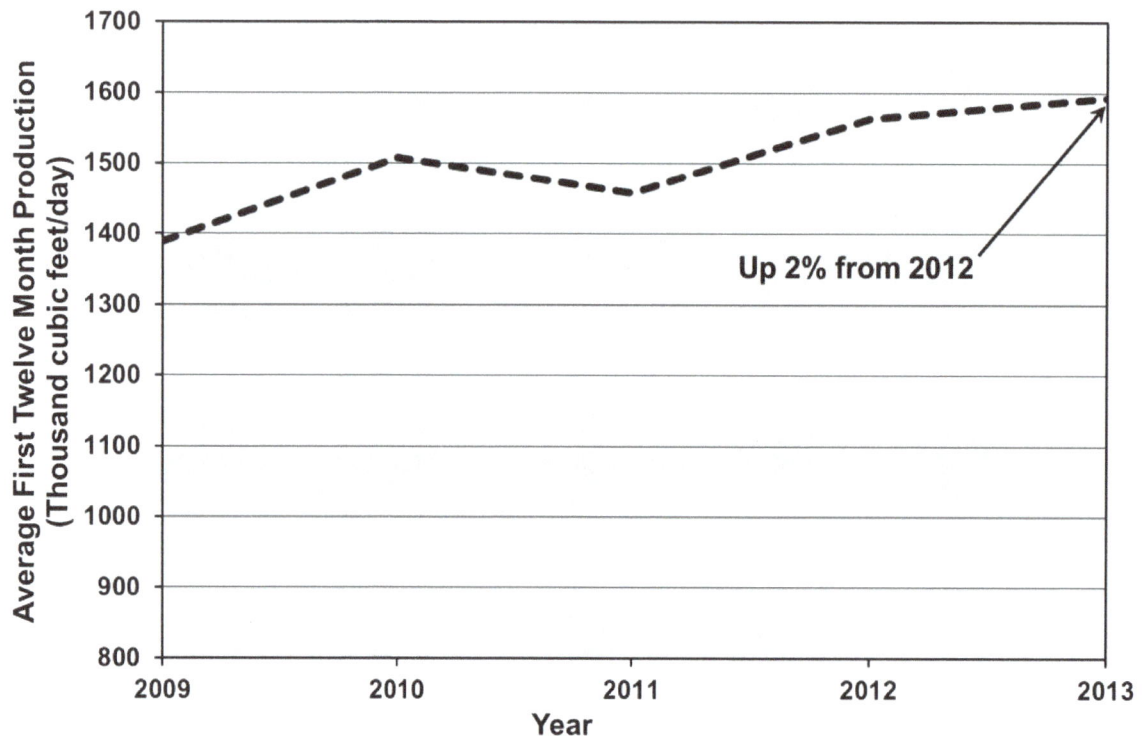

Figure 3-49. Average first-year production rates for Fayetteville gas wells, 2009 to 2013.[76]

Average well quality rose slightly in the most recent year.

[76] Data from Drillinginfo retrieved August 2014.

Another measure of well quality is cumulative production and well life. Nearly 8% of the wells that have been drilled in the Fayetteville are no longer productive. Figure 3-50 illustrates the cumulative production of these shut-down wells over their lifetime. At a mean lifetime of 31 months and a mean cumulative production of 0.34 billion cubic feet, most of these wells would be economic losers.

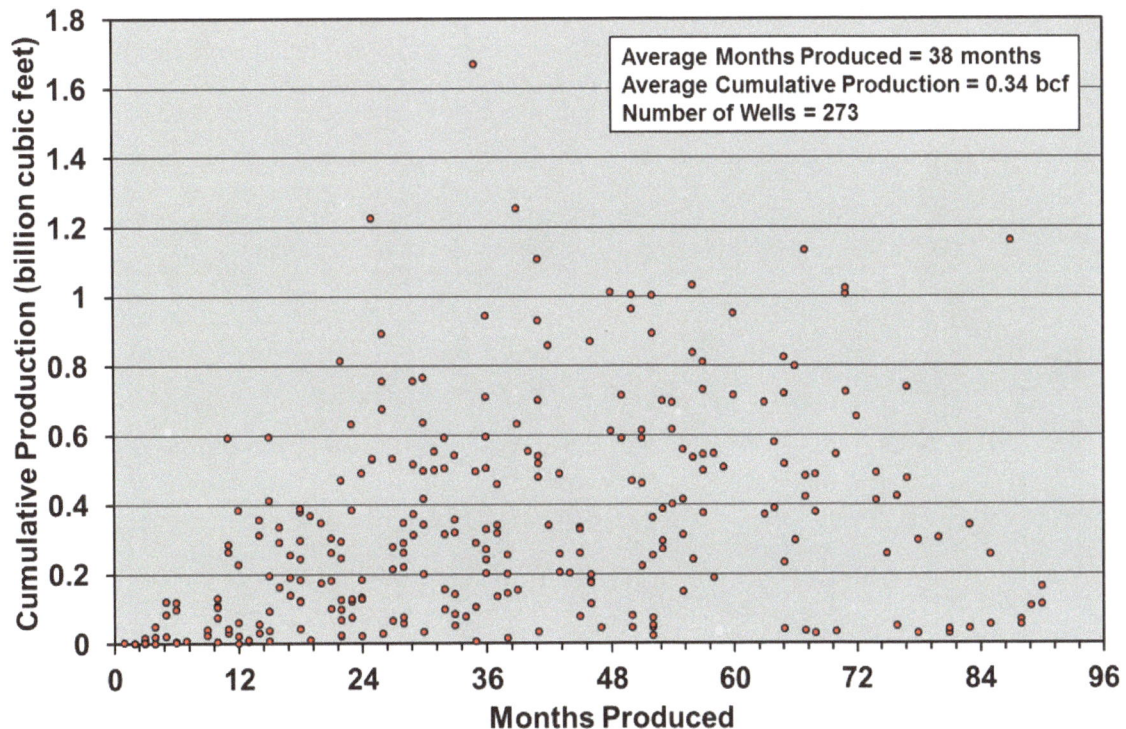

Figure 3-50. Cumulative gas production and length of time produced for Fayetteville wells that were not producing as of February 2014.[77]

These well constitute nearly 8% of all wells drilled; most would be economic failures, given the mean life of 38 months and average cumulative production of 0.34 billion cubic feet when production ended.

---

[77] Data from Drillinginfo retrieved August 2014.

Figure 3-51 illustrates the cumulative production of all wells that were producing in the Fayetteville in March 2014. Roughly 6% of the wells have produced more than 2 billion cubic feet over a relatively short lifespan and are clearly economic, however 57% have yet to produce 1 billion cubic feet. The average well has produced 0.99 billion cubic feet over a lifespan of 44 months. Just 5% of these wells are more than 7 years old.

The lifespan of wells is another key parameter as many operators assume a minimum well life of 30 years and longer, though this is conjectural given the lack of long term production data.

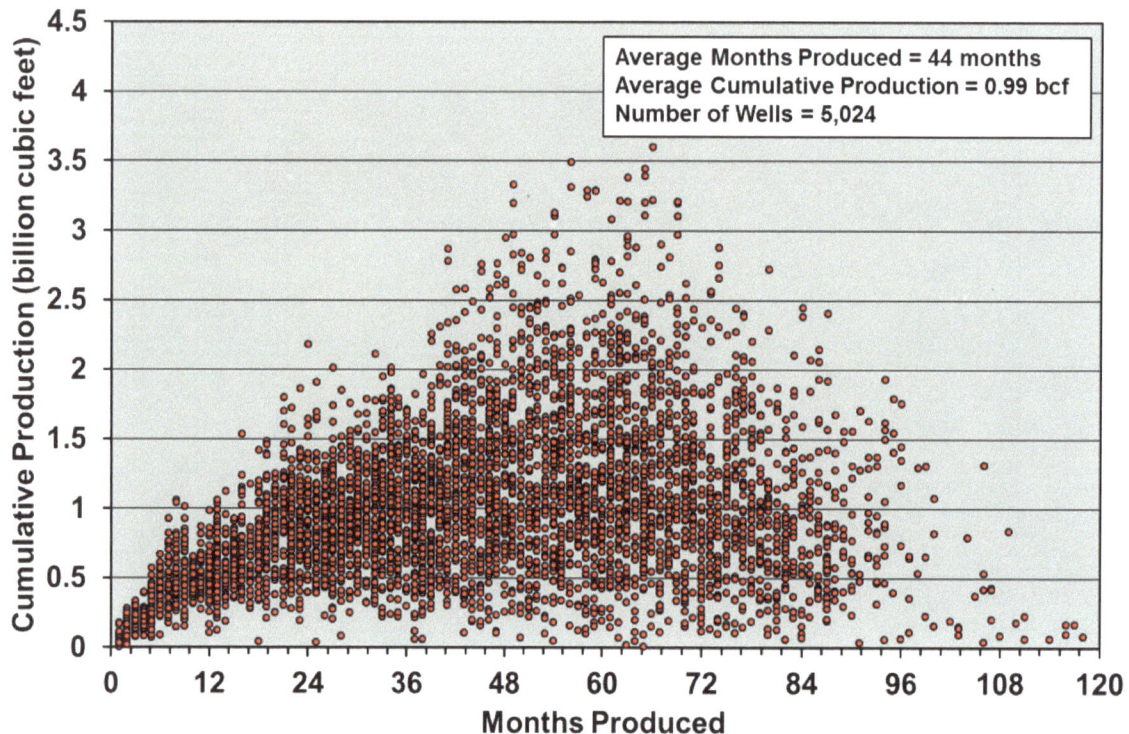

Figure 3-51. Cumulative gas production and length of time produced for Fayetteville wells that were producing as of March 2014.[78]

These constitute 92% of all wells drilled. Very few wells are greater than seven years old, with a mean age of 44 months and a mean cumulative recovery of 0.99 billion cubic feet.

[78] Data from Drillinginfo retrieved August 2014.

Cumulative production of course depends on how long a well has been producing, so looking at young wells in not necessarily a good indication of how much gas these wells will produce over their lifespan (although production is heavily weighted to the early years of well life). A measure of well quality, independent of age, is initial productivity (IP). Figure 3-52 illustrates the average daily output over the first six months of production for all wells in the Fayetteville play (six-month IP). Again, as with cumulative production, there are a few exceptional wells—5% produced more than 3 million cubic feet per day (MMcf/d)—but the average for all wells drilled since 2009 is 1.73 MMcf/d. Figure 3-45 illustrates the distribution of IPs in map form.

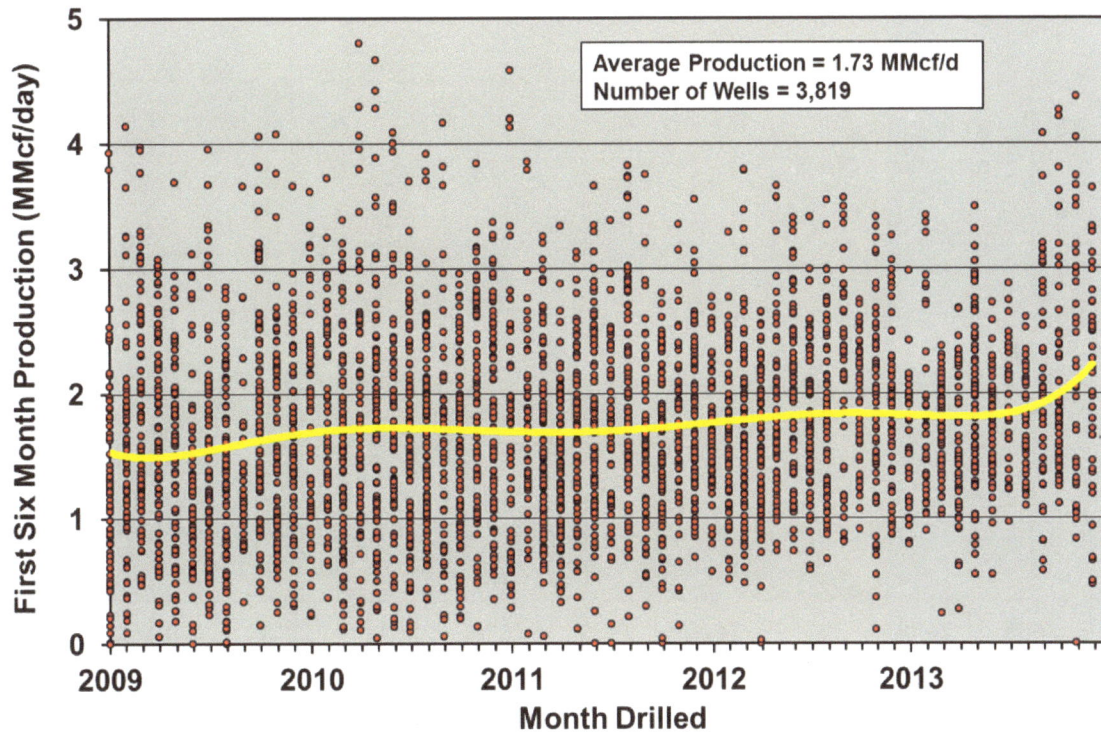

Figure 3-52. Average gas production over the first six months for all wells drilled in the Fayetteville play, 2009 to 2014.[79]

Although there are a few exceptional wells, the average well produced 1.73 million cubic feet per day over this period. The trend line indicates mean productivity over the time period.

---

[79] Data from Drillinginfo retrieved August 2014.

Different counties in the Fayetteville display different well quality characteristics, which are critical in determining the most likely production profile in the future. Figure 3-53, which illustrates production over time by county, shows that, as of May 2014, the top two counties produced 53% of the total, the top four produced 93%, and the remaining 6 counties produced just 7%. All counties are below peak production except Cleburne.

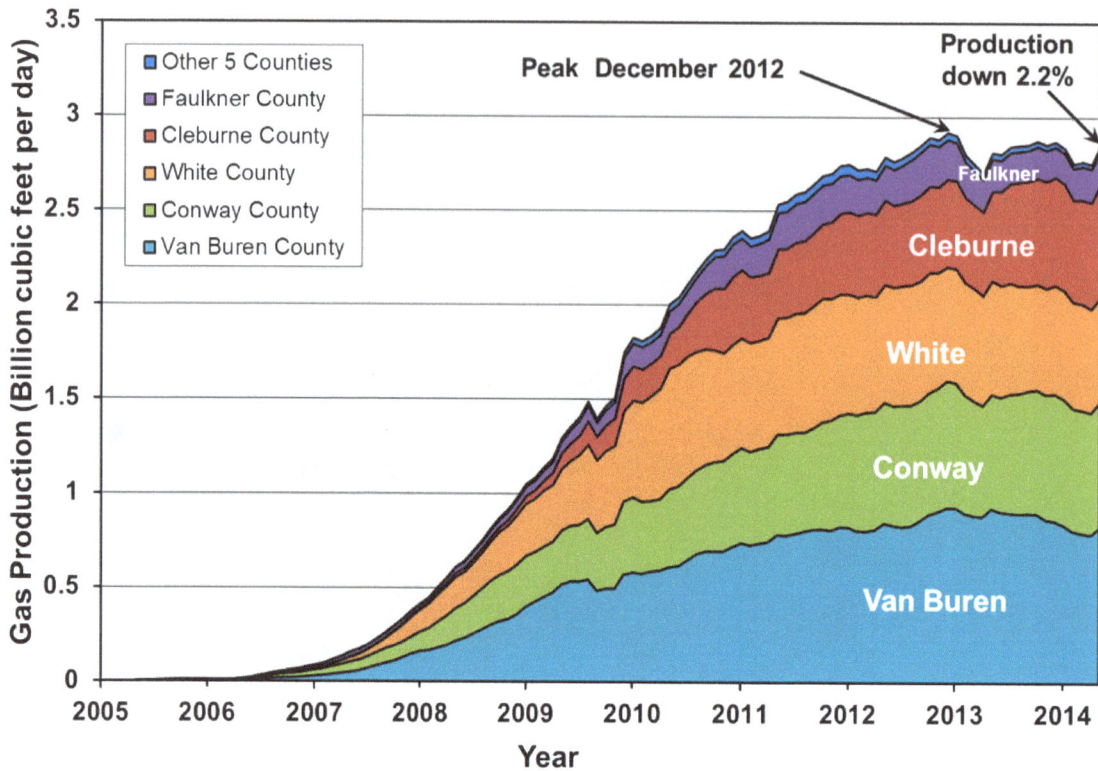

Figure 3-53. Gas production by county in the Fayetteville play, 2005 through 2014.[80]
The top four counties produced 93% of production in May 2014.

---

[80] Data from Drillinginfo retrieved August 2014. Three-month trailing moving average.

The same trend holds in terms of cumulative production since the field commenced. As illustrated in Figure 3-54, the top two counties have produced 54% of the gas and the top four have produced 92%. All of the counties except Cleburne have peaked, although with increased drilling rates some could conceivably resume production growth. Production in the top county—Van Buren—is down 11% from peak and production in other counties outside of the top four is down from 21 to 56%.

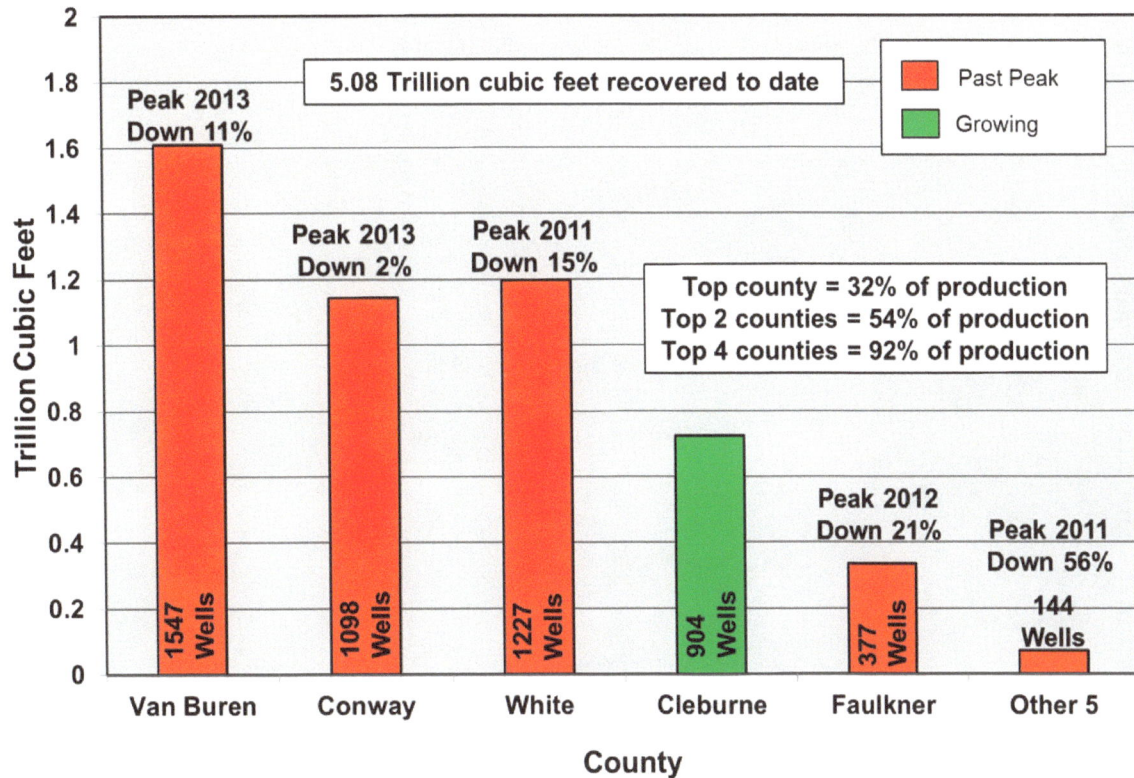

Figure 3-54. Cumulative gas production by county in the Fayetteville play through 2014.[81]

The top four counties have produced 92% of the 5.08 trillion cubic feet of gas produced to date.

---

[81] Data from Drillinginfo retrieved August 2014.

Operators are highly sensitive to the economic performance of the wells they drill, which typically cost on the order of $2.4 million each[82], not including leasing costs and other expenses. The areas of highest quality—the "core" or "sweet spots"—have now been well defined. Figure 3-55 illustrates average well decline curves by county which are a measure of well quality. Initial well productivities (IPs) are more closely grouped than in the Barnett; however, the top producing counties—Van Buren and Conway, which are both in decline—are somewhat better than the overall Fayetteville average and are significantly better than counties outside the top five. There are still a significant number of locations to drill wells in the top producing counties, although the overall play area of the Fayetteville is much smaller than plays like the Barnett and is dwarfed by the Marcellus.[83]

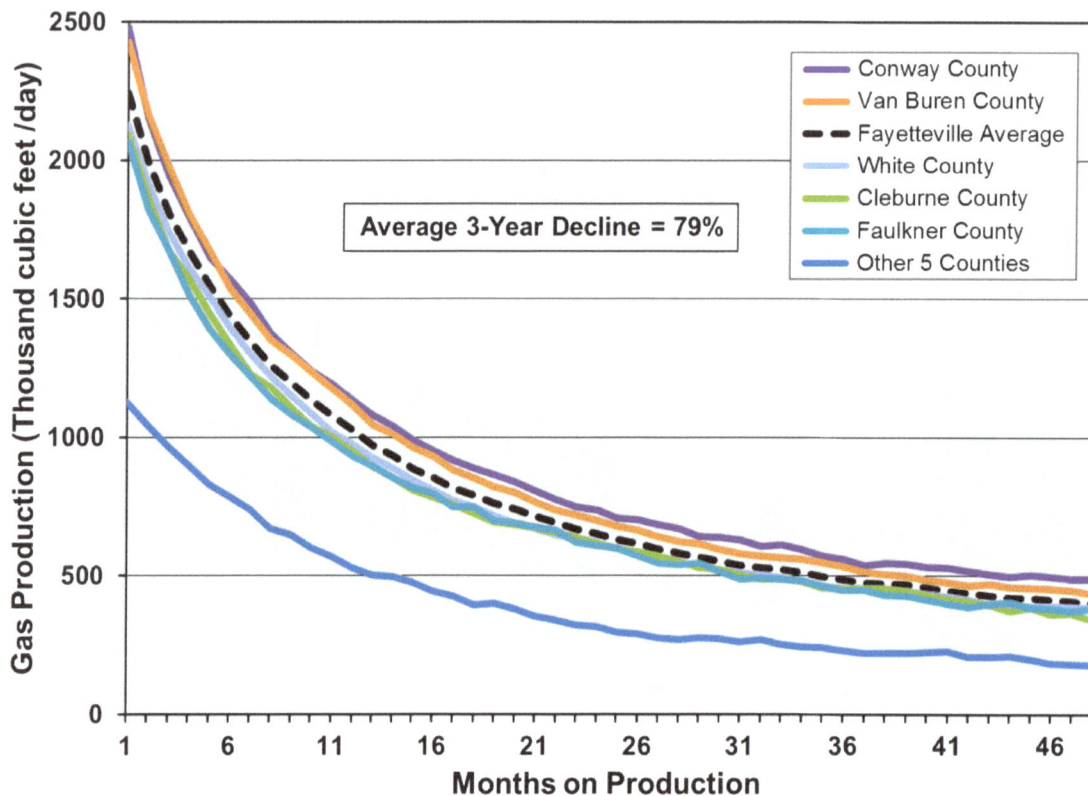

Figure 3-55. Average horizontal gas well decline profiles by county for the Fayetteville play.[84]

The low productivity outside of the top five counties seriously limits expansion of the play.

Another measure of well quality is "estimated ultimate recovery" or EUR—the amount of gas a well will recover over its lifetime. To be clear no one knows what the average lifespan of a Fayetteville well is, given that few of them are more than seven years old (see Figure 3-50 and Figure 3-51), and some 8% of wells drilled have ceased production at an average age of about three years. Operators fit hyperbolic and/or exponential curves to data such as presented in Figure 3-55, assuming well life spans of 30-50 years (as is typical for conventional wells) by comparison to conventional wells, but so far this is speculation given the nature of the extremely low permeability reservoirs and the completion technologies used in the Fayetteville.

---

[82] Southwest Energy, 2014, "Fayetteville Shale," http://www.swn.com/operations/pages/fayettevilleshale.aspx.
[83] EIA, *Assumptions to the Annual Energy Outlook 2014*, http://www.eia.gov/forecasts/aeo/assumptions/pdf/oilgas.pdf.
[84] Data from Drillinginfo retrieved August 2014.

Nonetheless, for comparative well quality purposes only, one can use the data in Figure 3-55, which exhibits steep initial decline with progressively more gradual decline rates, and assume a constant terminal decline rate thereafter to develop a theoretical EUR.

Figure 3-56 illustrates theoretical EURs by county for the Fayetteville, for comparative purposes of well quality. These range from 1.02 to 2.43 billion cubic feet per well, which is somewhat higher than the 0.84 to 1.44 billion cubic feet assumed by the EIA.[85] The range of EURs in the top five counties is fairly small, but all are roughly double the outlying counties which will serve to limit expansion of the play in future. The steep initial well production declines mean that well payout, if it is achieved, comes in the first few years of production, as between 55% and 62% of an average well's lifetime production occurs in the first four years.

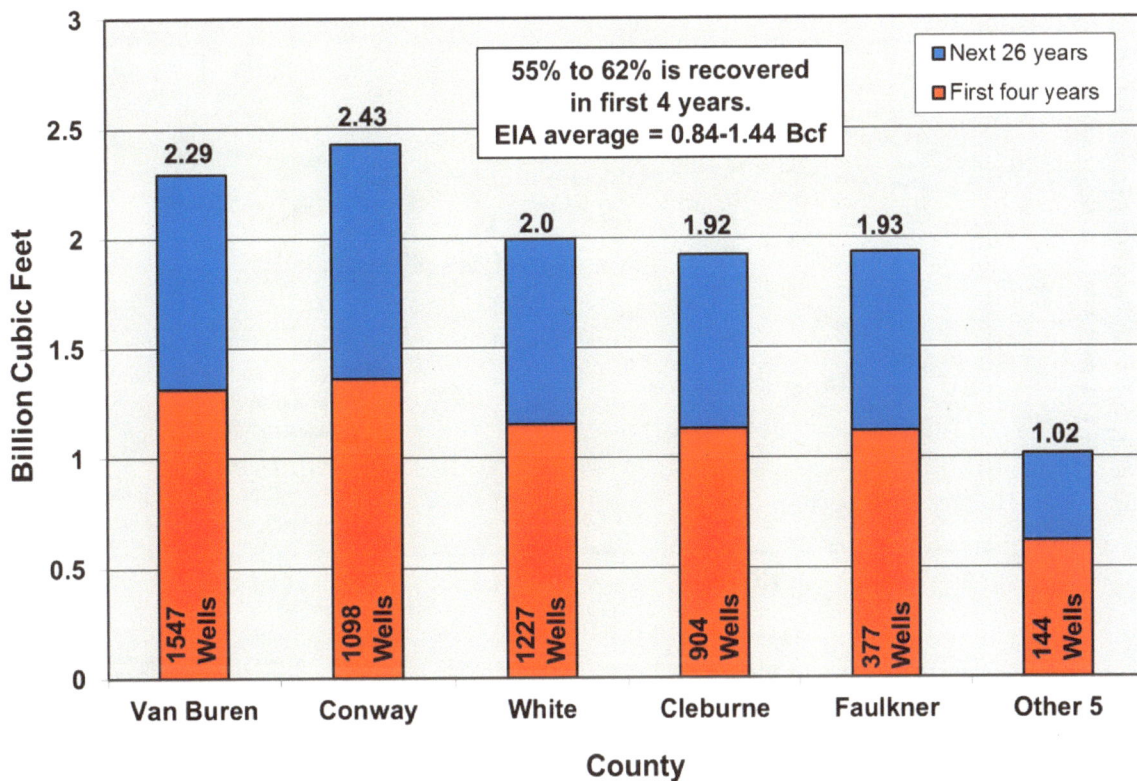

Figure 3-56. Estimated ultimate recovery of gas per well by county for the Fayetteville play.[86]

EURs are based on average well decline profiles (Figure 3-55) and a terminal decline rate of 15%. These are for comparative purposes only as it is highly uncertain if wells will last for 30 years. The steep decline rates mean that most production occurs early in well life.

---

[85] EIA, *Assumptions to the Annual Energy Outlook 2014*, http://www.eia.gov/forecasts/aeo/assumptions/pdf/oilgas.pdf.
[86] Data from Drillinginfo retrieved August 2014.

Well quality can also be expressed as the average rate of production over the first year of well life. If we know both the rate of production in the first year of the average well and the field decline rate, we can calculate the number of wells that need to be drilled each year to offset field decline in order to maintain production. Figure 3-57 illustrates the average first year production rate of wells by county over time. As noted earlier, average well quality for the play is up 2% in 2013 and four of the top five counties are flat to slightly rising. Van Buren County—the top producer—is declining, and no wells were drilled in 2013 outside of the top five counties, hence an estimate for that year was not possible.

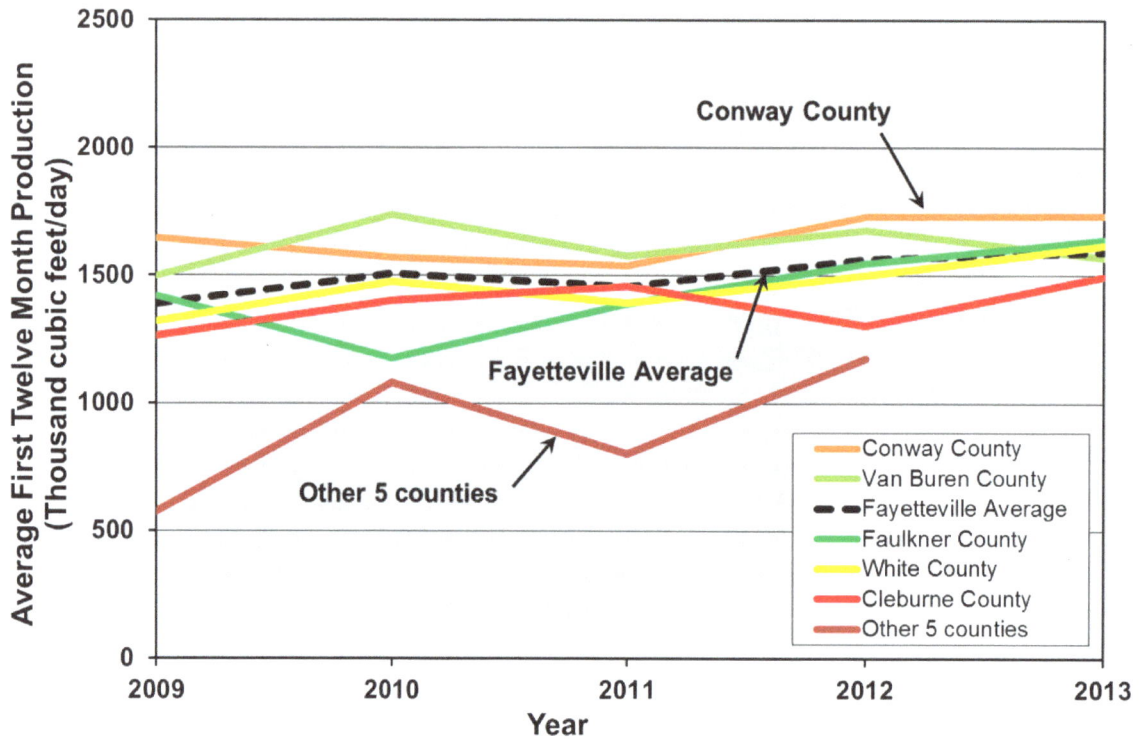

© Hughes GSR Inc, 2014 (data from Drillinginfo, August, 2014)

Figure 3-57. Average first-year gas production rates of wells by county in the Fayetteville play, 2009 to 2013.[87]

Well quality is flat to slightly rising in four counties and declining in Van Buren County which is the top producer. There were no wells drilled in 2013 outside of the top five counties so no estimate was possible for that year.

[87] Data from Drillinginfo retrieved August 2014.

3.3.3.4 Number of Wells

The fourth key fundamental is the number of wells that can ultimately be drilled in the Fayetteville play. The EIA has estimated the total play area as 2,904 square miles, including 2,132 in the "central" and 772 in the "west" area, and suggests this can be drilled at a well density of eight per square mile, for a total of 23,232 wells. In fact, the "west" area of the EIA has limited prospectivity—most wells there have ceased production—and drilling in areas outside the top five counties has ceased as of 2014. A close look at the drilling data limits the overall play area to 2,150 square miles, even allowing for 525 square miles of prospective area outside of the top five counties, for a total well count of 17,230 when the play is completely developed. As 5,297 wells have already been drilled this leaves 11,933 yet-to-drill wells.

Table 3-3 breaks down the number of yet-to-drill wells by county along with other critical parameters used for determining the future production rates of the Fayetteville play.

Parameter	County						Total
	Cleburne	Conway	Faulkner	Van Buren	White	Other 5	
Production May 2014 (Bcf/d)	0.60	0.66	0.18	0.83	0.56	0.02	2.85
% of Field Production	21.20	23.27	6.25	28.93	19.53	0.83	100.00
Cumulative Gas (Tcf)	0.72	1.14	0.33	1.61	1.20	0.07	5.08
Cumulative Liquids (MMBBL)	0.00	0.00	0.00	0.00	0.00	0.00	0.00
Number of Wells	904	1098	377	1547	1227	144	5297
Number of Producing Wells	848	1015	322	1441	1168	120	4914
Average EUR per well (Bcf)	1.92	2.43	1.93	2.29	2.00	1.02	2.10
Field Decline (%)	34.64	37.02	37.02	27.22	26.12	31.32	34.02
3-Year Well Decline (%)	78	78	78	79	79	80	79
Peak Year	Rising	2013	2012	2013	2011	2011	2012
% Below Peak	N/A	2	21	11	15	56	2.2
Average First Year Production in 2013 (Mcf/d)	1496	1734	1641	1571	1616	1174	1592
New Wells Needed to Offset Field Decline	140	142	40	143	90	6	610
Area in square miles	553	556	647	712	1034	3500	7002
% Prospective	70	50	30	50	40	15	31
Net square miles	387	278	194	356	414	525	2153
Well Density per square mile	2.34	3.95	1.94	4.35	2.97	0.27	2.46
Additional locations to 8/sq. Mile	2193	1126	1176	1301	2082	4056	11933
Population	25970	21273	113237	17295	77076	N/A	N/A
Total Wells 8/sq. Mile	3097	2224	1553	2848	3309	4200	17230
Total Producing Wells 8/sq. Mile	3041	2141	1498	2742	3250	4176	16847

Table 3-3. Parameters for projecting Fayetteville production, by county.

Area in square miles under "Other" is estimated.

A recent in-depth study of the Fayetteville by the Bureau of Economic Geology at the University of Texas (UT) at Austin takes a more conservative view.[88] Although they assign a study area of 2,737 square miles, they exclude 20% of "partly drained" portions and 60% of undrilled portions from consideration, given uncertainties about surface access and prospectivity. At the time of the 2011 data cutoff used in that study, 1,252 square miles had been tested by drilling, leaving 1,485 square miles undrilled—which leaves a net developable area of 1,596 square miles. The UT study assumes in its base case that a total of 10,117 wells will be drilled by 2030, which leaves just 4,820 yet-to-drill wells by 2030 given the 5,297 wells drilled as of May 2014.

---

[88] Browning et al., 2014, *Oil and Gas Journal*, "Study develops Fayetteville Shale reserves, production forecast," http://www.beg.utexas.edu/info/docs/Fayetteville%20Shale%20OGJ%20article.pdf.

3.3.3.5 Rate of Drilling

Given known well- and field-decline rates, well quality by area, and the number of available drilling locations, the most important parameter in determining future production levels is the rate of drilling—the fifth key fundamental. Figure 3-58 illustrates the historical drilling rates in the Fayetteville. Drilling rates peaked in 2011 at just over 800 wells per year and have fallen to current levels of about 500 wells per year. Current drilling rates are close to the 600 wells per year required to maintain production at current levels, hence production is maintaining a slowly downward trending plateau.

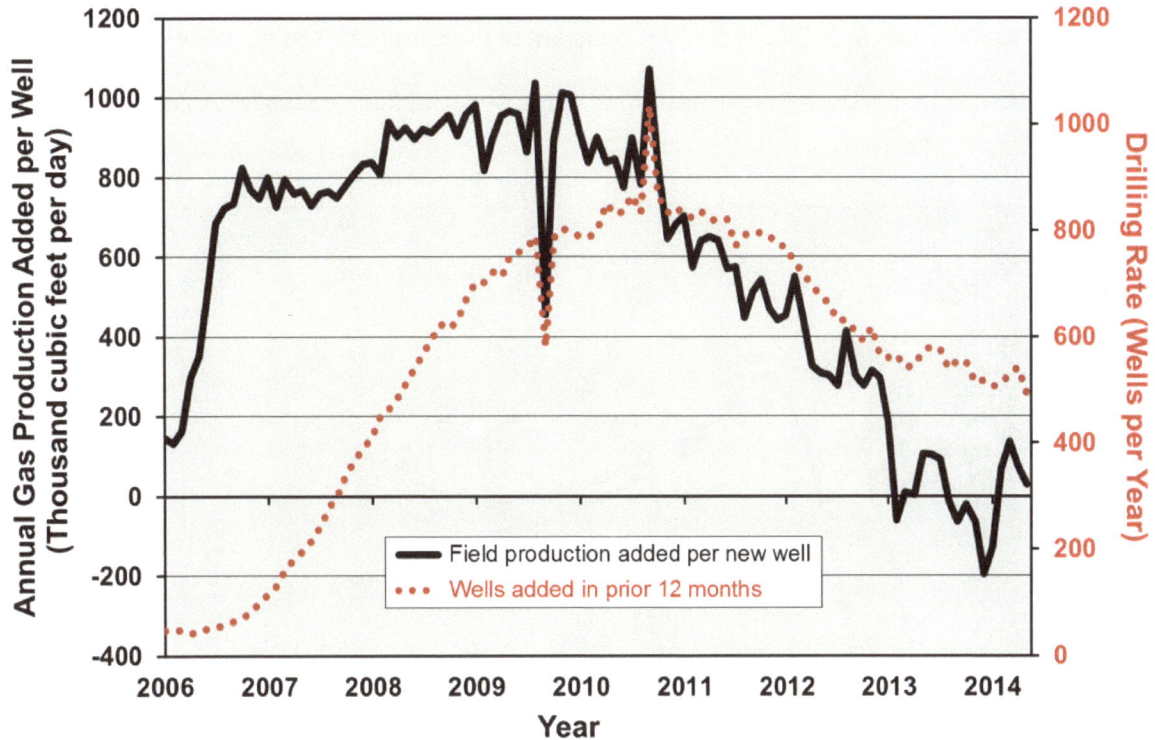

Figure 3-58. Annual gas production added per new horizontal well and annual drilling rate in the Fayetteville play, 2006 through 2014.[89]

Drilling rate peaked in 2010 and is now slightly below the level needed to keep production flat.

---

[89] Data from Drillinginfo retrieved August 2014. Three-month trailing moving average.

3.3.3.6   Future Production Scenarios

Based on the five key fundamentals outlined above, several production projections for the Fayetteville play were developed to illustrate the effects of changing the rate of drilling. Figure 3-59 illustrates the production profiles of three drilling rate scenarios if 100% of the prospective play area is drillable at eight wells per square mile. These scenarios are:

1. MOST LIKELY RATE scenario: Drilling remains at the current rate of 500 wells per year, then gradually declines to 300 wells per year.

2. EXISTING RATE scenario: Drilling remains constant at the current rate of 500 wells per year.

3. HIGH RATE scenario: Drilling increases to 750 wells per year, then gradually declines to 500 wells per year.

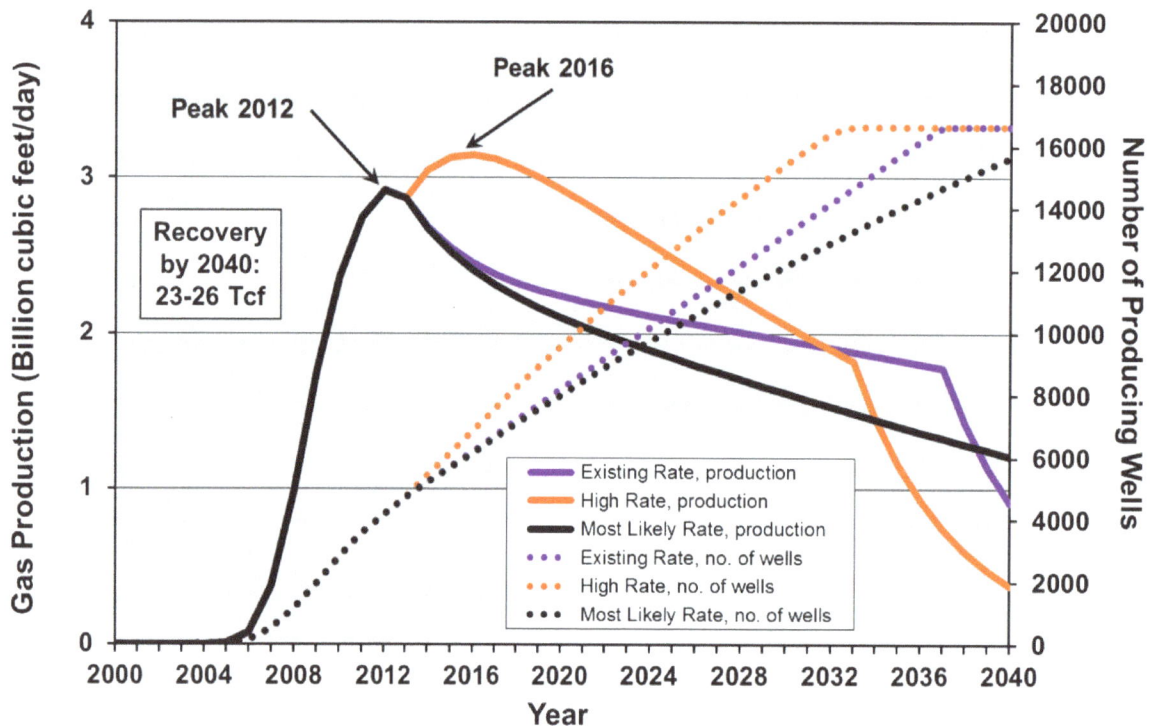

Figure 3-59. Three drilling rate scenarios of Fayetteville gas production (assuming 100% of the area is drillable eight wells per square mile).[90]

"Most Likely Rate" scenario: drilling holds at 500 wells/year, declining to 300 wells per year.
"Existing Rate" scenario: drilling holds constant at 500 wells/year.
"High Rate" scenario: drilling increases to 750 wells/year, declining to 500 wells/year.

[90] Data from Drillinginfo retrieved August 2014.

The drilling rate scenarios have the following results:

1. MOST LIKELY RATE scenario: The rate of drilling declines as the inventory of drilling locations is used up and drilling moves into outlying areas. Total gas recovery by 2040 would be 22.8 trillion cubic feet and drilling would continue beyond 2040.

2. HIGH RATE scenario: The rate of drilling increases by 50% immediately and production would increase to a new peak in 2016.. This scenario is considered unlikely unless there is a marked increase in gas price in the very near future. Total gas recovery by 2040 would be 26 trillion cubic feet and drilling would end in 2033.

3. EXISTING RATE scenario: Drilling continues at 500 wells per year until locations run out; this scenario is also considered unlikely given the decline in well quality in later years as drilling moves into lower productivity counties. Total gas recovery by 2040 would be 24.9 trillion cubic feet and drilling would end in 2037.

Total recovery of 22.8 trillion cubic feet by 2040 in the "Most Likely Rate" scenario is more than four times what has been recovered so far in the Fayetteville. In the "High Rate" scenario as much as 26 trillion cubic feet could be recovered; however, production rates would be far below those projected by the EIA for the Fayetteville play.

3.3.3.7 Comparison to EIA Forecast

Figure 3-60 illustrates the EIA's projection for Fayetteville production through 2040 compared to the "Most Likely Rate" scenario. The EIA projects a recovery by 2040 of 41.5 Tcf to meet its reference case forecast, and projects a new peak of the play in 2036 at a level far higher than the late-2012 peak. This represents the recovery of 98% of proved reserves[91] and unproved resources.[92] Furthermore, the EIA projects that production in 2040 will be much higher than the 2012 peak, suggesting that vastly more gas will be recovered beyond 2040. This strains credibility to the limit. How can all the proved and unproved resources and reserves be extracted and still have production near all-time highs in 2040?

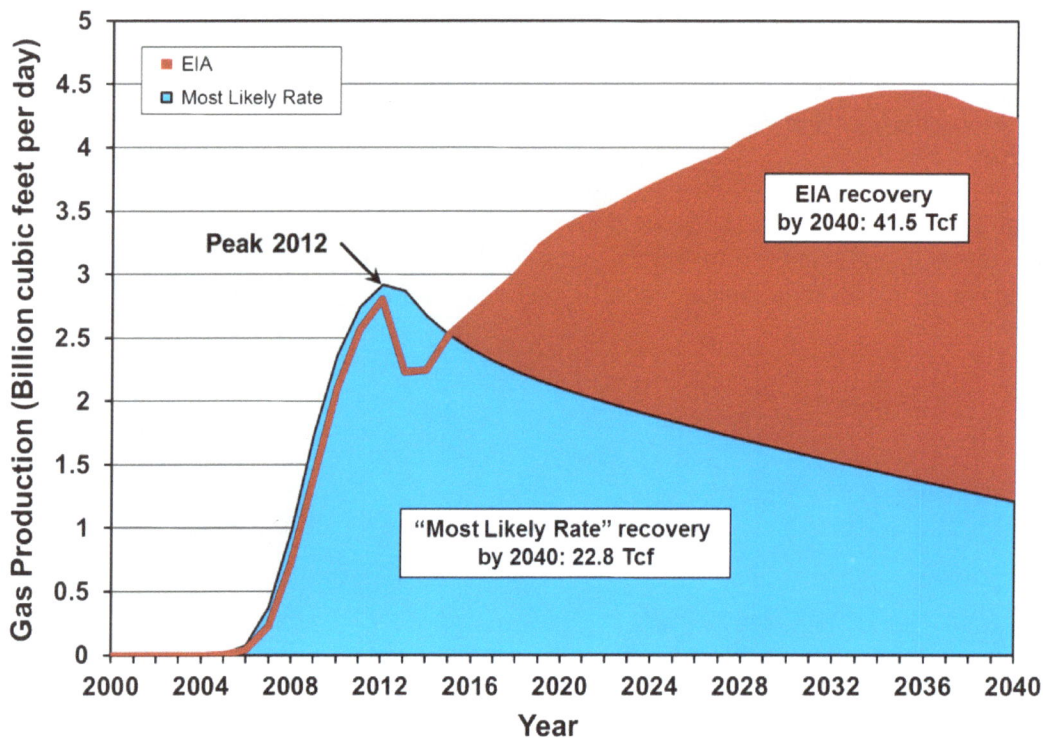

Figure 3-60. "Most Likely Rate" scenario of Fayetteville gas production compared to the EIA reference case, 2000 to 2040.[93]

The EIA assumes the Fayetteville will reach a new all-time high by 2036, produce 98% of proved reserves and unproved resources by 2040, and presumably produce a great deal more gas in the post-2040 period. The EIA forecast is made on a "dry gas" basis, whereas the "Most Likely Rate" scenario forecast is made on a "raw gas" basis.

---

[91] EIA, 2014, "Principal shale gas plays: natural gas production and proved reserves, 2011-12,"
http://www.eia.gov/naturalgas/crudeoilreserves/excel/table_4.xls.
[92] EIA, *Assumptions to the Annual Energy Outlook 2014*, http://www.eia.gov/forecasts/aeo/assumptions/pdf/oilgas.pdf.
[93] EIA, *Annual Energy Outlook 2014*, unpublished tables from AEO 2014 provided by the EIA.

3.3.3.8   Fayetteville Play Analysis Summary

Several things are clear from this analysis:

1. Drilling rates have fallen somewhat in the Fayetteville due to gas prices, but are still remarkably high likely due to the relatively low cost of wells compared to other plays.

2. High well- and field-decline rates mean a continued high rate of drilling is required to maintain, let alone increase, production. The Fayetteville field decline rate of 34% is in the lower range for shale gas plays. Current drilling rates of 500 wells per year are slightly below the level required to maintain current production levels. Maintaining production at current levels would require the investment of $1.4 billion per year for drilling (assuming $2.4 million per well). Future production profiles are most dependent on drilling rate and, to a lesser extent, on the number of drilling locations (i.e., greatly increasing the number of drilling locations would not change the production profile nearly as much as changing the drilling rate). Growing production in the Fayetteville would require considerably higher gas prices to justify higher drilling rates.

3. Increasing current drilling rates by 50% could reverse the current production decline and raise production to a new peak, at 3.15 Bcf/d, in 2016, which is 10% higher than current levels. Cumulative recovery by 2040 in this high drilling rate scenario would be increased by 14% but would still be only 63% of that projected by the EIA in its reference case.

4. The projected recovery of 22.8 Tcf by 2040 in the "Most Likely Rate" scenario represents four times as much gas as has been recovered so far from the Fayetteville, and is more optimistic than the "base case" estimated ultimate recovery of 18.2 Tcf projected by the Bureau of Economic Geology at the University of Texas.[94] Both are significantly less than the 41.5 Tcf projected by the EIA in its reference case forecast.

5. This report's projections are optimistic in that they assume the capital will be available for the drilling treadmill that must be maintained. They also assume that 100% of the prospective area is drillable. This is not a sure thing as drilling in the poorer quality parts of the play will require higher gas prices to be economic. Failure to increase current drilling rates will result in a steeper drop off in production.

6. Nearly three times the current number of wells will need to be drilled to meet the production projection of the "Most Likely Rate" scenario by 2040.

7. The EIA projection for future Fayetteville gas production included in its reference case forecast for AEO 2014,[95] which forecasts recovery of 98% of proved reserves plus unproved resources by 2040, strains credibility to the limit. It is highly unlikely to be realized, especially at the gas prices the EIA forecasts.[96]

---

[94] Browning et al., 2014, *Oil and Gas Journal*, "Study develops Fayetteville Shale reserves, production forecast," http://www.beg.utexas.edu/info/docs/Fayetteville%20Shale%20OGJ%20article.pdf.

[95] EIA, *Annual Energy Outlook 2014*, unpublished tables from AEO 2014 provided by the EIA.

[96] EIA, *Annual Energy Outlook 2014*, http://www.eia.gov/forecasts/aeo.

3.3.4   Woodford Play

The EIA forecasts recovery of 23.8 Tcf of gas from the Woodford play by 2040. The analysis of actual production data presented below suggests that this forecast is somewhat—but not significantly—higher than the data suggest, although the forecast production profile is improbable.

The Woodford play in Oklahoma is primarily a shale gas play, for although parts of it are liquids rich, 92% of the energy produced from it in mid-2014 was natural gas. It is a complex play, comprising parts of the Anadarko Basin on the west, the Arkoma Basin on the east, the Chautauqua Platform in the central and northern portions, and the Oklahoma- and Ouachita-folded belts in the south and southeast. Figure 3-61 illustrates the distribution of wells as of early 2014. Since 2005 over 3,600 wells have been drilled, of which 3,062 were producing at the time of writing. The play covers parts of 31 counties although 70% of production is concentrated in five counties.

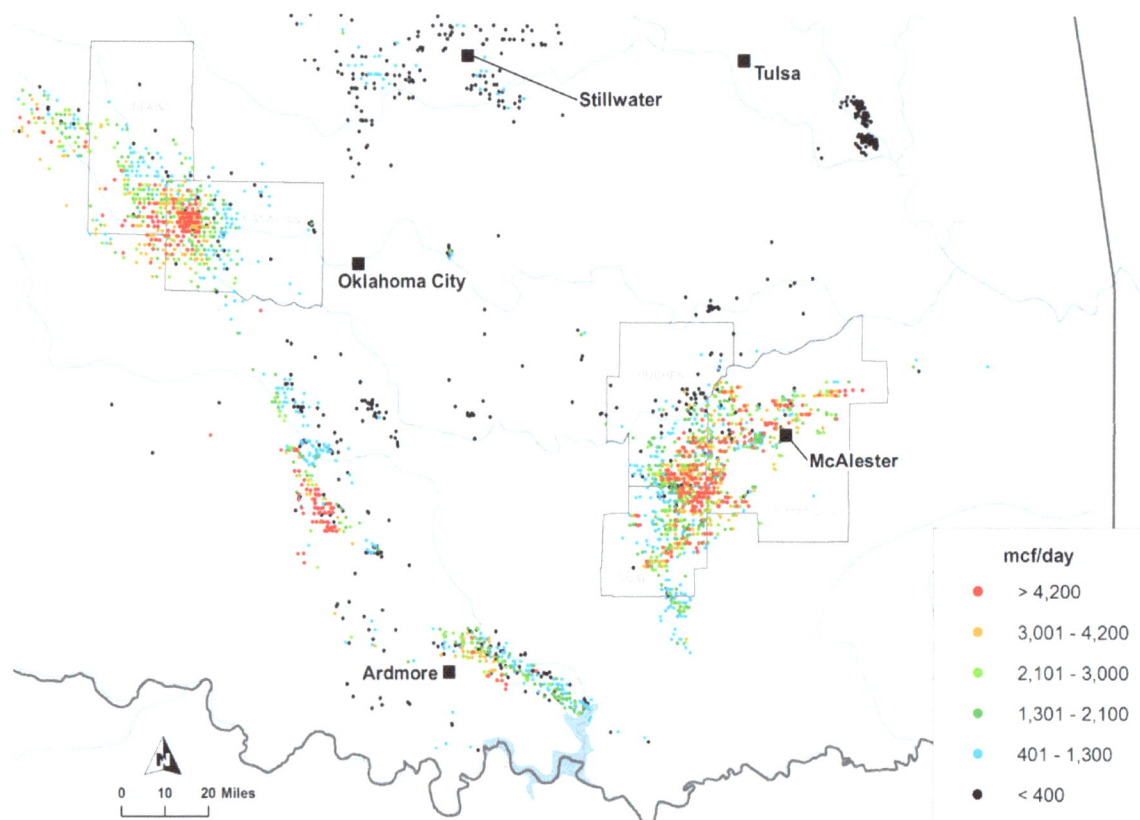

Figure 3-61. Distribution of wells in the Woodford play as of early 2014, illustrating highest one-month gas production (initial productivity, IP).[97]

Well IPs are categorized approximately by percentile; see Appendix.

[97] Data from Drillinginfo retrieved August 2014.

Production in the Woodford peaked at nearly 1.9 billion cubic feet per day in June 2013 as illustrated in Figure 3-62.

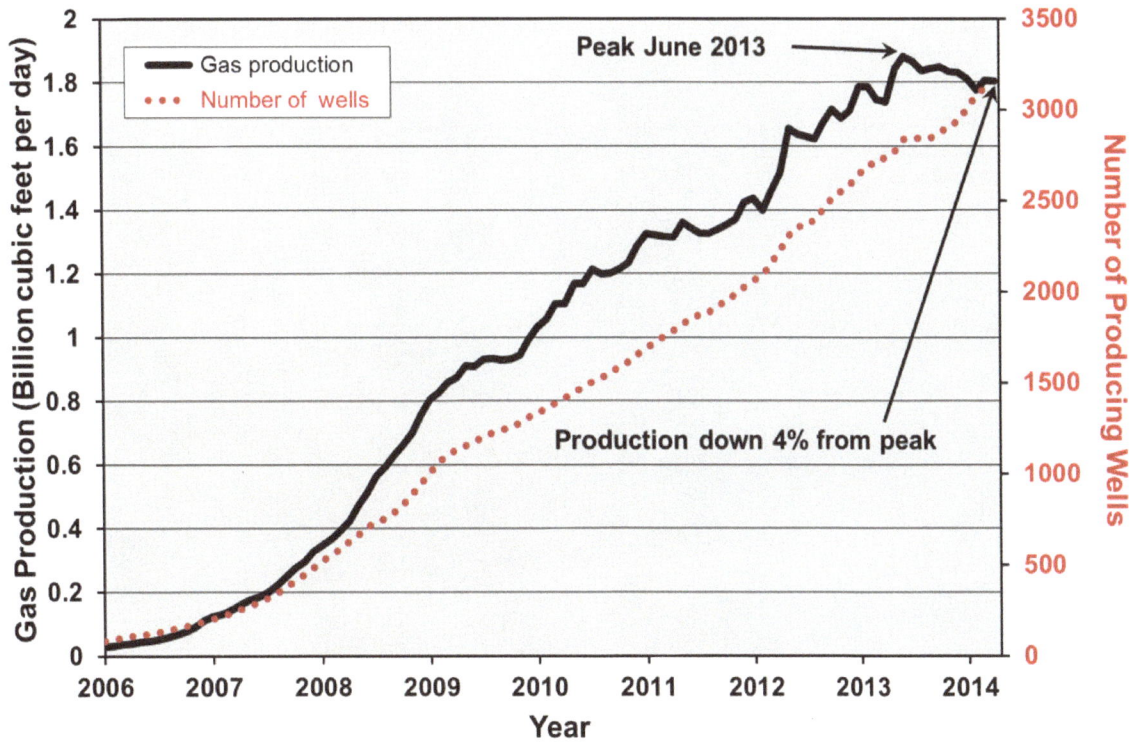

Figure 3-62. Woodford play shale gas production and number of producing wells, 2006 to 2014.[98]

Gas production data are provided on a "raw gas" basis.

---

[98] Data from Drillinginfo retrieved September 2014. Three-month trailing moving average.

Although some 14% of producing wells in the Woodford are vertical/directional, 98% of current production is from horizontal fracked wells as illustrated in Figure 3-63. The rate of drilling peaked at more than 600 wells per year in 2010 but has since fallen to less than the roughly 405 wells per year required to keep production flat at current production levels. Very few vertical/directional wells are being drilled today—the future of the play lies in drilling horizontal fracked wells.

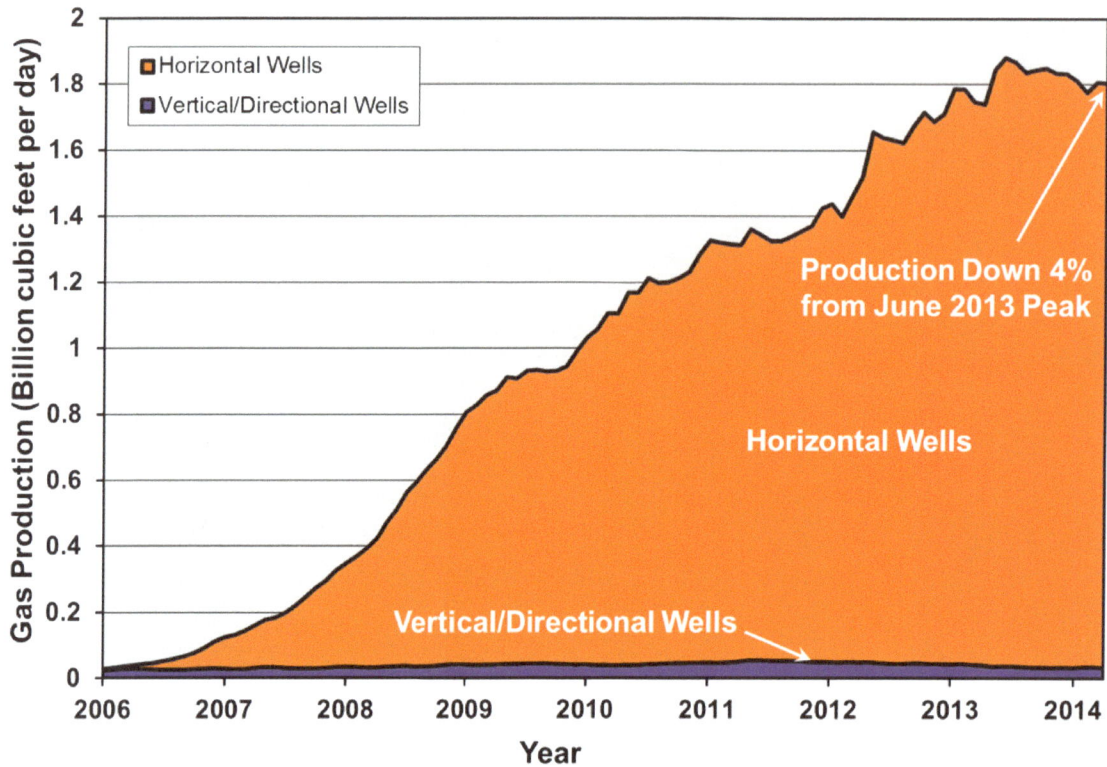

Figure 3-63. Gas production from the Woodford play by well type, 2006 to 2014.[99]

---

[99] Data from Drillinginfo retrieved September 2014. Three-month trailing moving average.

3.3.4.1 Well Decline

The first key fundamental in determining the life cycle of Woodford production is the *well decline rate*. Woodford wells exhibit high decline rates in common with all shale plays. Figure 3-64 illustrates the average decline rate of Woodford horizontal and vertical/directional wells. Decline rates are steepest in the first year and are progressively less in the second and subsequent years. The decline rate over the first three years of average well life is 74%, which is at the low end of typical shale plays. As can be seen, vertical/directional wells have much lower productivity than horizontal wells and hence are being phased out.

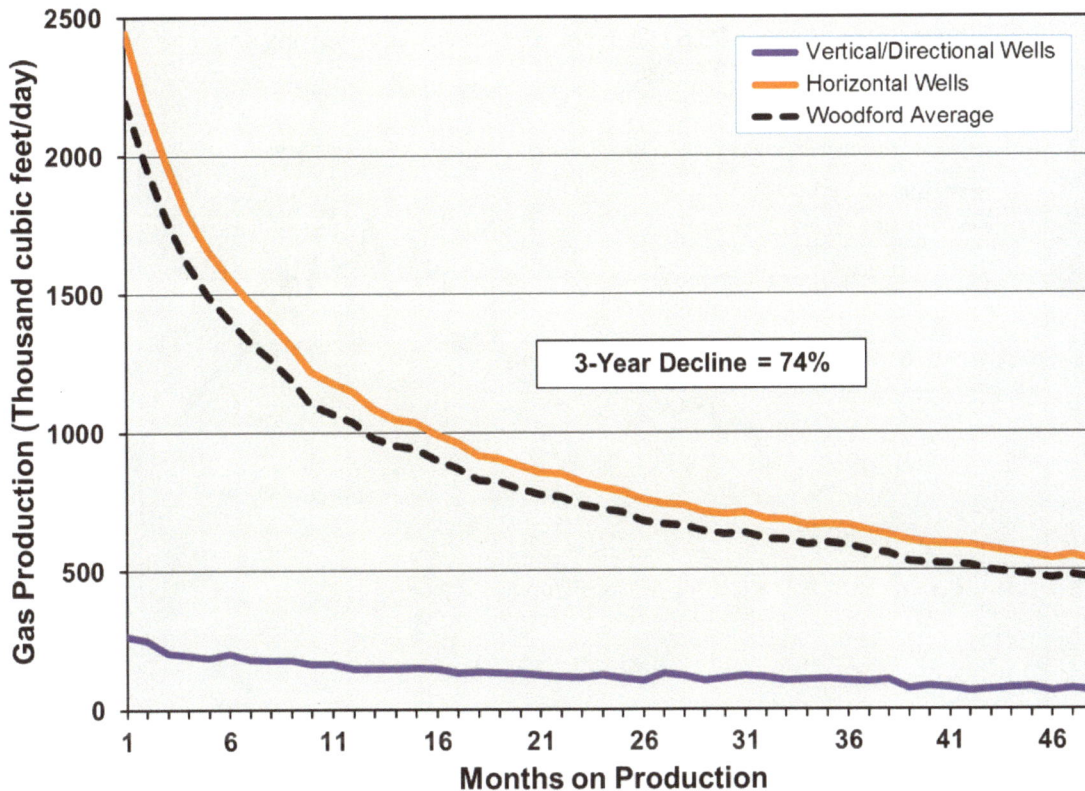

Figure 3-64. Average decline profile for gas wells in the Woodford play.[100]
Decline profile is based on all shale gas wells drilled since 2009.

---

[100] Data from Drillinginfo retrieved September 2014.

3.3.4.2 Field Decline

A second key fundamental is the overall *field decline rate*, which is the amount of production that would be lost in the Woodford in a year without more drilling. Figure 3-65 illustrates production from the 2,600 horizontal wells drilled prior to 2013 (horizontal wells only are considered as very few vertical/directional wells are being drilled). The first-year decline rate is 34%. This is lower than the well decline rate as the field decline is made up of both new wells declining at high rates and older wells declining at lesser rates. It's also at the low end of field decline rates observed in shale plays. Assuming new wells will produce in their first year at the average first-year rates observed for wells drilled in 2013, 405 new wells each year would be required to offset field decline at current production levels. At an average cost of $9 million per well,[101] this would represent a capital input of about $3.6 billion per year, exclusive of leasing and other infrastructure costs, just to keep production flat at 2013 levels.

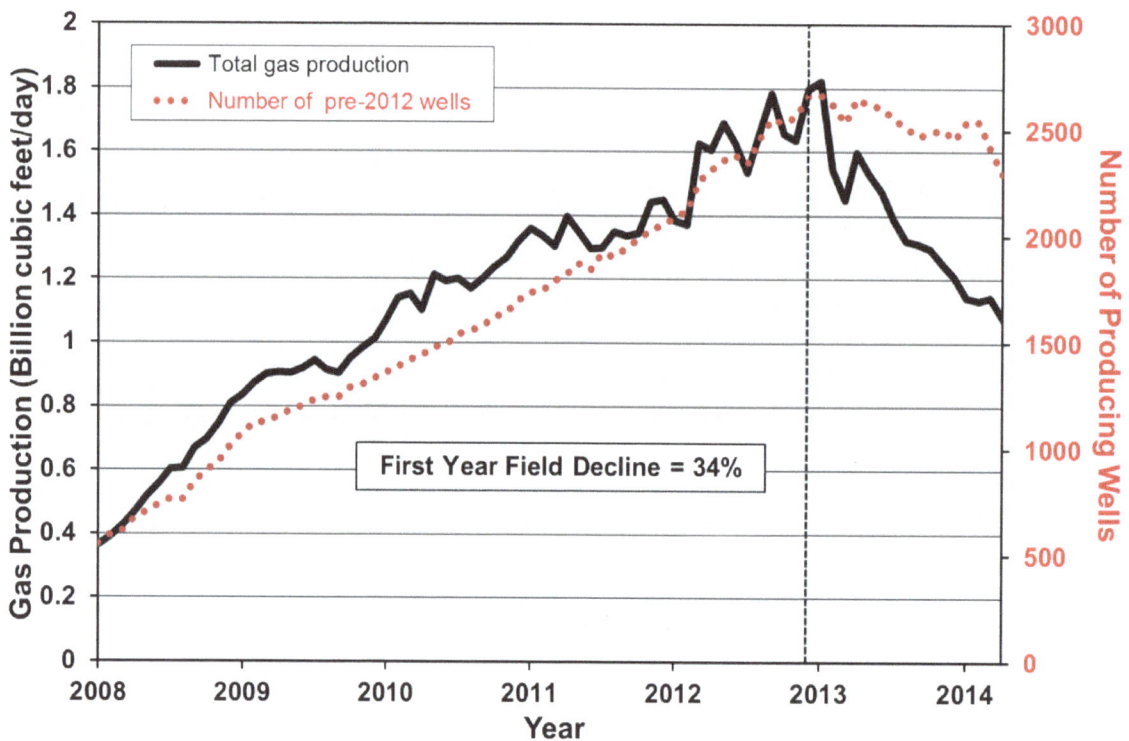

Figure 3-65. Production rate and number of horizontal shale gas wells drilled in the Woodford play prior to 2013, 2008 to 2014.[102]

This defines the field decline for the Woodford play which is 34% per year (only production from horizontal wells is analyzed as few vertical/directional wells are likely to be drilled in the future).

---

[101] Mason, Richard, June 27, 2013, "Targeting Oklahoma's Ubiquitous Woodford Shale," http://www.ugcenter.com/Woodford/Targeting-Oklahomas-Ubiquitous-Woodford-Shale_118127.
[102] Data from Drillinginfo retrieved September 2014.

3.3.4.3 Well Quality

The third key fundamental is the *average well quality* in the Woodford by area and its trend over time. Petroleum engineers tell us that technology is constantly improving, with longer horizontal laterals, more frack stages per well, more sophisticated mixtures of proppants and other additives in the frack fluid injected into the wells, and higher-volume frack treatments. This has certainly been true over the past few years, along with multi-well pad drilling which has reduced well costs. It is, however, approaching the limits of diminishing returns, and improvements in average well quality are non-existent in the Woodford. The average first year production rate of Woodford wells is down 24% from what it was in 2010, as illustrated in Figure 3-66. This is clear evidence that geology is winning out over technology, as drilling moves into lower-quality locations, as investigated further below, although some of the decline may be related to moves into more liquids-rich parts of the play.

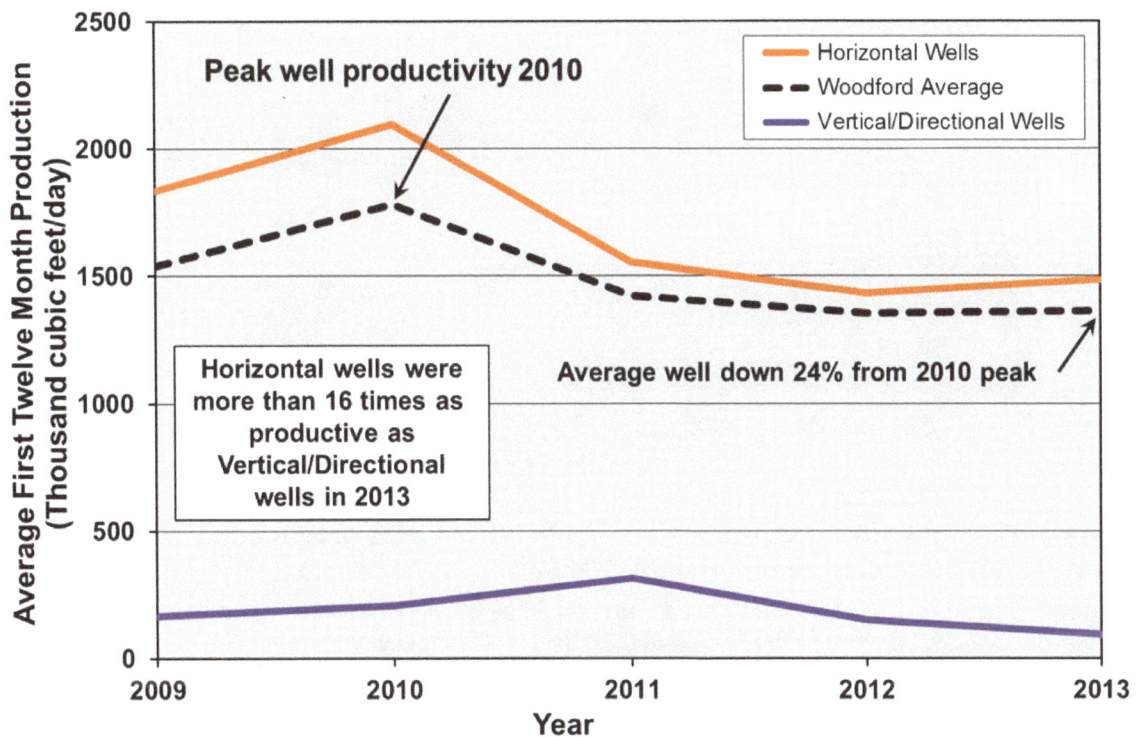

Figure 3-66. Average first-year production rates for Woodford horizontal and vertical/directional gas wells from 2009 to 2013.[103]

Average well quality has fallen by 24% from 2010, a clear indication that geology is trumping technology in this shale play.

---

[103] Data from Drillinginfo retrieved September 2014.

Another measure of well quality is cumulative production and well life. Ten percent of the wells that have been drilled in the Woodford are no longer productive. Figure 3-67 illustrates the cumulative production of these shut-down wells over their lifetime. At a mean lifetime of 32 months and a mean cumulative production of 0.26 billion cubic feet, these wells would in large part be economic losers.

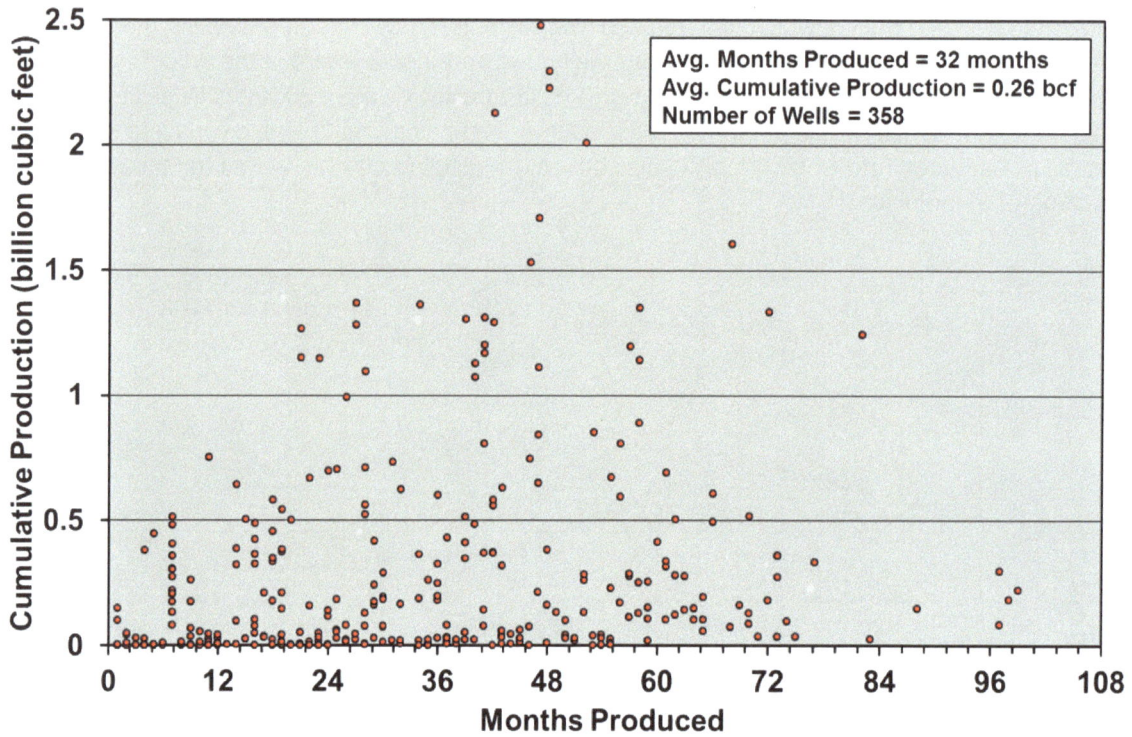

Figure 3-67.Cumulative gas production and length of time produced for Woodford wells that were not producing as of February 2014.

These well constitute 10% of all wells drilled; most would be economic failures, given the mean life of 32 months and average cumulative production of 0.26 billion cubic feet when production ended.[104]

---

[104] Data from Drillinginfo retrieved September 2014.

Figure 3-68 illustrates the cumulative production of all horizontal wells that were producing in the Woodford as of March 2014. Although it can be seen that there are a few very good wells that recovered large amounts of gas in the first few years, and undoubtedly were great economic successes, the average well had produced just 0.92 billion cubic feet over a lifespan averaging 42 months. Just 3% of these wells are more than 8 years old.

The lifespan of wells is another key parameter, as many operators assume a minimum well life of 30 years and longer, though this is conjectural given the lack of data and the significant number of wells that have been shut down after less than 8 years.

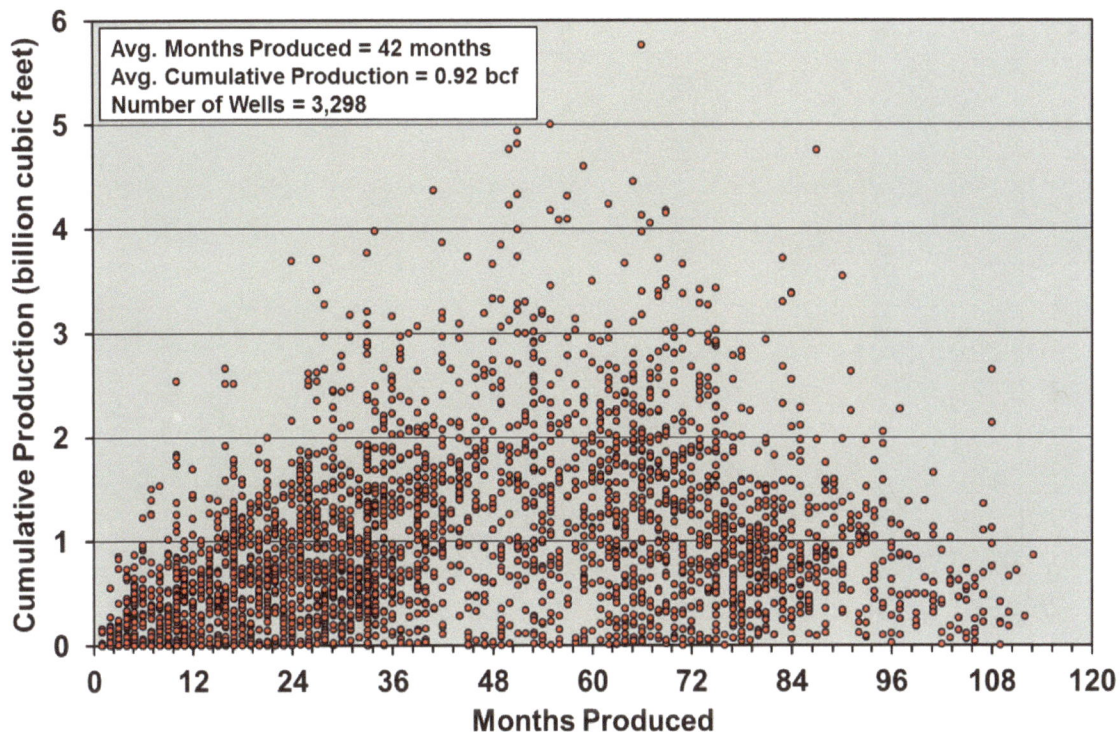

Figure 3-68. Cumulative gas production and length of time produced for Woodford wells that were producing as of March 2014.[105]

These well constitute 90% of all wells drilled. Very few wells are greater than eight years old, with a mean age of 42 months and a mean cumulative recovery of 0.92 billion cubic feet.

---

[105] Data from Drillinginfo retrieved September 2014.

Cumulative production of course depends on how long a well has been producing, so looking at young wells in not necessarily a good indication of how much gas these wells will produce over their lifespan (although production is heavily weighted to the early years of well life). A measure of well quality independent of age is initial productivity (IP). Figure 3-69 illustrates the average daily output over the first six months of production for all wells in the Woodford play (six-month IP). Again, as with cumulative production, there are a few exceptional wells—5% produced more than 4 million cubic feet per day (MMcf/d)—but the average for all wells drilled since 2005 is just 1.41 MMcf/d. Figure 3-61 illustrates the distribution of IPs in map form.

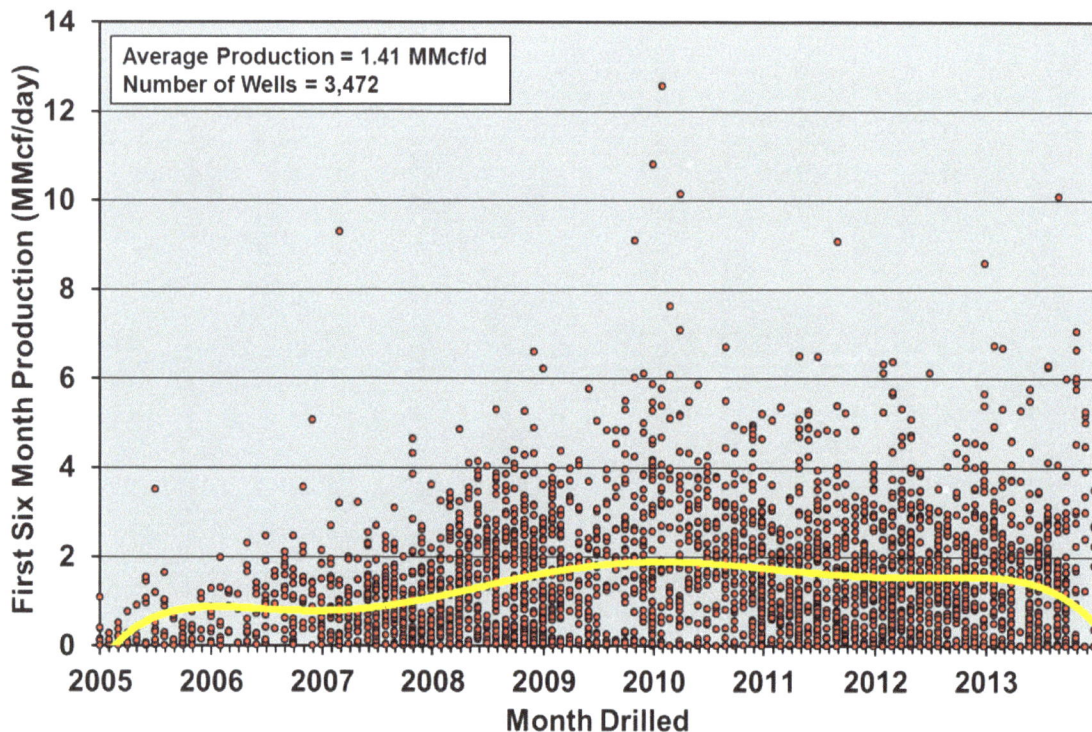

Figure 3-69. Average gas production over the first six months for all wells drilled in the Woodford play, 2005 to 2014.[106]

Although there are a few exceptional wells, the average well produced 1.41 million cubic feet per day over this period. The trend line indicates mean productivity over time.

[106] Data from Drillinginfo retrieved September 2014.

Different counties in the Woodford display markedly different well quality characteristics, which are critical in determining the most likely production profile in the future. Figure 3-70, which illustrates production over time by county, shows that as of April 2014, the top two counties produced 45% of the total, the top five produced 69%, and the remaining 26 counties produced 31%.

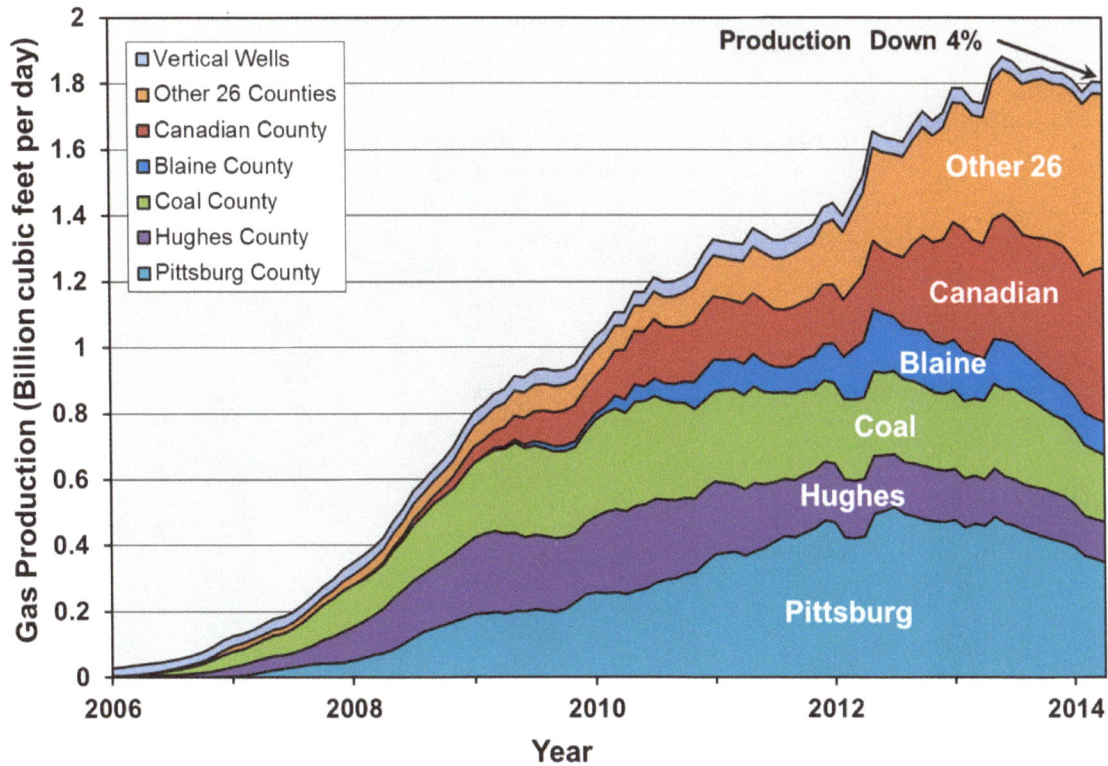

Figure 3-70. Gas production by county in the Woodford play, 2006 through 2014.[107]

The top five counties produced 69% of production in April 2014.

---

[107] Data from Drillinginfo retrieved September 2014. Three-month trailing moving average.

The same trend holds in terms of cumulative production since the field commenced. As illustrated in Figure 3-71, the top two counties have produced 49% of the gas and the top five have produced 85%. Production in four of the top five counties peaked in 2010 to 2012 and is down sharply. Production is growing in Canadian County and is flat in the 26 counties outside the top five, which tend to be richer in liquids and are the focus of drilling in a period of low priced gas. An increase in the rate of drilling given higher gas prices could temporarily halt and perhaps reverse declines in those counties that have peaked.

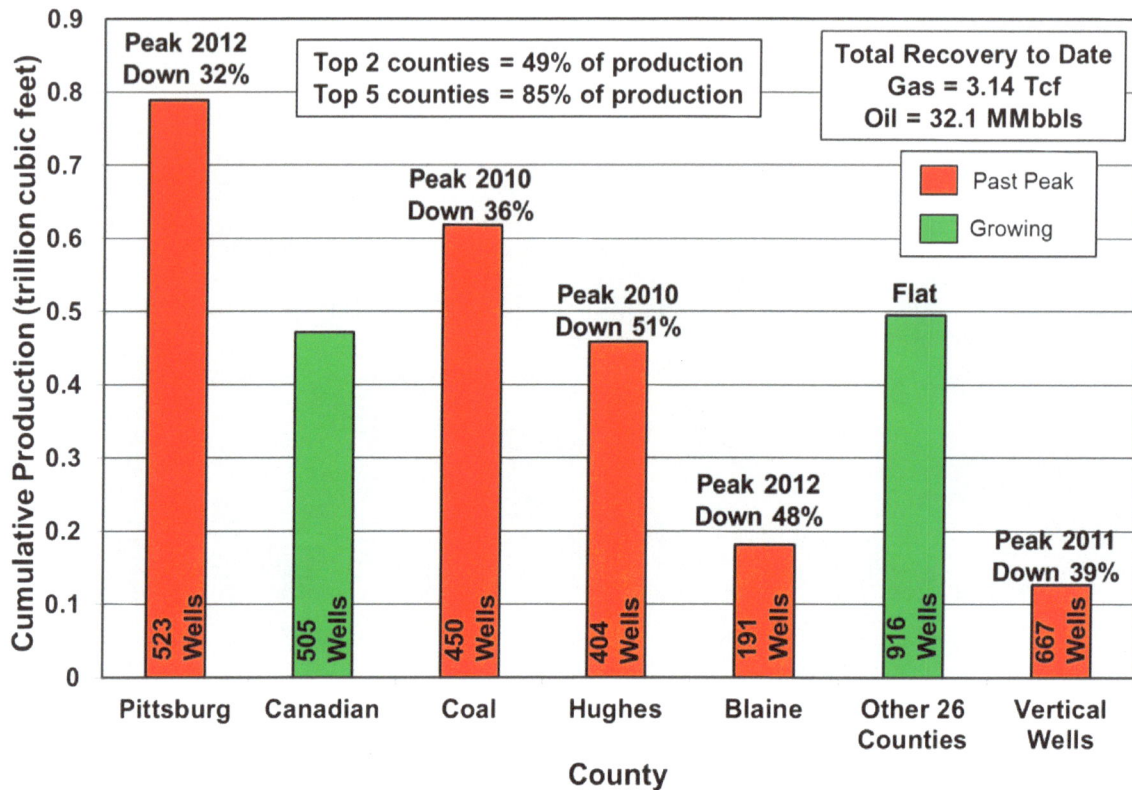

Figure 3-71. Cumulative gas production by county in the Woodford play through 2014.[108]
The top five counties have produced 85% of the 3.14 trillion cubic feet of gas produced to date.

[108] Data from Drillinginfo retrieved September 2014.

The Woodford also produces limited amounts of natural gas liquids and oil. With the exception of Canadian County, most liquids production is not within the top five counties but is located in the central and northern portions, as illustrated in Figure 3-72. Some 32 million barrels of liquids have been produced since 2005, and, given low gas prices, has improved economics and driven drilling to counties where liquids can be produced.

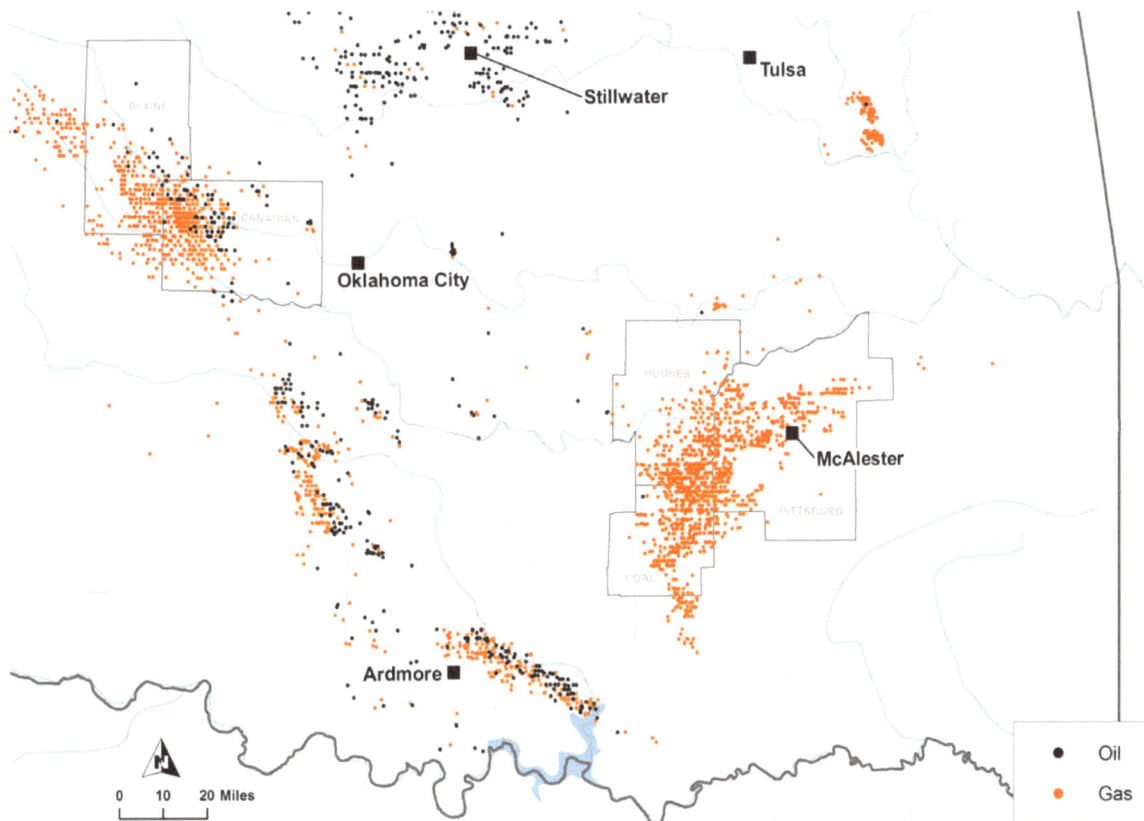

Figure 3-72. Distribution of gas and oil wells in Woodford play as of early 2014.[109]
Liquids production from wells classified as "oil" occurs mainly in the central and northern portions of the play.

[109] Data from Drillinginfo retrieved August 2014.

Figure 3-73 illustrates liquids production in the Woodford by county. In the big picture liquids production from the Woodford is relatively insignificant, for although it has grown significantly since 2005 it still amounted to less than 8% of the energy produced from the Woodford play in early 2014. In fact, liquids production has fallen more than 30% from a peak of 38,000 barrels per day reached in June 2013.

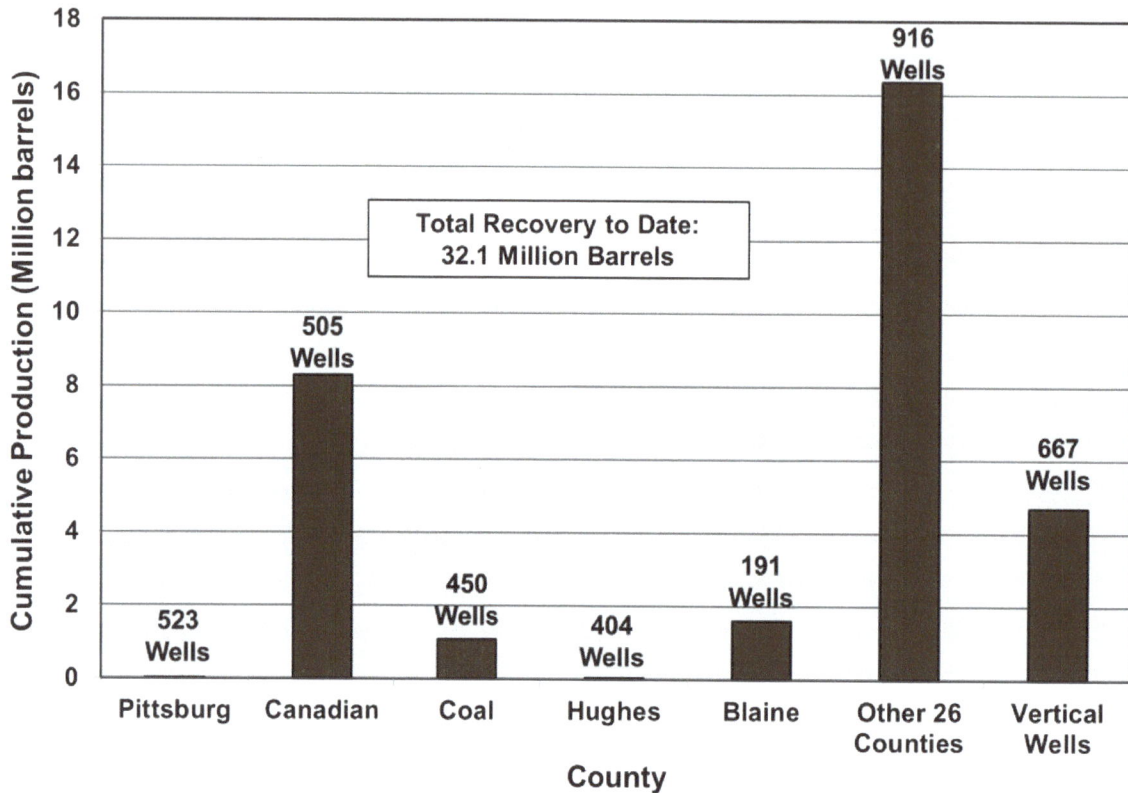

Figure 3-73. Cumulative liquids production by county in the Woodford play through 2014.[110]

Canadian and the "other 26" counties account for 77% of the 32 million barrels produced to date.

[110] Data from Drillinginfo retrieved September 2014.

Operators are highly sensitive to the economic performance of the wells they drill, which typically cost on the order of $9 million or more each, not including leasing costs and other expenses.[111] The areas of highest quality—the "core" or "sweet spots"—have now been well defined. Figure 3-74 illustrates average horizontal well decline curves by county which are a measure of well quality (recognizing that future gas production from the Woodford will be dominantly from horizontal, not vertical, wells). Initial well productivities (IPs) from Pittsburg, Coal and Hughes counties are significantly higher than Canadian, Blaine and the "other 26" counties, although the latter benefit from significant liquids production which improves economics. Notwithstanding the higher productivity of wells in the top counties, production has fallen between 32% and 52% from peak in four of the top five counties—a function of low gas prices, expensive wells, and available drilling locations. Halting production declines even temporarily in these counties will require significantly higher gas prices.

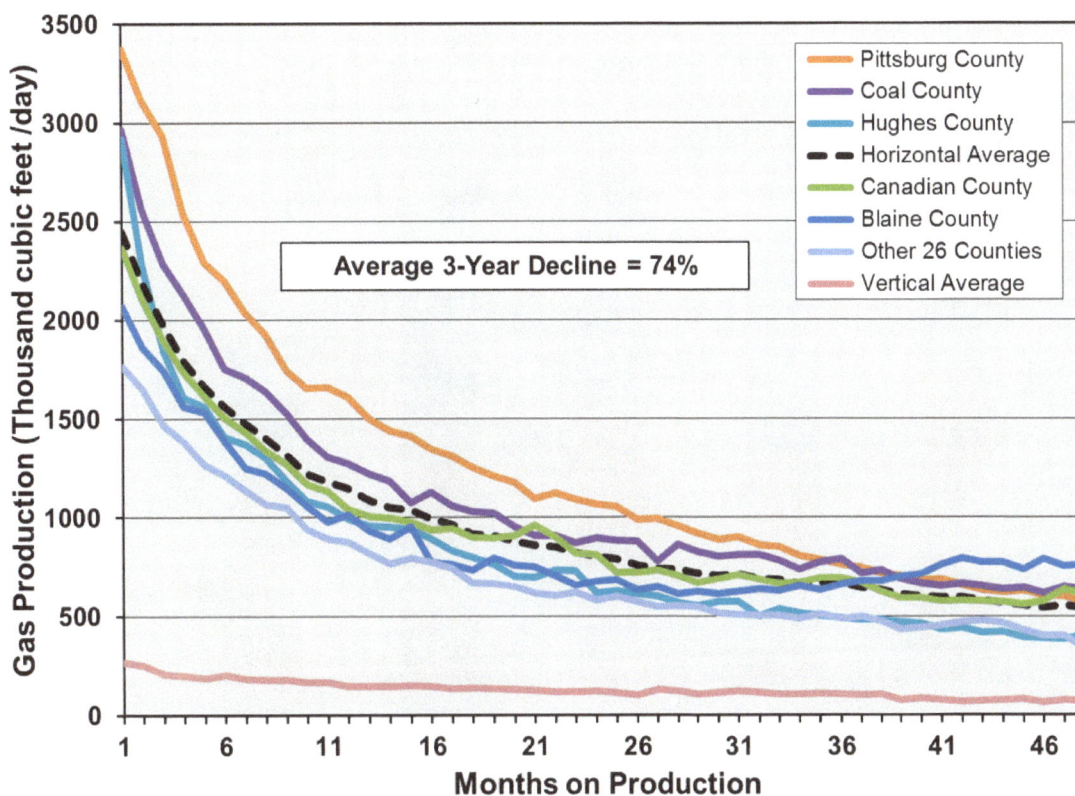

Figure 3-74. Average horizontal gas well decline profiles by county for the Woodford play.[112]

The top two counties, which have produced much of the gas in the Woodford, are clearly superior.

Another measure of well quality is "estimated ultimate recovery" or EUR—the amount of gas a well will recover over its lifetime. To be clear, no one knows what the lifespan of an average Woodford well is, given that few of them are more than eight years old (see Figure 3-67 and Figure 3-68), and some 10% of horizontal wells drilled have ceased production at an average age of less than three years. Operators fit hyperbolic and/or exponential curves to data such as presented in Figure 3-74, assuming well life spans of

---

[111] Mason, Richard, June 27, 2013, "Targeting Oklahoma's Ubiquitous Woodford Shale," http://www.ugcenter.com/Woodford/Targeting-Oklahomas-Ubiquitous-Woodford-Shale_118127.
[112] Data from Drillinginfo retrieved September 2014.

30-50 years (as is typical for conventional wells), but so far this is speculation given the nature of the extremely low permeability reservoirs and the completion technologies used in the Woodford. Nonetheless, for comparative well quality purposes only, one can use the data in Figure 3-74, which exhibits steep initial decline with progressively more gradual decline rates, and assume a constant terminal decline rate thereafter to develop a theoretical EUR.

Figure 3-75 illustrates theoretical EURs by county for the Woodford for comparative purposes of well quality. These range from 1.95 to 3.19 billion cubic feet per well, which are somewhat higher than the 1.18 to 1.51 billion cubic feet assumed by the EIA.[113] The steep initial well production declines mean that well payout, if it is achieved, comes in the first few years of production, as between 45% and 60% of an average well's lifetime production occurs in the first four years.

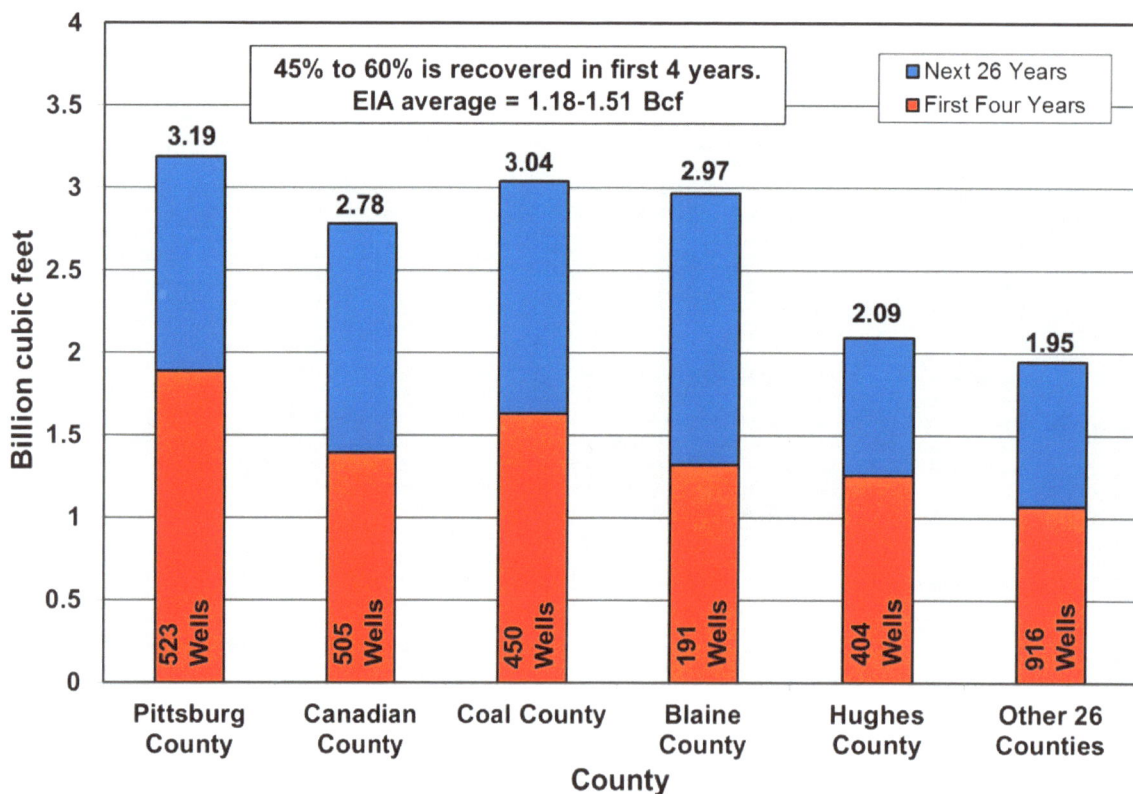

Figure 3-75. Estimated ultimate recovery of gas per well by county for the Woodford play.[114]

EURs are based on average well decline profiles (Figure 3-74) and a terminal decline rate of 15%. These are for comparative purposes only as it is highly uncertain if wells will last for 30 years. The steep decline rates mean that most production occurs early in well life.

Well quality can also be expressed as the average rate of production over the first year of well life. If we know both the rate of production in the first year of the average well and the field decline rate, we can calculate the number of wells that need to be drilled each year to offset field decline in order to maintain production. Figure 3-76 illustrates the average first-year production rate of wells in the Woodford by county. With the

---

[113] EIA, *Assumptions to the Annual Energy Outlook 2014*, http://www.eia.gov/forecasts/aeo/assumptions/pdf/oilgas.pdf.
[114] Data from Drillinginfo retrieved September 2014.

exception of Pittsburg County, which had its peak rate in 2011, all counties experienced peak rates in 2010 and on average the play is down 24% since then. In the past two years average productivity has been flat, including significant improvement in Coal County and continued decline in Blaine and Pittsburg counties. This reflects both a lack of improvement from better technology as well as a move into liquids rich-parts of the play which in general have somewhat lower gas productivities.

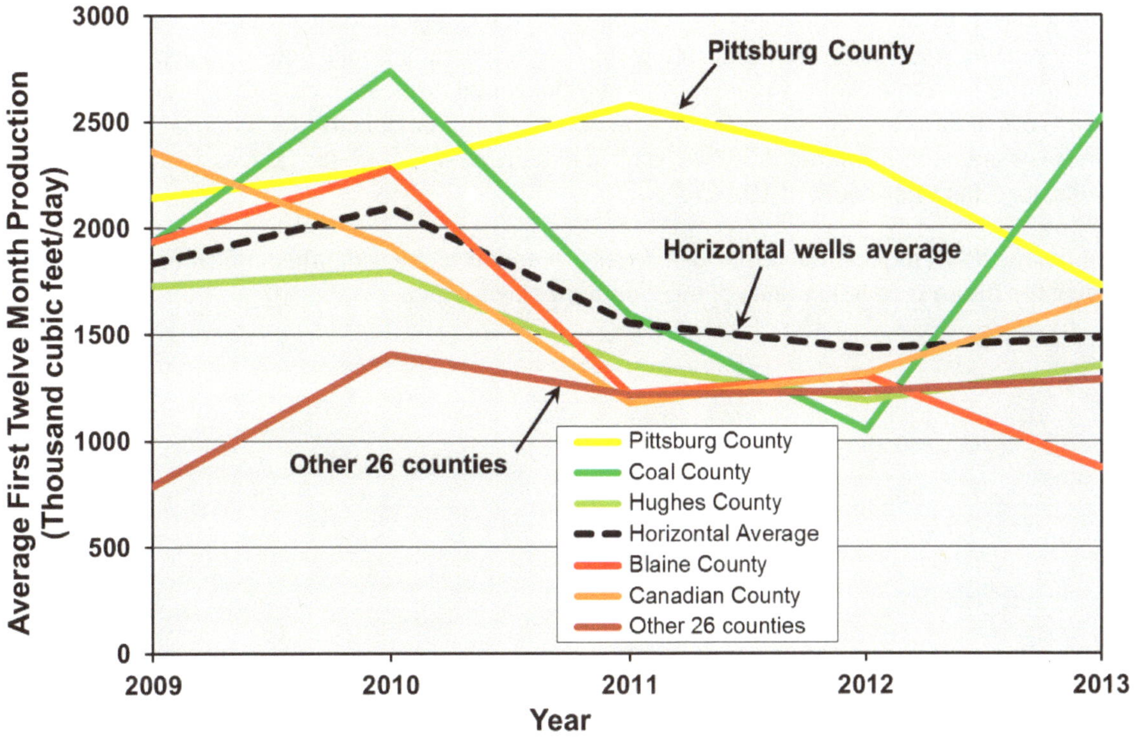

Figure 3-76. Average first-year gas production rates of wells by county for the Woodford play, 2009 to 2013.[115]

Well quality is down 24% on average from 2010, notwithstanding a recent increase in Coal County.

---

[115] Data from Drillinginfo retrieved September 2014.

3.3.4.4 Number of Wells

The fourth key fundamental is the number of wells that can ultimately be drilled in the Woodford play. A careful review of the top five counties suggests a prospective area of 2,358 square miles within them. The EIA has estimated the total play area at 4,246 square miles,[116] which leaves 1,888 prospective square miles outside the top five counties. This appears to be overly optimistic, given the distribution of production outlined in Figure 3-61, but for the sake of argument is assumed to be correct. The EIA further assumes that between 4 and 8 wells can be drilled per square mile, for an average well density of 4.6 wells per square mile.[117] The existing well density over this area is 0.84 wells per square mile (including vertical wells), and 0.7 including only horizontal wells. Assuming that only horizontal wells will be drilled in future, and given that vertical wells are already at a density of 0.14 per square mile, a final density of 4.5 horizontal wells per square mile is assumed. Given that 3,656 wells have already been drilled, that leaves 16,118 horizontal yet-to-drill wells, for a final well count of 19,107.

Table 3-4 breaks down the number of yet-to-drill wells by county along with other critical parameters used for determining the future production rates of the Woodford play.

---

[116] EIA, *Assumptions to the Annual Energy Outlook 2014*, http://www.eia.gov/forecasts/aeo/assumptions/pdf/oilgas.pdf.
[117] EIA, *Assumptions to the Annual Energy Outlook 2014*, http://www.eia.gov/forecasts/aeo/assumptions/pdf/oilgas.pdf.

Parameter	County						Total
	Blaine County	Canadian County	Coal County	Hughes County	Pittsburg County	Other 26 Counties	
Production April 2014 (Bcf/d)	0.10	0.47	0.20	0.12	0.35	0.53	1.77
% of Field Production	6	26	11	7	20	30	100
Cumulative Gas (Tcf)	0.18	0.47	0.62	0.46	0.79	0.50	3.01
Cumulative Liquids (MMbbl)	1.61	8.31	1.09	0.04	0.00	16.35	27.41
Number of Wells	191	505	450	404	523	916	2989
Number of Producing Wells	171	451	423	361	481	745	2632
Average EUR per well (Bcf)	2.09	2.78	3.04	2.97	3.19	1.95	2.64
Field Decline (%)	38.1	46.5	14.1	17.9	28.4	40.3	32.7
3-Year Well Decline (%)	63	74	79	86	83	81	78
Peak Year	2012	Rising	2010	2010	2012	Flat	2012
% Below Peak	48	N/A	36	51	32	N/A	4
Average First Year Production in 2013 (Mcf/d)	875	1673	2522	1354	1728	1290	1486
New Wells Needed to Offset Field Decline	29	170	27	19	43	170	405
Area in square miles	929	900	518	807	1306	10000	14460
% Prospective	50	60	70	50	45	19	29
Net square miles	465	540	363	404	588	1888	4246
Well Density per square mile	0.41	0.94	1.24	1.00	0.89	0.49	0.70
Additional locations to 4.5/sq. Mile	1899	1925	1182	1412	2122	7579	16118
Population	11943	115541	5925	14003	45837	N/A	N/A
Total Wells 4.5/sq. Mile	2090	2430	1632	1816	2645	8495	19107
Total Producing Wells 4.5/sq. Mile	2070	2376	1605	1773	2603	8324	18750

Table 3-4. Parameters for projecting Woodford production, by county.

Area in square miles under "Other" is estimated.

3.3.4.5   Rate of Drilling

Given known well- and field-decline rates, well quality by area, and the number of available drilling locations, the most important parameter in determining future production levels is the rate of drilling—the fifth key fundamental. Figure 3-77 illustrates the historical drilling rates in the Woodford. Horizontal drilling rates peaked in January 2013 at 601 wells per year and have fallen to current levels of less than 300 wells per year. Current drilling rates are somewhat less than the roughly 400 wells per year required to maintain current production, hence production is gradually declining.

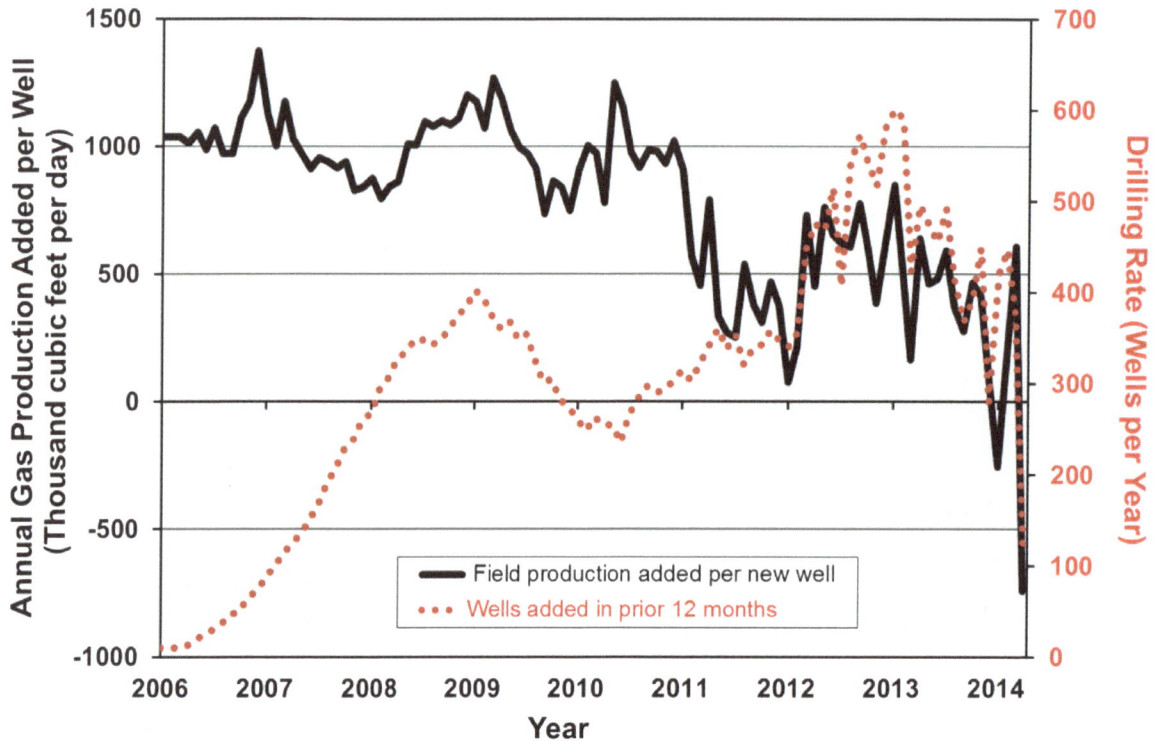

Figure 3-77. Annual production added per new horizontal well and annual drilling rate in the Woodford play, 2006 through 2014.[118]

Drilling rate peaked in January 2013 and is now somewhat below the level needed to keep production flat, hence each new well now only serves to slow the overall production decline of the play.

---

[118] Data from Drillinginfo retrieved September 2014. Three-month trailing moving average.

3.3.4.6   Future Production Scenarios

Based on the five key fundamentals outlined above, several production projections for the Woodford play were developed to illustrate the effects of changing the rate of drilling. Figure 3-78 illustrates the production profiles of three drilling rate scenarios if 100% of the prospective play area is drillable at 4.5 horizontal wells per square mile. These scenarios are:

1. MOST LIKELY RATE scenario: Drilling increases somewhat to 400 wells per year, then gradually declines to 300 wells per year.

2. LOW RATE scenario: Drilling remains at 300 wells per year, then gradually declines to 250 wells per year.

3. HIGH RATE scenario: Drilling roughly doubles to 550 wells per year, then gradually declines to 300 wells per year.

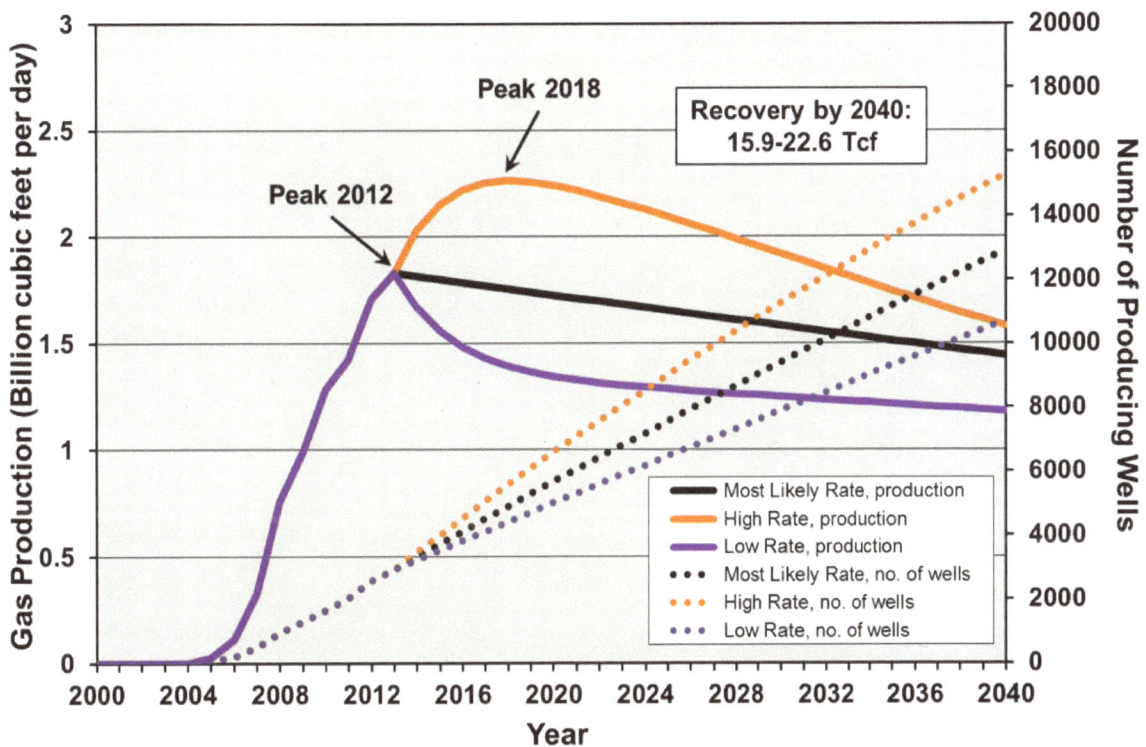

Figure 3-78. Three drilling rate scenarios of Woodford gas production (assuming 100% of the area is drillable at 4.5 horizontal wells per square mile).[119]

"Most Likely Rate" scenario: drilling increases to 400 wells/year, declining to 300 wells per year.
"Low Rate" scenario: drilling continues at 300 wells/year, declining to 250 wells/year.
"High Rate" scenario: drilling increases to 550 wells/year, declining to 300 wells/year.

The drilling rate scenarios have the following results:

1. MOST LIKELY RATE scenario: The drilling rate increases somewhat from current levels on strengthening gas prices, and then gradually declines as lower quality parts of the play are drilled.

---

[119] Data from Drillinginfo retrieved September 2014.

Total gas recovery by 2040 would be 19.1 trillion cubic feet and drilling would continue beyond 2040.

2. LOW RATE scenario: Drilling would continue at current rates. Total gas recovery by 2040 would be 15.9 trillion cubic feet and drilling would continue beyond 2040.

3. HIGH RATE scenario: Nearly doubling drilling rates would reverse decline and production would grow to a new peak in 2018. Total gas recovery by 2040 would be 22.6 trillion cubic feet and drilling would continue beyond 2040.

The recovery of 19.1 trillion cubic feet by 2040 in the "Most Likely" drilling rate scenario, and the recovery of 22.6 trillion cubic feet in the "High" drilling rate scenario, are somewhat less but reasonably close to the recovery of 23.8 trillion cubic feet assumed by the EIA. The "Most Likely" drilling rate scenario would see the recovery of more than six times as much gas as has been recovered to date (3.01 Tcf).

3.3.4.7   Comparison to EIA Forecast

Figure 3-79 illustrates the EIA's projection for Woodford production through 2040 compared to the "Most Likely Rate" scenario. Although the total recovery is not that different, the EIA has underestimated actual recovery through 2014 and assumes that production rate will ramp to a new peak in 2026 some 36% higher than the peak in 2012, and maintain production at levels considerably higher than today through 2040.[120] This implies the recovery of 82% of the proved reserves[121] and unproved resources[122] that the EIA assigns to the Woodford play. Although this seems highly optimistic, the EIA forecast for the Woodford is more restrained than its estimates for most other major shale plays.

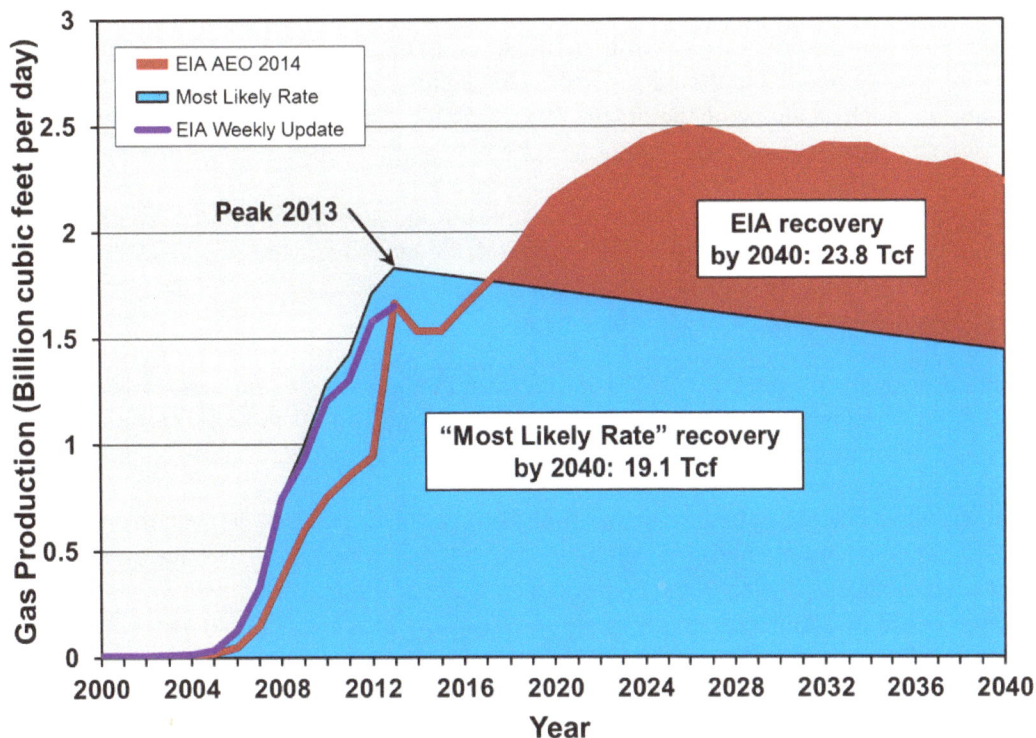

Figure 3-79. "Most Likely Rate" scenario of Woodford gas production compared to the EIA reference case, 2000 to 2040.[123]

The EIA assumes the Woodford will reach a new all-time high by 2026, and maintain production at considerably higher than present levels through 2040. The EIA forecast is made on a "dry gas" basis, whereas the "Most Likely Rate" scenario forecast is made on a "raw gas" basis. Also shown are the EIA's Woodford gas production statistics from its *Natural Gas Weekly Update*,[124] which contradict the early years of its AEO 2014 forecast.

---

[120] EIA, *Annual Energy Outlook 2014*, unpublished tables from AEO 2014 provided by the EIA.
[121] EIA, 2014, "Principal shale gas plays: natural gas production and proved reserves, 2011-12," http://www.eia.gov/naturalgas/crudeoilreserves/excel/table_4.xls.
[122] EIA, *Assumptions to the Annual Energy Outlook 2014*, http://www.eia.gov/forecasts/aeo/assumptions/pdf/oilgas.pdf.
[123] EIA, *Annual Energy Outlook 2014*, unpublished tables from AEO 2014 provided by the EIA.
[124] EIA, *Natural Gas Weekly Update*, retrieved October 2014, http://www.eia.gov/naturalgas/weekly.

3.3.4.8   Woodford Play Analysis Summary

Several things are clear from this analysis:

1. Drilling rates have fallen in the Woodford due to gas prices, and drilling has moved to liquids-rich parts of the play.

2. High well- and field-decline rates mean a continued high rate of drilling is required to maintain, let alone increase, production. Current drilling rates of about 300 wells per year are somewhat below the level of about 400 wells per year required to maintain production, which would require the investment of $3.6 billion per year for drilling (assuming $9 million per well). Future production profiles are most dependent on drilling rate and, to a lesser extent, on the number of drilling locations (i.e., greatly increasing the number of drilling locations would not change the production profile nearly as much as changing the drilling rate). Maintaining or growing gas production in the Woodford would require considerably higher gas prices to justify higher drilling rates.

3. Doubling current drilling rates could reverse the current production decline and raise production to a new peak in the 2018 timeframe, but would increase cumulative recovery only by 19% by 2040 and wouldn't change the ultimate recovery of the play. Increasing drilling rates effectively recovers the gas sooner making the supply situation worse later.

4. The projected recovery of 19.1 Tcf by 2040 in the "Most Likely Rate" scenario, is somewhat less than the 23.8 Tcf projected by the EIA in its reference case forecast. The EIA forecast of the Woodford rising to a new production peak in 2026 at significantly higher rates than today is improbable.

5. This report's projections are optimistic in that they assume the capital will be available for the drilling treadmill that must be maintained. They also assume that 100% of the prospective area is drillable. This is not a sure thing as drilling in the poorer quality parts of the play will require considerably higher gas prices to be economic. Failure to maintain drilling rates will result in a steeper drop off in production.

6. More than triple the current number of wells will need to be drilled to meet the production projection of the "Most Likely Rate" scenario by 2040.

7. The EIA projection for future Woodford gas production included in its reference case forecast for AEO 2014[125] is highly optimistic in that it forecasts the current production decline will be reversed and rise to a new peak in 2026 at a level 36% higher than the 2012 peak of the play, and then maintain production through 2040 at levels far higher than today. This is highly unlikely to be realized, especially at the gas prices the EIA forecasts.[126]

---

[125] EIA, *Annual Energy Outlook 2014*, unpublished tables from AEO 2014 provided by the EIA.
[126] EIA, *Annual Energy Outlook 2014*, http://www.eia.gov/forecasts/aeo.

3.3.5   Marcellus Play

The Marcellus play is now the largest and fastest growing shale gas play in the U.S. Production growth in the Marcellus has more than compensated for declines in other plays. It is also the largest play in terms of areal extent, stretching from New York State to southern West Virginia and west to Ohio, although most production comes from Pennsylvania. Figure 3-80 illustrates the distribution of wells as of mid-2014. Over 10,700 wells have been drilled to date of which 7,006 were producing at the time of writing. Of these, more than 7,900 are in Pennsylvania, 5,302 of which were producing in mid-2014. There is a large backlog of drilled but not connected wells (also indicated in Figure 3-80), believed to be over two thousand in number. This is a function of the rate of drilling and the relative youth of the play; most of these wells will be connected over time as pipeline infrastructure catches up.

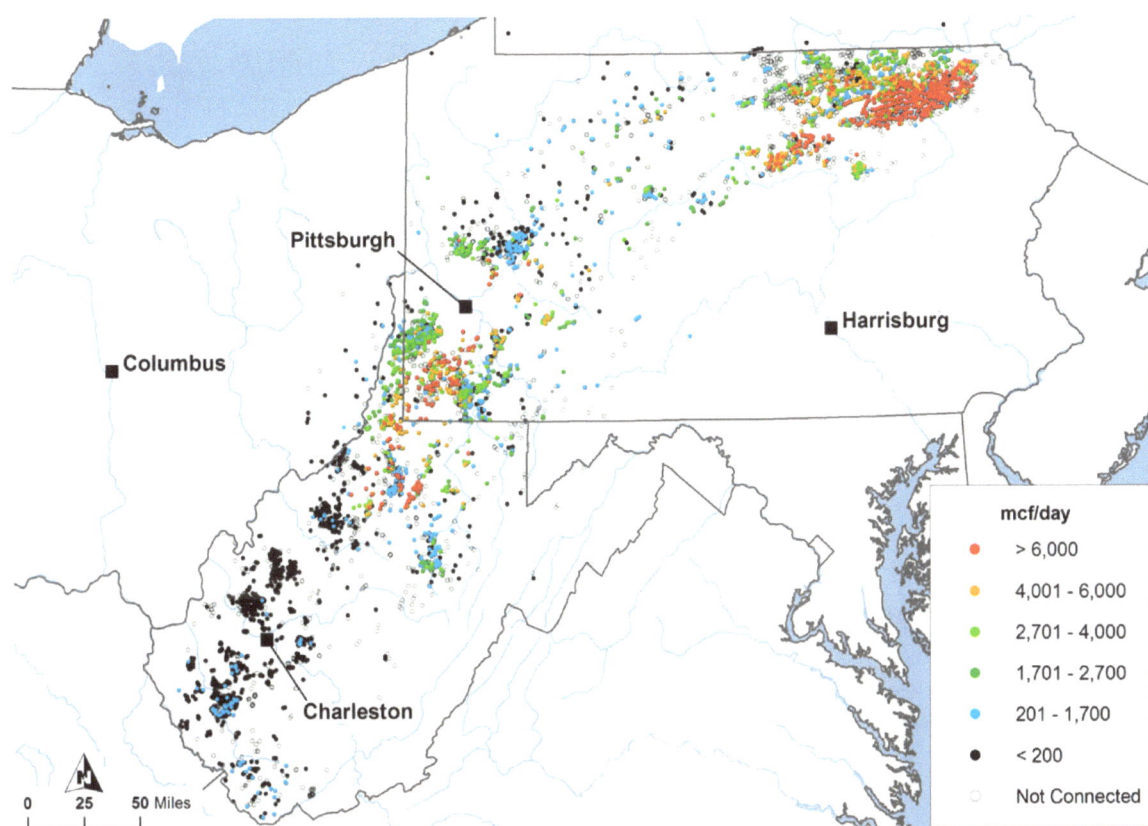

Figure 3-80. Distribution of wells in Marcellus play as of mid-2014, illustrating highest one-month gas production (initial productivity, IP).[127]

Well IPs are categorized approximately by percentile; see Appendix.

[127] Data from Drillinginfo retrieved September 2014.

Production from the Marcellus exceeded 12 billion cubic feet per day in June 2014 as illustrated in Figure 3-81. More than 91% of production came from Pennsylvania with most of the remainder from West Virginia. Ohio and New York State production is negligible. Over 98% of Pennsylvania production is from horizontal fracked wells, whereas 22% of production in West Virginia came from vertical/directional wells. The rate of drilling grew to a maximum of more than 1,500 wells per year in mid-2012 through 2013 and has now fallen to about 1,300 per year. Drilling rates are still well above the approximately 1,000 wells per year required to keep production flat at current production levels, so production will keep rising.

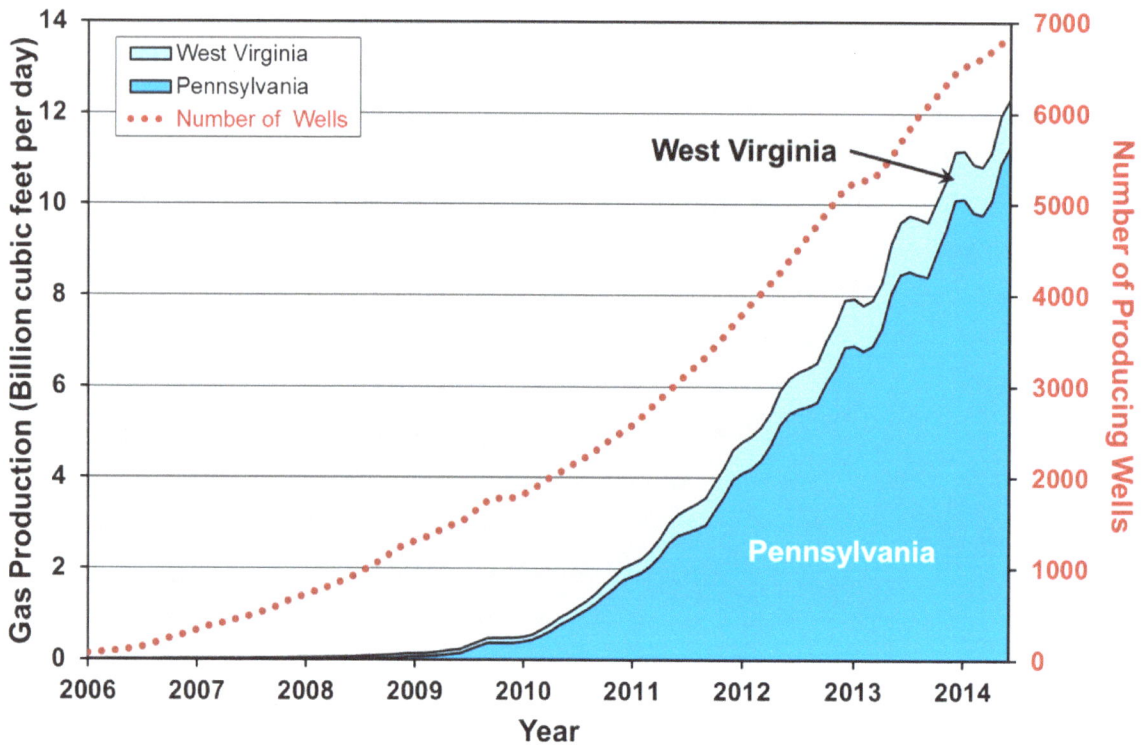

Figure 3-81. Marcellus play shale gas production, differentiating between Pennsylvania and West Virginia, and number of producing wells, 2006 to 2014.[128]

Gas production data are provided on a "raw gas" basis.

---

[128] Data from Drillinginfo retrieved September 2014. Three-month trailing moving average.

Vertical wells played a significant role in the early development of the Marcellus play in West Virginia and still produce some oil and gas, although new wells are predominantly horizontal. Although there are some legacy vertical wells in Pennsylvania, virtually all new drilling is horizontal. The distribution of horizontal and vertical/directional wells in the play is illustrated in Figure 3-82.

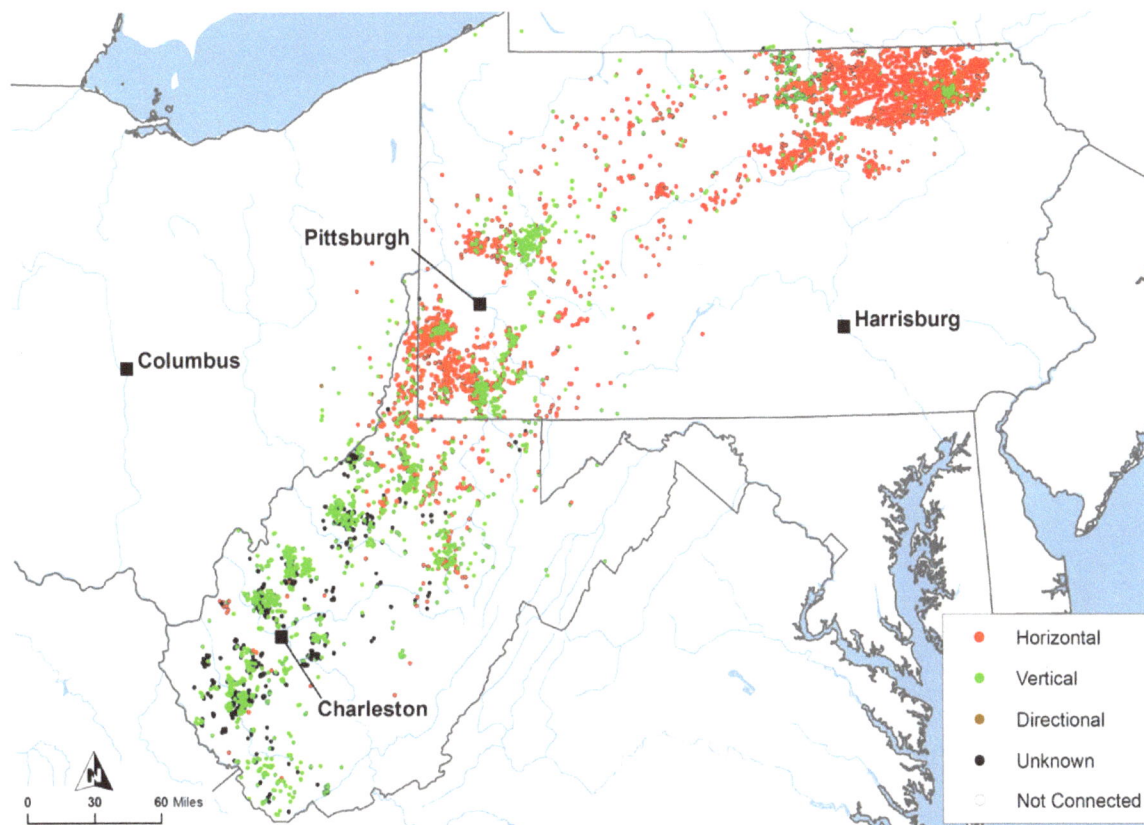

Figure 3-82. Distribution of wells in Marcellus play categorized by drilling type as of mid-2014.[129]

Development began with vertical and directional wells before expanding to largely horizontal drilling at present.

---

[129] Data from Drillinginfo retrieved September 2014.

Cumulative gas recovery by well type in Pennsylvania and West Virginia is illustrated in Figure 3-83. Although vertical/directional wells make up 23% of currently producing wells, they have produced less than 4% of the gas. There will be few if any additional vertical/directional wells drilled in the Marcellus play—future production growth will rely on horizontal fracked wells.

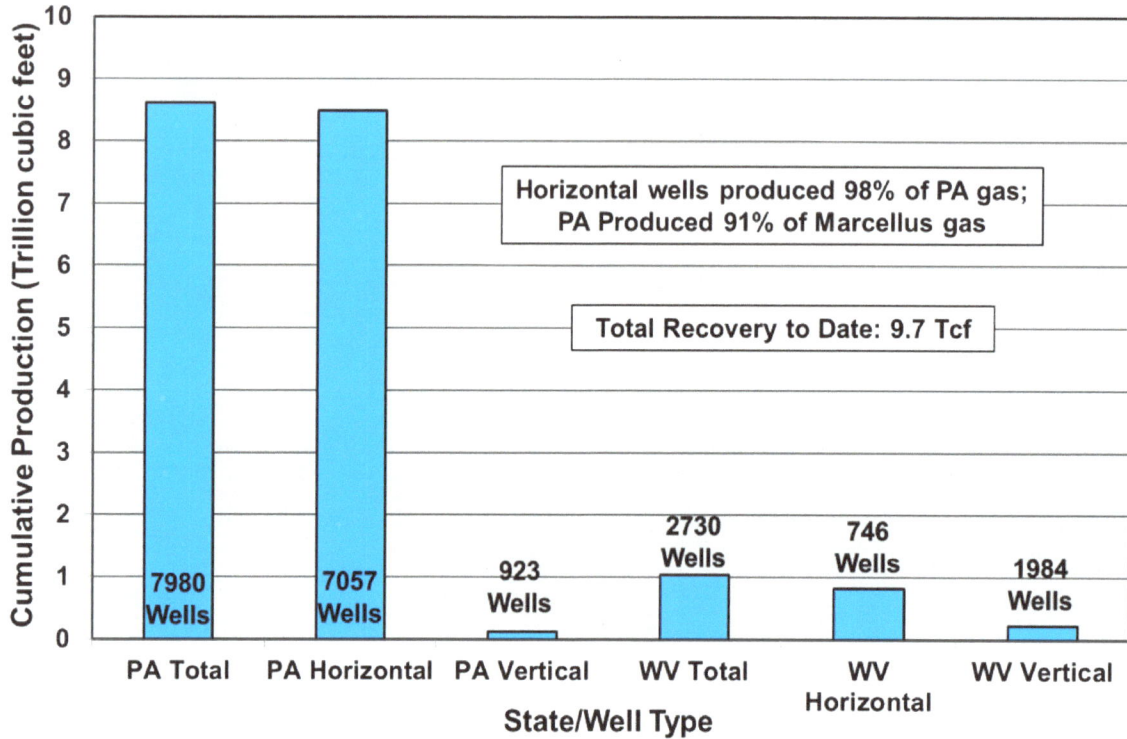

Figure 3-83. Cumulative gas production in the Marcellus play by well type and state, 2000 to 2014.[130]

The well count includes all producing wells as well as those drilled but not producing, either because they are not connected to pipelines or have ceased production.

---

[130] Data from Drillinginfo retrieved September 2014.

3.3.5.1 Well Decline

The first key fundamental in determining the life cycle of Marcellus production is the *well decline rate*. Marcellus wells exhibit high decline rates in common with all shale plays. Figure 3-84 illustrates the average decline rate of the most recent Marcellus horizontal and vertical/directional wells by state. Decline rates are steepest in the first year and are progressively less in the second and subsequent years. The decline rates over the first three years of average well life range between 74% and 82%, which is on the lower end of the range for most shale plays. As can be seen, vertical/directional wells have much lower productivity than horizontal wells and hence are being phased out.

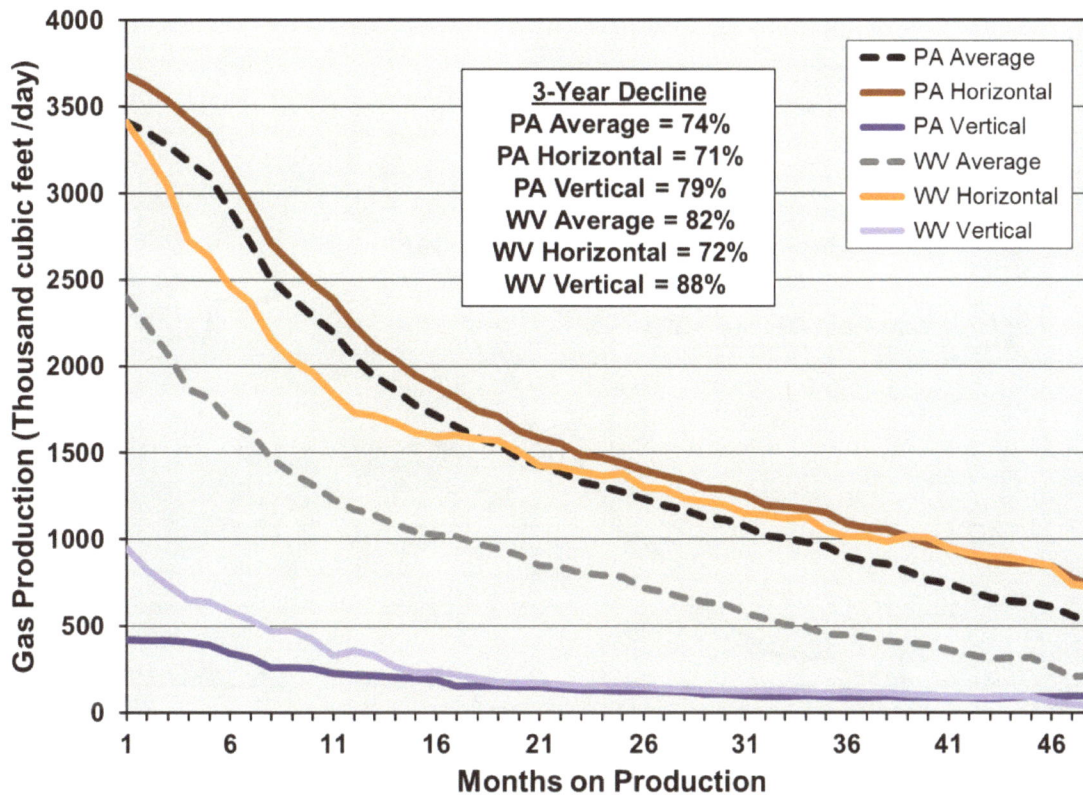

Figure 3-84. Average decline profile for horizontal and vertical/directional gas wells in the Marcellus play, by state.[131]

Decline profile is based on all shale gas wells drilled since 2009.

---

[131] Data from Drillinginfo retrieved September 2014.

3.3.5.2   Field Decline

A second key fundamental is the overall *field decline rate*, which is the amount of production that would be lost in a year without more drilling. Figure 3-85 illustrates production from the 3,500 horizontal wells drilled prior to 2013 in Pennsylvania. The first-year decline rate is 32%, which is on the low end of field decline rates observed for shale plays. Assuming new wells will produce in their first year at the average first-year rates observed for wells drilled in 2013, approximately 1,000 new wells each year would be required to offset field decline at current production levels. At an average cost of $5 million per well, this would represent a capital input of about $5 billion per year, exclusive of leasing and other infrastructure costs, to keep production flat at mid-2014 levels.

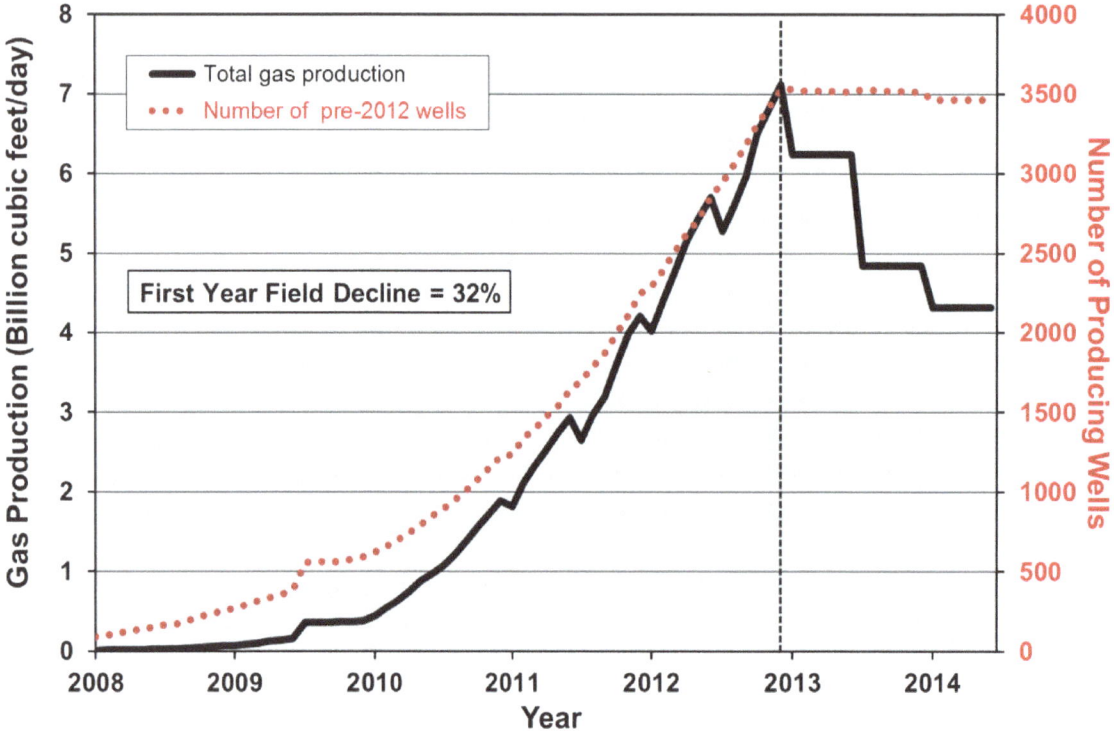

Figure 3-85. Production rate and number of horizontal shale gas wells drilled in the Marcellus play in Pennsylvania prior to 2013, 2008 to 2014.[132]

This defines the field decline for the Marcellus play which is 32% per year (horizontal wells will be responsible for virtually all future production). The stepped nature of the production curve is due to the fact that Pennsylvania releases data in six month chunks, not on a monthly basis.

[132] Data from Drillinginfo retrieved September 2014.

3.3.5.3 Well Quality

The third key fundamental is the *average well quality* by area and its trend over time. Petroleum engineers tell us that technology is constantly improving, with longer horizontal laterals, more frack stages per well, more sophisticated mixtures of proppants and other additives in the frack fluid injected into the wells, and higher-volume frack treatments. This has certainly been true over the past few years, along with multi-well pad drilling which has reduced well costs. In the Marcellus, well quality is continuing to grow strongly, suggesting that better technology is having an effect, along with a better understanding of the reservoir and the location of sweet spots. The average first-year production rate of Marcellus wells over time is illustrated in Figure 3-86.

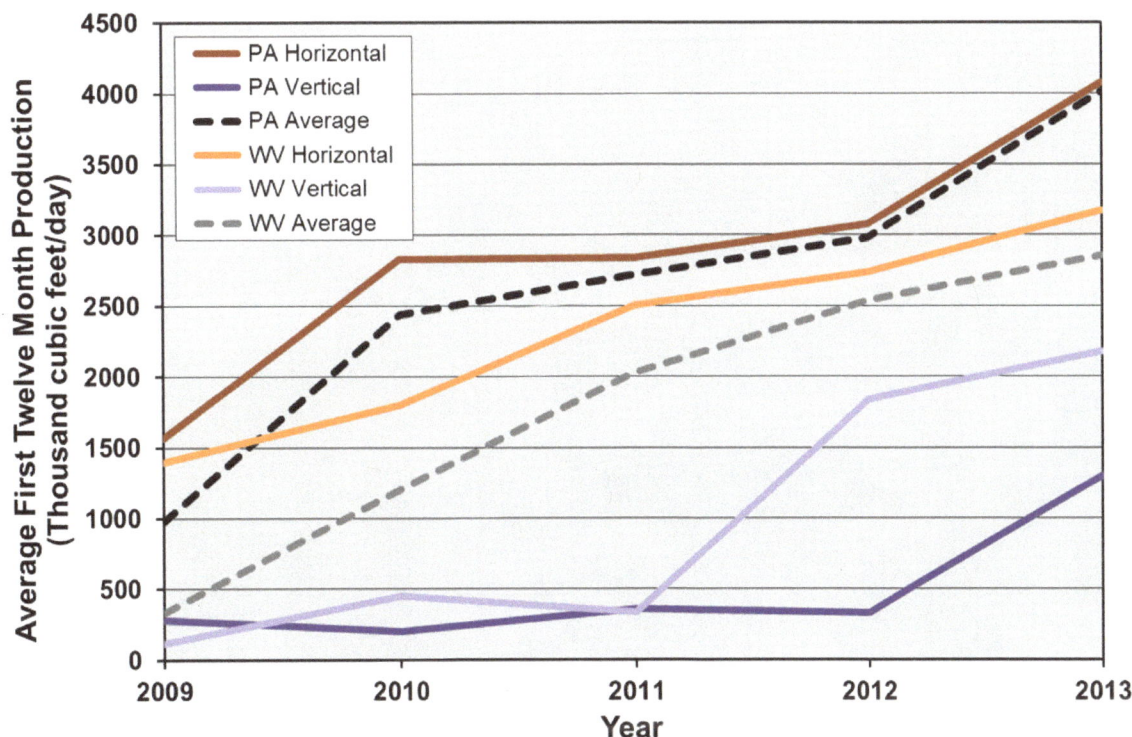

Figure 3-86. Average first-year production rates for Marcellus horizontal and vertical/directional gas wells by state, 2009 to 2013.[133]

Average well quality has increased substantially as better technology is applied and drilling is focused on the sweet spots.

---

[133] Data from Drillinginfo retrieved September 2014.

Another measure of well quality is cumulative production and well life. Figure 3-87 illustrates the cumulative production of all horizontal wells that were producing in the Pennsylvania Marcellus as of June 2014 (Pennsylvania is focused on as it has generally higher quality wells and more than 90% of Marcellus production). Although it can be seen that there are a few very good wells that recovered large amounts of gas in the first few years, and undoubtedly were great economic successes—7% of wells had recovered more than 4 billion cubic feet after less than 5 years—the average well had produced just 1.56 billion cubic feet over a lifespan averaging 28 months. Less than 6% of these wells are more than 5 years old.

The lifespan of wells is another key parameter as many operators assume a minimum life of 30 years and longer—this is conjectural at this point given the lack of long term well performance data.

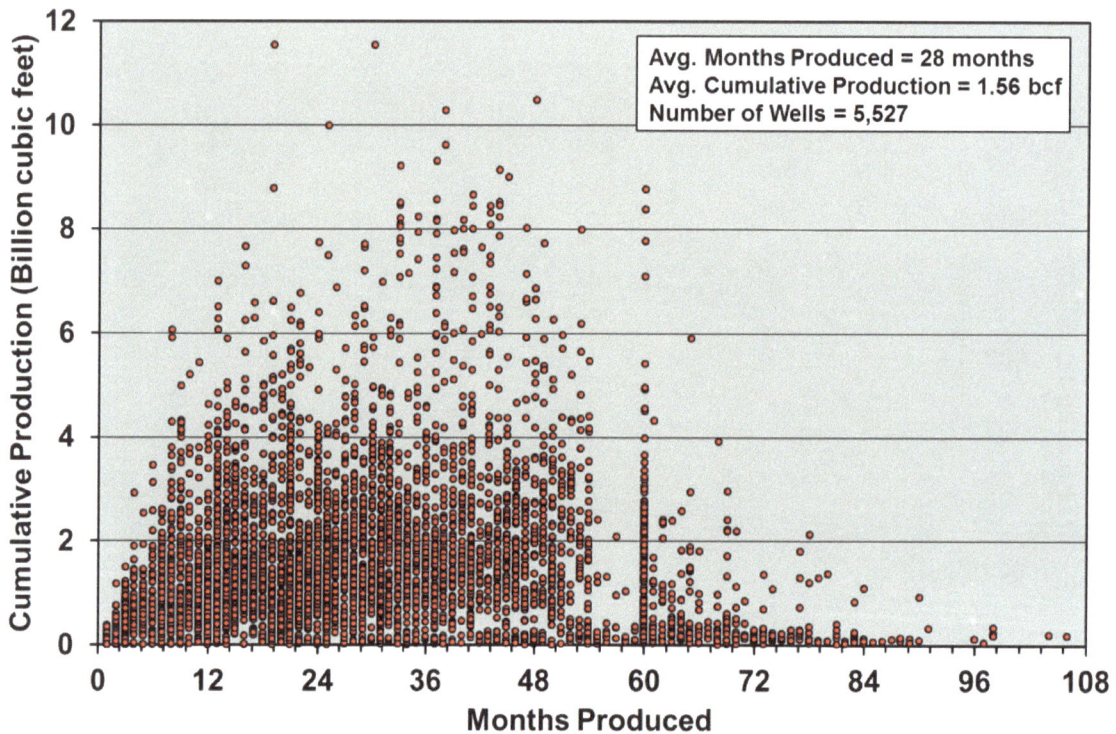

Figure 3-87. Cumulative gas production and length of time produced for wells in the Marcellus play in Pennsylvania.

Few wells are greater than five years old, with a mean age of 28 months and a mean cumulative recovery of 1.56 billion cubic feet.[134]

---

[134] Data from Drillinginfo retrieved September 2014.

Cumulative production of course depends on how long a well has been producing, so looking at young wells is not necessarily a good indication of how much gas these wells will produce over their lifespan (although production is heavily weighted to the early years of well life). A measure of well quality independent of age is initial productivity (IP) which is often focused on by operators. Figure 3-88 illustrates the average daily output over the first six months of production for all wells in the Pennsylvania portion of the Marcellus play (six month IP). The IPs are higher than most other shale plays—averaging 3.45 million cubic feet per day (MMcf/d) for all wells over the 2010 to 2014 period—and are trending upward, through both better technology and concentration of drilling in sweet spots. Again, as with cumulative production, there are a few exceptional wells—4% produced more than 10 MMcf/d—although the average of the most recent wells was about 5 MMcf/d overall. Figure 3-82 illustrates the distribution of IPs in map form illustrating the concentration of drilling in sweet spots.

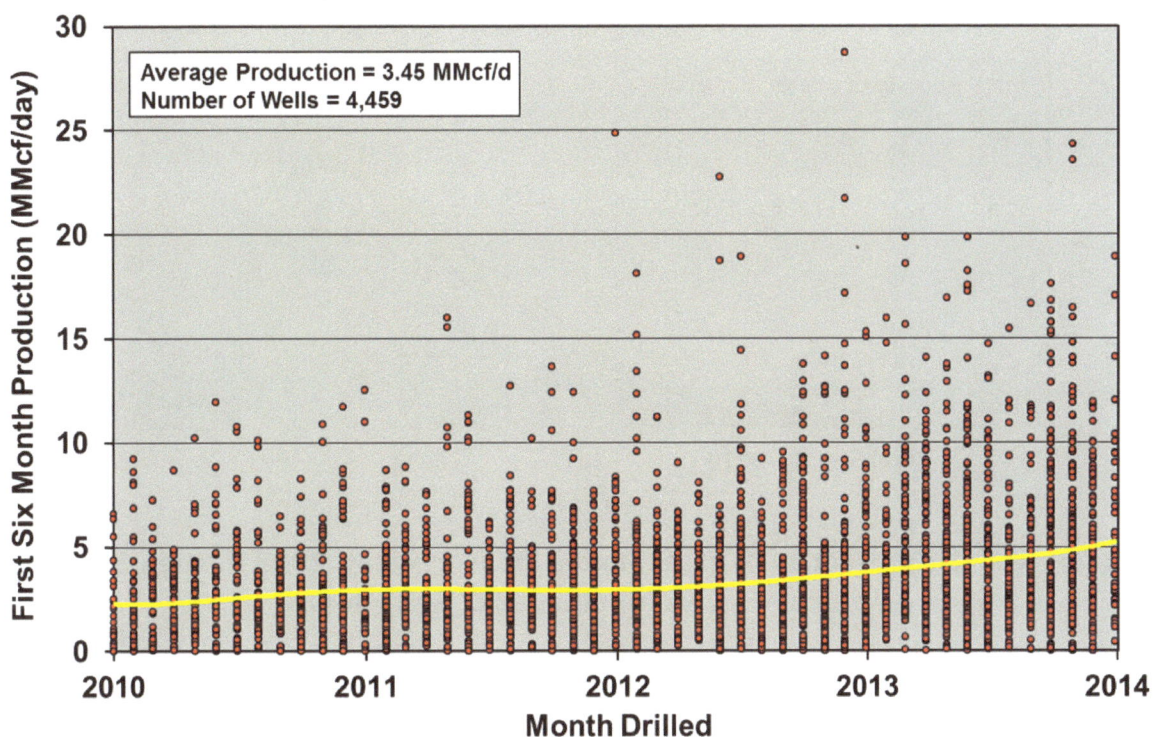

Figure 3-88. Average gas production over the first six months for all wells drilled in the Marcellus play of Pennsylvania, 2010 to 2014.[135]

Although there are a few exceptional wells, the average well produced 3.45 MMcf/d over the 2010 to 2014 period, with the most recent wells producing 5 MMcf/d. The trend line indicates mean productivity over time.

[135] Data from Drillinginfo retrieved September 2014.

Different counties in the Marcellus display markedly different well quality characteristics which are critical in determining the most likely production profile in the future. Figure 3-89, which illustrates production over time by county and state, shows that in June 2014, two counties in Pennsylvania produced 41% of all Marcellus gas and the top six Pennsylvania counties produced 76%.

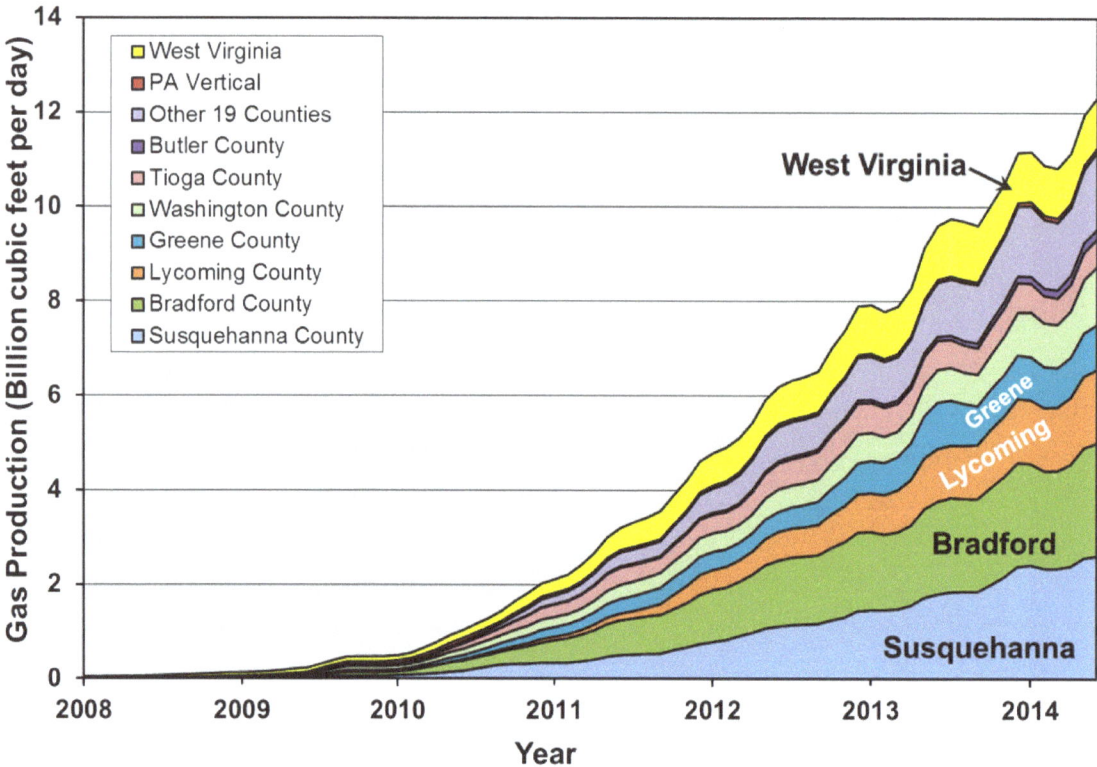

Figure 3-89. Gas production by county in the Marcellus play, 2008 through 2014.[136]

The top six Pennsylvania counties produced 76% of production in June 2014.

---

[136] Data from Drillinginfo retrieved September 2014. Three-month trailing moving average.

The location of sweet spots is a function of the combination of many geological characteristics, including depth, thickness, organic matter content, thermal maturity, lithological characteristics allowing fractures to propagate, and the presence of natural fracture systems. Despite the widespread nature of the Marcellus, two sweet spots have been defined that produce the bulk of the gas. The northeast Pennsylvania sweet spot, centered in Susquehanna and Bradford counties, is illustrated with IPs in Figure 3-90, and the southwest Pennsylvania/West Virginia sweet spot, centered on Washington and Greene counties, is illustrated in Figure 3-91. Berman and Pettinger provide an in-depth discussion of the variation in quality of the Marcellus and the price of gas required to be profitable in various areas; they conclude that relatively little commercial gas exists in southern New York State.[137]

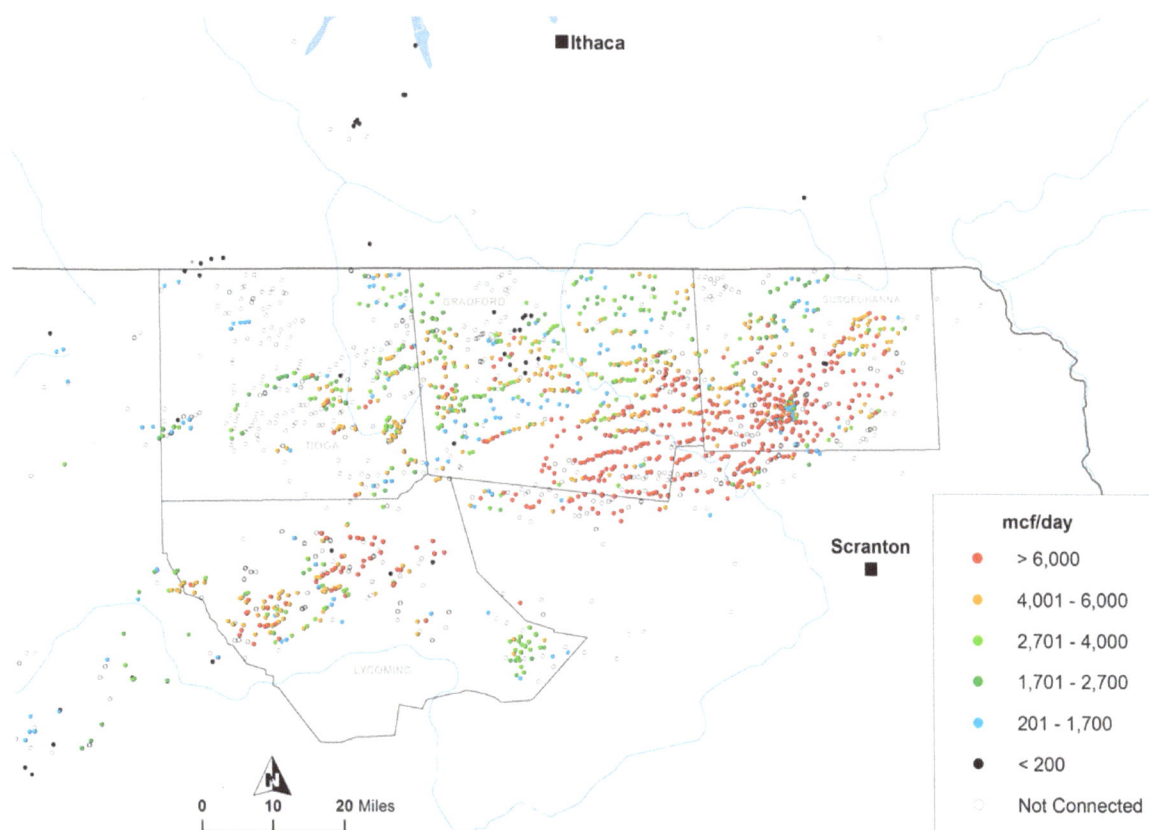

Figure 3-90. Distribution of wells in the northeast Pennsylvania sweet spot of the Marcellus play, illustrating highest one-month gas production (initial productivity, IP).[138]
Bradford and Susquehanna counties produced 41% of all Marcellus gas in June 2014.

[137] Berman, A.E. and Pettinger, L, 2014, *Resource Assessment of Potentially Producible Natural Gas Volumes from the Marcellus Shale, State of New York*, http://www.lwvny.org/advocacy/natural-resources/hydrofracking/2014/Marcellus-Resource-Assessment-NY_0414pdf.pdf.
[138] Data from Drillinginfo retrieved September 2014.

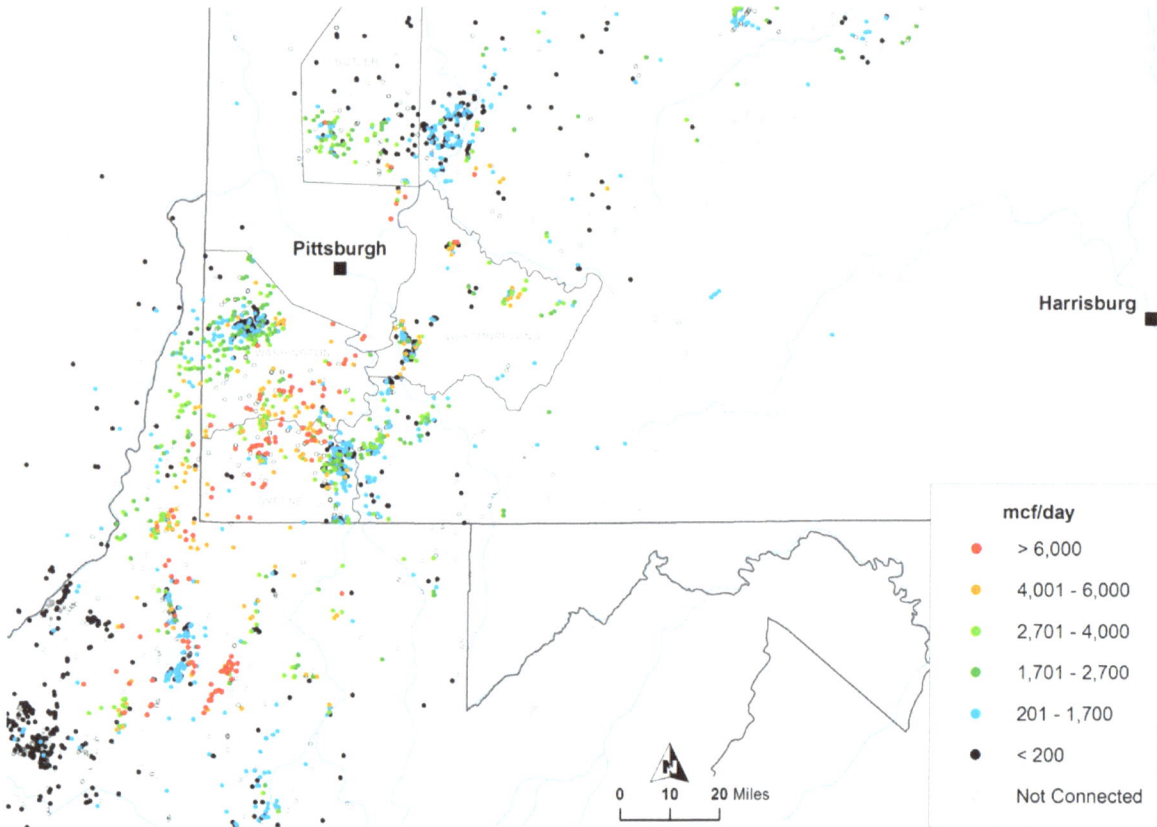

Figure 3-91. Distribution of wells in the southwest Pennsylvania / northern West Virginia sweet spot of the Marcellus play, illustrating highest one-month gas production (initial productivity, IP). [139]

Washington and Greene counties along with northern West Virginia produce most of the liquids associated with Marcellus gas.

---

[139] Data from Drillinginfo retrieved September 2014.

Cumulative production since the field commenced is also concentrated in the sweet spots. As illustrated in Figure 3-92, the top two counties have produced 40% of the gas and the top six have produced 75%. Production in most counties is growing although Greene and Tioga counties in Pennsylvania, and the state of West Virginia in general, are down somewhat from peak production.

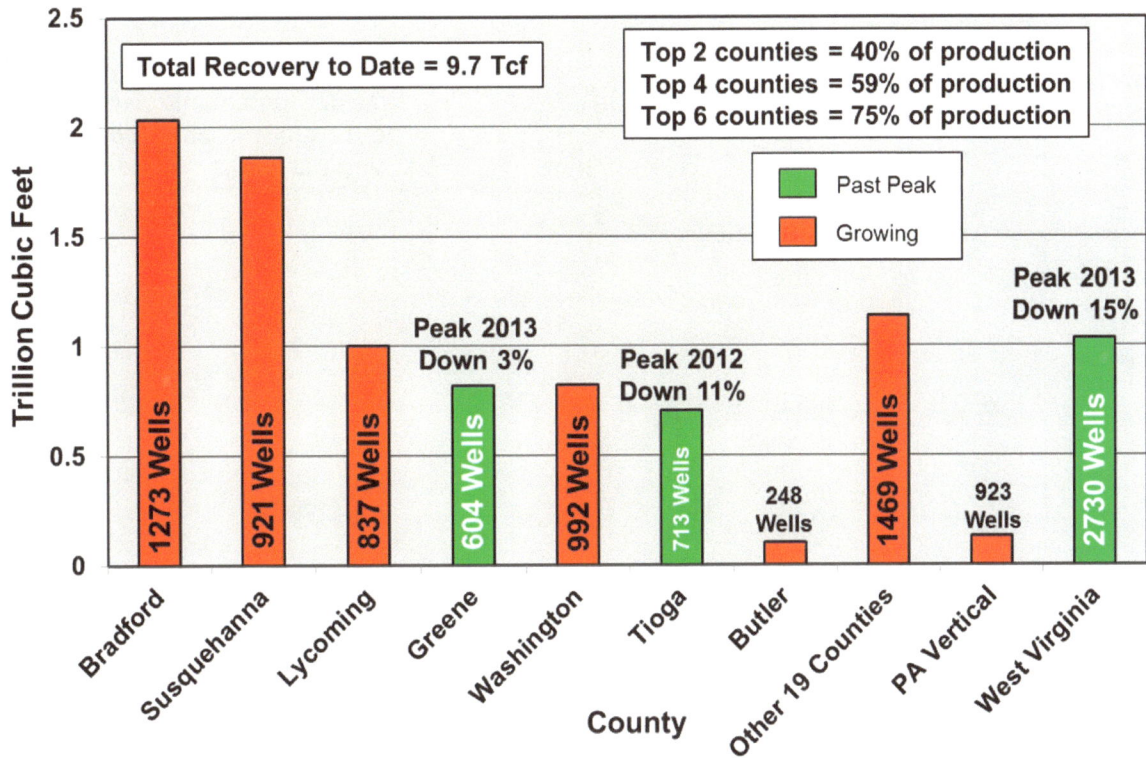

Figure 3-92. Cumulative gas production by county in the Marcellus play through June 2014.[140]

The top six counties have produced 75% of the 9.7 trillion cubic feet of gas produced to date. Greene and Tioga counties in Pennsylvania as well as West Virginia are below peak production, but all other areas are rising.

---

[140] Data from Drillinginfo retrieved September 2014.

The Marcellus also produces limited amounts of natural gas liquids and oil. Most liquids production is in Washington County in southwestern Pennsylvania and in northern West Virginia, as illustrated in Figure 3-93. Although more than 13 million barrels of liquids have been produced since 2005, in the big picture liquids production from the Marcellus is relatively insignificant.

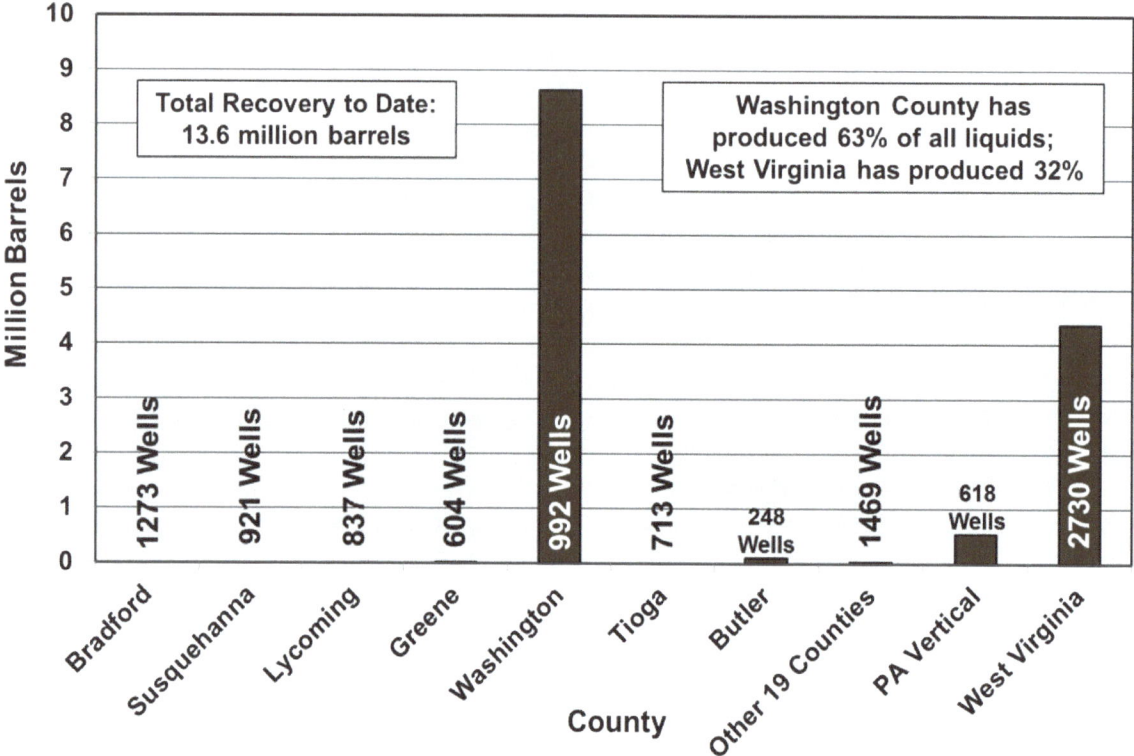

Figure 3-93. Cumulative liquids production by county in the Marcellus play through 2014.[141]

Production is concentrated in southwest Pennsylvania and northern West Virginia.

---

[141] Data from Drillinginfo retrieved September 2014.

Operators are highly sensitive to the economic performance of the wells they drill, which typically cost in the order of $6 million or more each, not including leasing costs and other expenses.[142] The areas of highest quality—the "core" or "sweet spots"—have now been well defined. Figure 3-94 illustrates average horizontal well decline curves by county, which are a measure of well quality (recognizing that future gas production from the Marcellus will be from horizontal, not vertical, wells). Initial well productivities (IPs) from Susquehanna County are more than double those of most other counties (excepting Bradford, Lycoming and Greene). The decline curves from the top four counties are all above the Marcellus average, hence these counties are attracting the bulk of the drilling and investment. Future drilling will have to focus more and more on lesser quality counties.

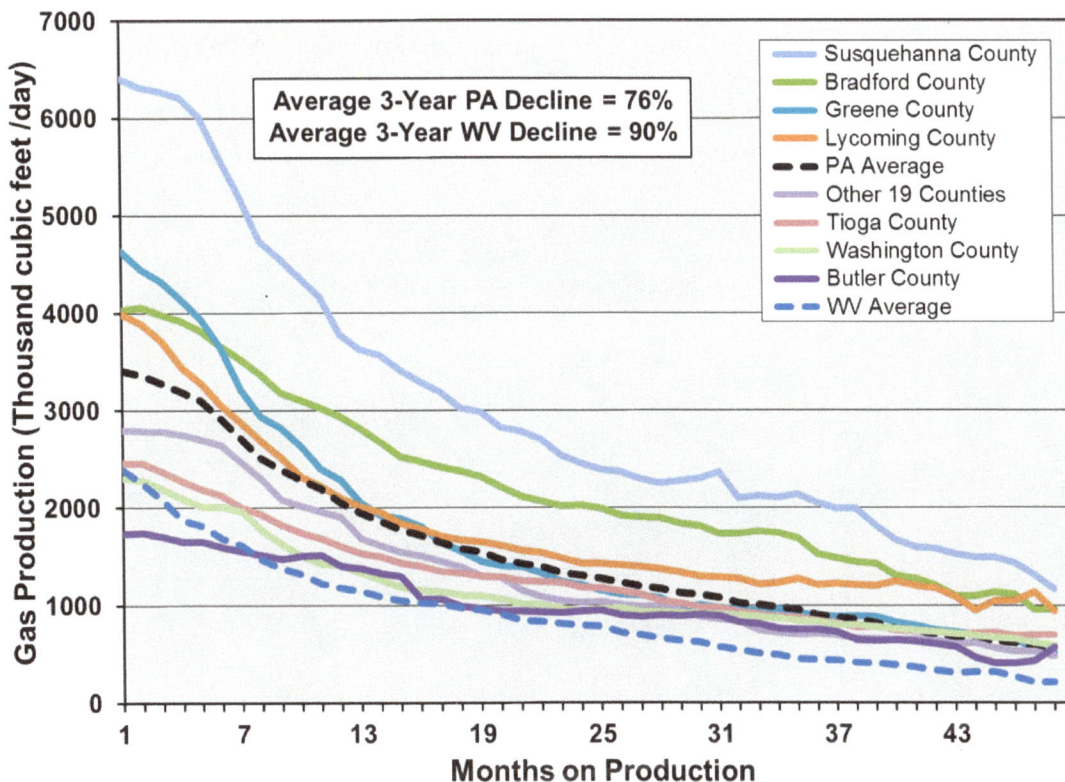

Figure 3-94. Average horizontal gas well decline profiles by county and state for the Marcellus play.[143]

The top four Pennsylvania counties, which have produced much of the gas in the Marcellus, are clearly superior.

Another measure of well quality is "estimated ultimate recovery" or EUR—the amount of gas a well will recover over its lifetime. Although to be clear no one knows what the lifespan of a Marcellus well is, given that few of them are more than five years old (see Figure 3-87 and Figure 3-88), EURs provide a useful metric to compare well quality between areas. Operators fit hyperbolic and/or exponential curves to data such as presented in Figure 3-94, assuming well life spans of 30-50 years (as is typical for conventional wells), but so far this is speculation given the nature of the extremely low permeability reservoirs and the completion technologies used in the Marcellus. Nonetheless, for comparative well quality purposes only, one can use the data in Figure 3-94, which exhibits steep initial decline with progressively more gradual decline rates, and assume a constant terminal decline rate thereafter to develop a theoretical EUR.

Figure 3-95 illustrates theoretical EURs by county in Pennsylvania for the Marcellus for comparative purposes of well quality. These range from 2.21 to 7.05 billion cubic feet per well, which are comparable to the 0.55 to 7.14 billion cubic feet assumed by the EIA.[144] The steep initial well production declines mean that well payout, if it is achieved, comes in the first few years of production, as between 63% and 72% of an average well's lifetime production occurs in the first four years.

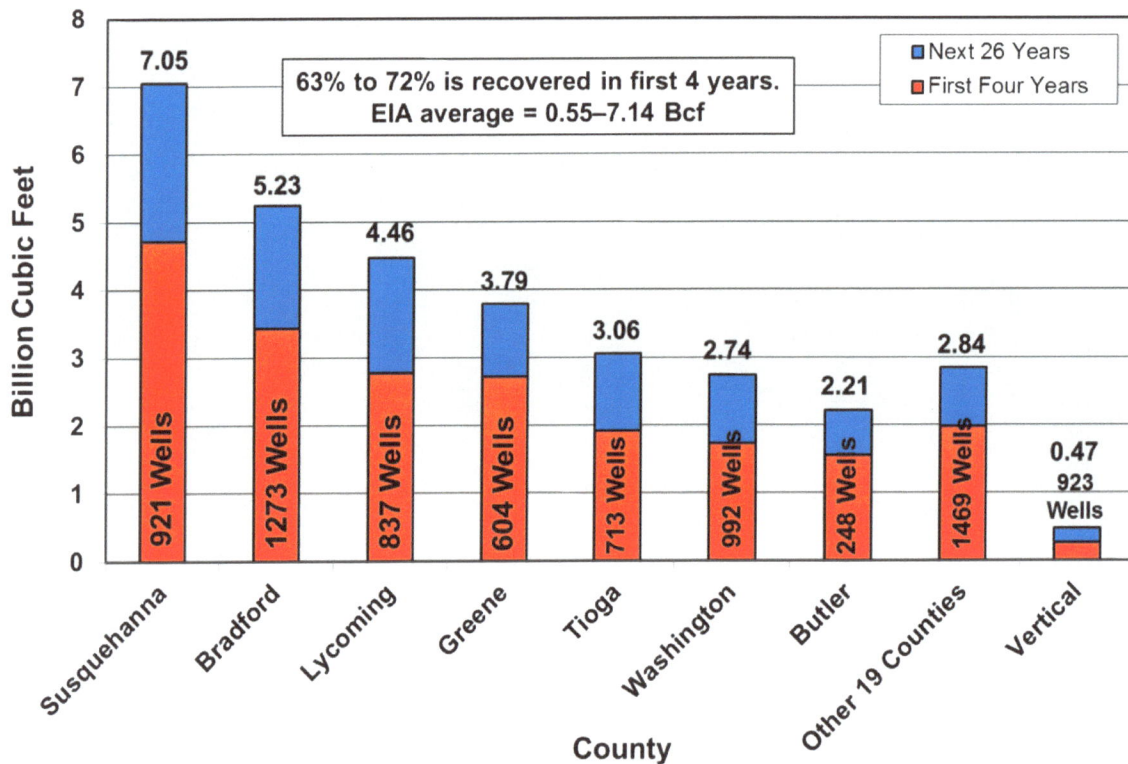

Figure 3-95. Estimated ultimate recovery of gas by county for the Marcellus play in Pennsylvania.[145]

EURs are based on average well decline profiles (Figure 3-94) and a terminal decline rate of 20%. These are for comparative purposes only as it is highly uncertain if wells will last for 30 years. The steep decline rates mean that most production occurs early in well life. The lowest 22 counties average less than half of the EUR of the top county, Susquehanna.

---

[144] EIA, July 2014, *Oil and Gas Supply Module*, http://www.eia.gov/forecasts/aeo/nems/documentation/ogsm/pdf/m063(2014).pdf.
[145] Data from Drillinginfo retrieved September 2014.

Well quality can also be expressed as the average rate of production over the first year of well life. If we know the rate of production in the first year of the average well, and the field decline rate, we can calculate the number of wells that need to be drilled each year to offset field decline in order to maintain production. Given that drilling is currently focused on the highest quality counties, the average first year production rate per well will fall as drilling moves into lower quality counties over time as the best locations are drilled off. As average well quality falls, the number of wells that must be drilled to offset field decline must rise, until the drilling rate can no longer offset decline and the field peaks.

Figure 3-96 illustrates the average first year production rate of wells by county. Average well quality has been rising in all areas through application of better technology—longer horizontal laterals, more frack stages, higher volumes of more sophisticated additives, and higher-volume frack treatments. The top three counties—Susquehanna, Greene and Bradford—are significantly higher than the average well productivity of the rest. Considering the large areal extent of the Marcellus play, relatively few wells have been drilled and thus there is still considerable room for more wells in the best areas. The current drilling rate of about 1,300 wells per year is above the 1,000 wells needed to offset field decline at current production levels, so Marcellus production will keep rising in the short to medium term as long as these drilling rates are maintained.

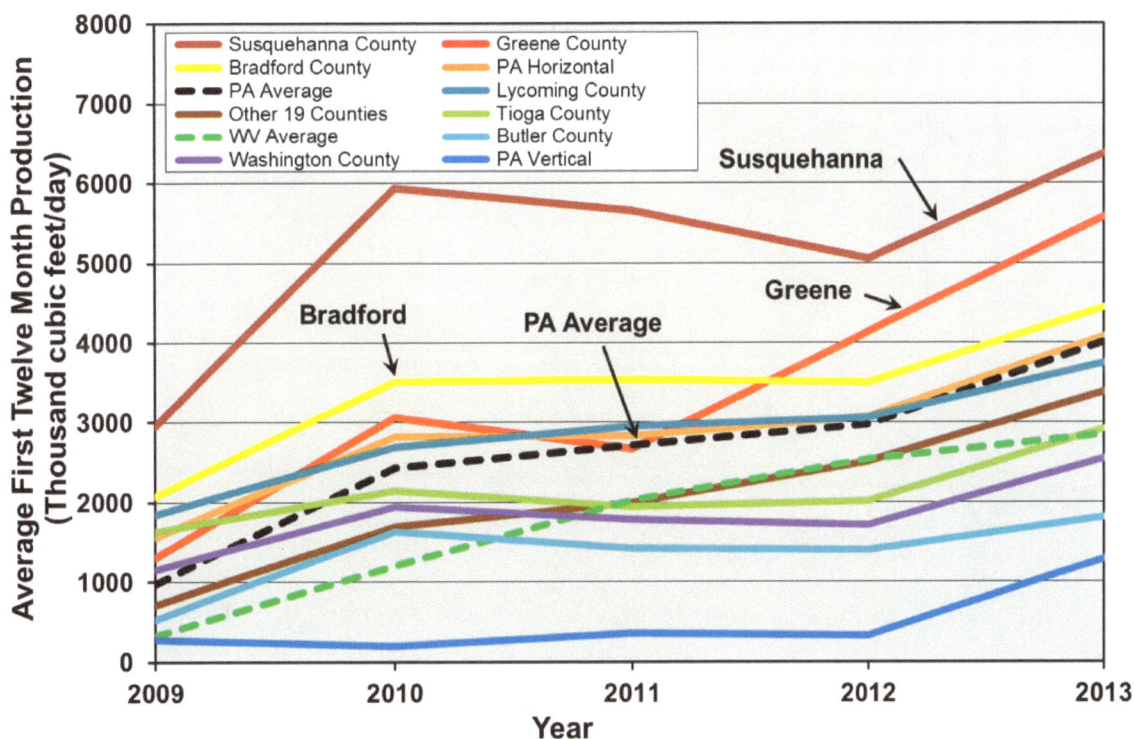

Figure 3-96. Average first-year gas production rates of wells by county in the Marcellus play, 2009 to 2013.[146]

Well quality is rising in most areas indicating that better technology—longer horizontal laterals and higher volume frack treatments—are improving productivity. First year production rate in the "other 19" counties, where more than half of the remaining drilling locations are found, is roughly half that of the top two counties.

[146] Data from Drillinginfo retrieved September 2014.

3.3.5.4 Number of Wells

A fourth critical parameter is the number of wells that can ultimately be drilled in the Marcellus play. The EIA estimates an area of 16,688 square miles for the "Marcellus Interior" and an additional 869 square miles for the "Marcellus Foldbelt" for a total of 17,566 square miles. They assign an average EUR of 1.59 Bcf to the former and 0.32 to the latter. They also include a "Marcellus Western" area of 2,684 square miles with an average EUR of 0.26 Bcf (which has less than 4% of total unproved resources). Assuming the EIA's estimates for the Marcellus interior and foldbelt regions are correct—and eliminating the low productivity western area due to its likely lack of economic viability—leaves a play area of 17,566 square miles. Using the EIA's estimate of 4.3 wells per square mile over this region, a total of 76,415 wells would be developed when the region is completely drilled off, or some ten times the current number of producing wells.

Given that Pennsylvania and West Virginia are relatively densely populated states, with some difficult topography, a more conservative estimate may be that only 80% of the remaining drilling locations are actually accessible to development—allowing for towns, cities, parks and other surface restrictions to development. In this case 63,274 wells would be drilled in total, or an additional 52,564 wells over what are currently in place.

Table 3-5 breaks down the number of yet-to-drill wells by county along with other critical parameters used for determining the future production rates of the Marcellus play.

Parameter	County											Total
	Bradford	Butler	Greene	Lycoming	Susquehanna	Tioga	Washington	Other 19	PA Vertical	PA Total	WV Total	
Production June 2014 (Bcf/d)	2.39	0.21	0.95	1.56	2.62	0.58	1.23	1.63	0.09	11.27	1.05	12.32
% of Field Production	19.40	1.73	7.71	12.67	21.23	4.72	9.98	13.26	0.75	91.46	8.54	100.00
Cumulative Gas (Tcf)	2.03	0.10	0.82	1.00	1.86	0.71	0.82	1.14	0.13	8.62	1.04	9.65
Cumulative Liquids (MMBBL)	0.00	0.09	0.00	0.00	0.00	0.00	8.63	0.03	0.53	9.29	4.35	13.64
Number of Wells	1273	248	604	837	921	713	992	1469	923	7980	2730	10710
Number of Producing Wells	896	142	416	615	662	485	734	862	490	5302	1704	7006
Average EUR per well (Bcf)	5.24	2.21	3.79	4.48	7.05	3.06	2.74	2.84	0.42	3.41	1.67	3.06
Field Decline (%)	25	31	48	37	33	33	32	26	30	32	29	32
3-Year Well Decline (%)	62	57	81	70	68	66	64	75	79	74	81	76
Average First Year Production in 2013 (Mcf/d)	4440	1823	5578	3750	6368	2924	2554	3390	1297	4012	2858	3932
New Wells Needed to Offset Field Decline	135	36	82	154	134	65	155	127	21	899	107	1003
Area in square miles	1161	795	578	1244	832	1137	861	19000	25608	25608	13656	39264
% Prospective	90	90	80	50	75	60	90	34	45	45	45	45
Net square miles	1045	716	462	622	624	682	775	6486	11412	11412	6145	17556
Well Density per square mile	1.22	0.35	1.31	1.35	1.48	1.05	1.28	0.23	0.08	0.70	0.44	0.61
Additional locations to 4.3/sq. Mile	3220	2829	1384	1838	1762	2220	2340	26420	0	42013	23692	65705
Population	62622	183862	38686	116111	43356	41981	207820	N/A	N/A	N/A	N/A	N/A
Total Wells 4.3/sq. Mile	4493	3077	1988	2675	2683	2933	3332	27889	923	49993	26422	76415
Producing Wells 4.3/sq. Mile	4116	2971	1800	2453	2424	2705	3074	27282	490	47315	25396	72711
Risked 80% Total Wells 4.3/sq. Mile	3849	2511	1711	2307	2331	2489	2864	22605	0	40667	21684	63274
Risked 80% Producing Wells 4.3/sq. Mile	3472	2405	1523	2085	2072	2261	2606	21998	0	38422	20658	59570

Table 3-5. Parameters for projecting Marcellus production, by county.

Area in square miles under "Other" is estimated. Wells by county are horizontal only; "Total" columns include both horizontal and vertical wells.

3.3.5.5   Rate of Drilling

Given known well- and field-decline rates, well quality by area, and the number of available drilling locations, the most important parameter in determining future production levels is the rate of drilling. Figure 3-97 illustrates the historical drilling rates in the Marcellus of Pennsylvania. Horizontal drilling rates in Pennsylvania peaked in 2013 at about 1,350 wells per year and have since fallen to current levels of about 1,200 wells per year. Coupled with drilling rates of 120 wells per year in West Virginia, current rates are about 1,320 wells per year. This is considerably higher that the approximately 1,000 wells per year needed to offset field decline at current production rates, hence Marcellus production will keep rising in the short to medium term as long as these drilling rates are maintained.

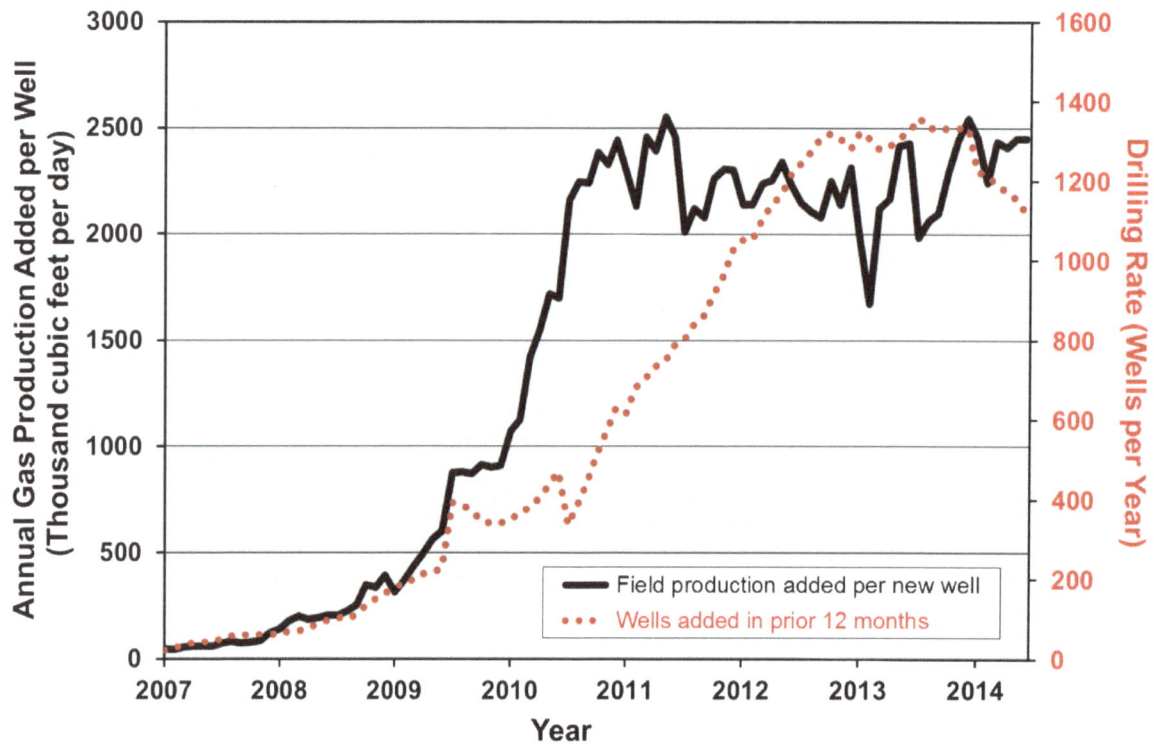

Figure 3-97. Annual gas production added per new horizontal well and annual drilling rate and in the Marcellus play, 2007 through 2014.[147]

Drilling rate peaked in 2013 but remains well above the level needed to offset field decline, hence production will continue to grow in the short to medium term.

[147] Data from Drillinginfo retrieved September 2014. Three-month trailing moving average.

3.3.5.6 Future Production Scenarios

Several drilling rate scenarios were used to develop production projections for the Marcellus play given the number of available drilling locations. Figure 3-98 illustrates the production profiles in Pennsylvania for three drilling rate scenarios if 80% of the prospective play area is drillable at 4.3 wells per square mile (for a total of 63,274 wells in the play with 40,677 of them in Pennsylvania). These scenarios are:

- MOST LIKELY RATE scenario: Assumes that drilling rate continues at current levels and then gradually declines to 800 wells per year as drilling moves into lower quality parts of the play.

- HIGH RATE scenario: Assumes that drilling will continue at current rates until all locations are drilled off.

- REDUCED RATE scenario: Assumes that wells will continue at current rates but decline more steeply to 200 wells per year as the last wells are drilled.

In all scenarios drilling continues through 2040 and beyond.

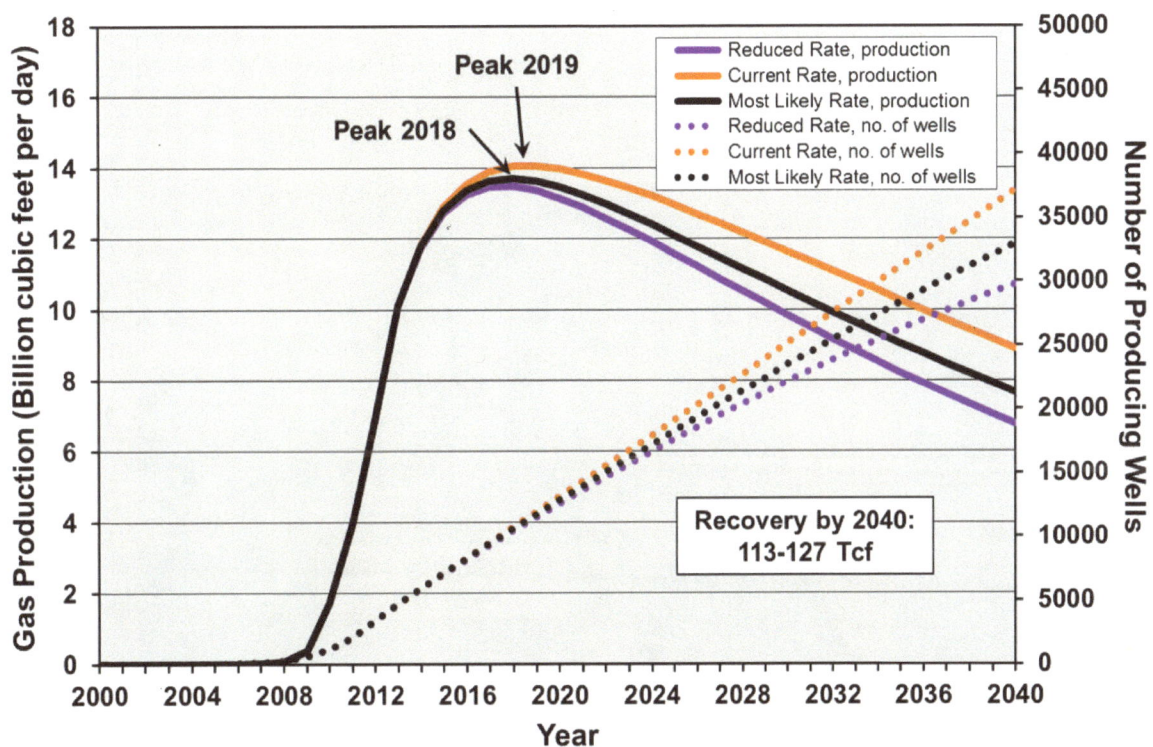

Figure 3-98. Three drilling rate scenarios of Marcellus gas production in Pennsylvania (assuming 80% of the area is drillable at 4.3 wells per square mile).[148]

"Most Likely Rate" scenario: drilling continues at 1,200 wells/year, declining to 800/year.
"High Rate" scenario: drilling continues at 1,200 wells/year.
"Reduced Rate" scenario: drilling continues at 1,200 wells/year, declining to 200/year.

[148] Data from Drillinginfo retrieved September 2014.

The drilling rate scenarios have the following results:

1. MOST LIKELY RATE scenario: Total gas recovery by 2040 would be 118.2 trillion cubic feet and drilling would continue beyond 2040. Peak production would occur in 2018.

2. HIGH RATE scenario: Total gas recovery by 2040 would be 127 trillion cubic feet and drilling would continue beyond 2040. Peak production would occur in 2019.

3. REDUCED RATE scenario: Total gas recovery by 2040 would be 113 trillion cubic feet and drilling would continue beyond 2040. Peak production would occur in 2017.

The recovery of between 113 and 127 trillion cubic feet, with 118.2 trillion cubic feet in the "Most Likely Rate" scenario by 2040, makes the Marcellus the most important shale gas play in the U.S. by a wide margin. Nonetheless, it peaks in the 2017-2019 timeframe followed by a long period of decline. If projected production from the Marcellus in West Virginia is included, production in the "Most Likely Rate" scenario will reach nearly 15 Bcf/d, with recovery of 129 trillion cubic feet by 2040 (assuming drilling is continued at the current rate in West Virginia of 120 wells per year) as illustrated in Figure 3-99 .

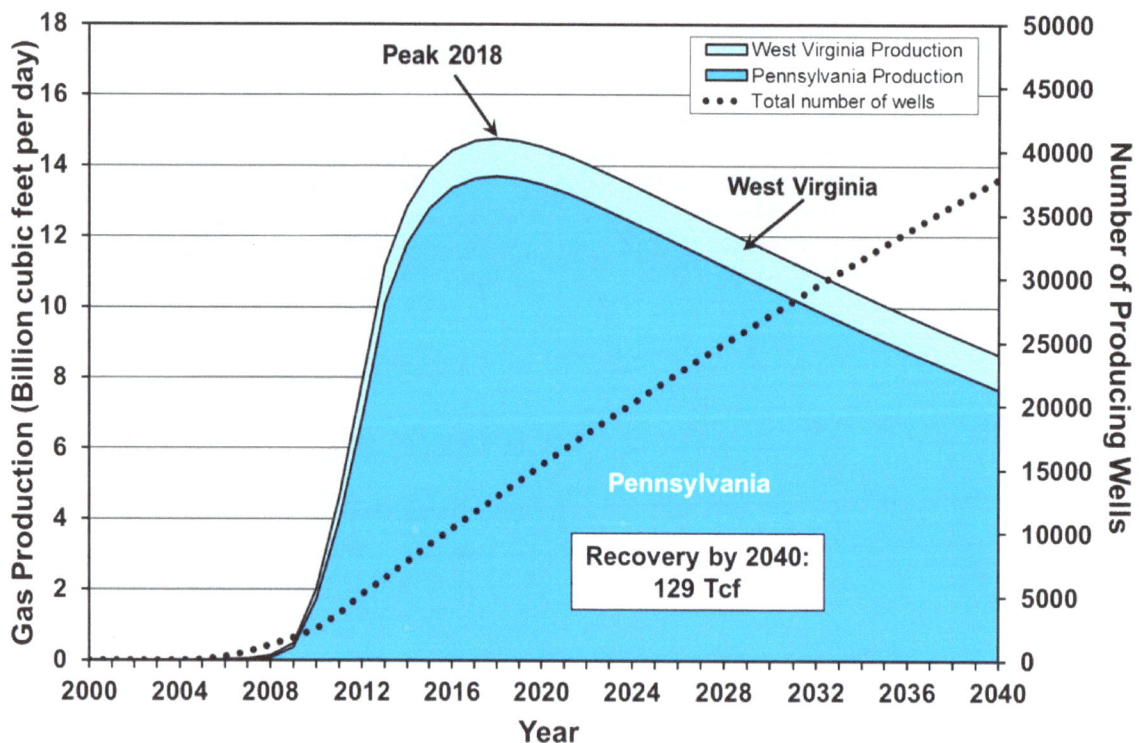

Figure 3-99. "Most Likely Rate" scenario of Marcellus gas production including both Pennsylvania and West Virginia.

Total recovery by 2040 of 129 Tcf is 13 times the amount of gas recovered to date. In this "Most Likely Rate" scenario, with the addition of West Virginia, drilling continues at 1,320 wells/year, declining to 920/year.

3.3.5.7   Comparison to EIA Forecast

Figure 3-100 illustrates the EIA's projection for Marcellus production through 2040 compared to the "Most Likely Rate" scenario. The EIA projects recovery by 2040 of 129 Tcf to meet its reference case forecast, which coincidentally is exactly the same quantity as projected in the "Most Likely Rate" scenario. The shape of the EIA production profile in its reference case, however, appears to underestimate past and current production—even compared to its own independent estimates (*Natural Gas Weekly Update* and *Drilling Productivity Report*[149])—and overestimate production in later years, beyond 2024. The EIA projects a peak in 2024 at 13.8 Bcf/d—lower than the 14.8 Bcf/d peak in 2018 in this report—and generally higher production in the post-2022 timeframe.

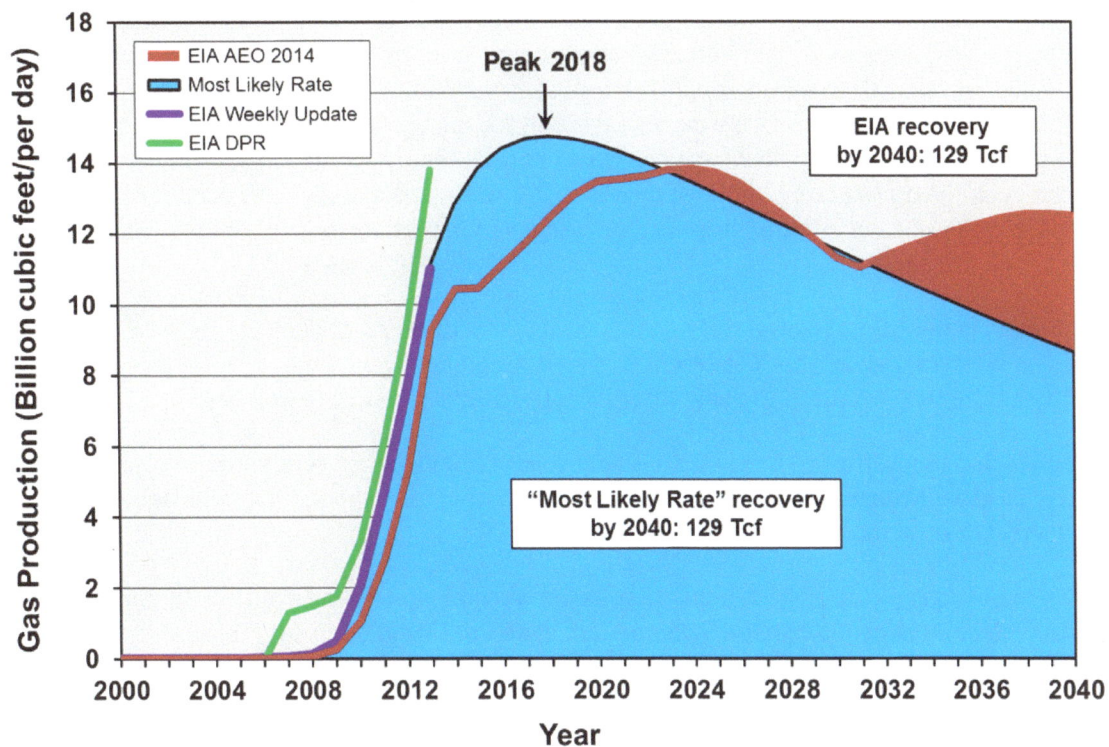

Figure 3-100. EIA reference case for Marcellus shale gas[150] vs. this report's "Most Likely Rate" scenario, 2000 to 2040.

The EIA underestimates past and current production compared to the "Most Likely Rate" scenario and its own independent estimates,[151] but overestimates production in later years. The EIA forecast is made on a "dry gas" basis, whereas the "Most Likely Rate" scenario forecast is made on a "raw gas" basis.

---

[149] EIA, *Natural Gas Weekly Update*, retrieved October 2014, http://www.eia.gov/naturalgas/weekly. EIA, *Drilling Productivity Report*, retrieved October 2014, http://www.eia.gov/petroleum/drilling.

[150] EIA, *Annual Energy Outlook 2014*, unpublished tables from AEO 2014 provided by the EIA.

[151] EIA, *Natural Gas Weekly Update*, retrieved October 2014, http://www.eia.gov/naturalgas/weekly. EIA, *Drilling Productivity Report*, retrieved October 2014, http://www.eia.gov/petroleum/drilling.

3.3.5.8 Marcellus Play Analysis Summary

Several things are clear from this analysis:

1. Marcellus production is growing strongly and drilling rates are sufficient to see continued growth through 2018. There is a significant backlog of wells drilled but not connected—estimated at over 2,000 wells—which will serve to maintain productive well additions in the near term even if rig count and new well drilling declines.

2. High well- and field-decline rates mean a continued high rate of drilling is required to maintain, let alone increase, production. Current drilling rates of 1,320 wells per year are considerably above the roughly 1,000 wells per year required to offset field decline at current production rates. Offsetting field decline requires an investment of $6 billion per year for drilling (assuming $6 million per well), not including leasing, infrastructure and operating costs. Future production profiles are most dependent on drilling rate and to a lesser extent on the number of drilling locations (i.e., greatly increasing the number of drilling locations would not change the production profile nearly as much as changing the drilling rate). Although drilling in the sweet spots is certainly economic at current prices, prices will have to increase to justify drilling in lower quality parts of the play when sweet spots are exhausted.

3. Production in the "Most Likely Rate" scenario will rise to 15 Bcf/d at peak in the 2018 timeframe followed by a gradual decline. The "High" drilling rate scenario would move this peak forward to 2019 at more than 15 Bcf/d. Drilling will continue in all scenarios until well beyond 2040.

4. The projected recovery of 129 Tcf by 2040 in the "Most Likely Rate" scenario, is the same as the EIA reference case. However, the EIA has underestimated near term production rates and overestimated production rates in the longer term.

5. These projections are optimistic in that they assume the capital will be available for the drilling treadmill that must be maintained to keep production up. This is not a sure thing as drilling in the poorer quality parts of the play will require higher gas prices to make it economic. Failure to maintain drilling rates will result in a lower production profile.

6. More than four times the current number of wells will need to be drilled by 2040 to meet production projections.

7. The projections in this report assume that of the total number of wells that could be drilled if 100% of the surface area was accessible for drilling at 4.3 wells per square mile, only 80% of the undrilled locations will be available, owing to surface land use. Any additional restrictions on land use would further limit the number of wells that could be drilled and result in lower production.

3.4 Major U.S. Tight Oil Plays with Significant Associated Shale Gas Production

Two tight oil plays which were analyzed in depth in Part 2 (Tight Oil) of this report also produce significant quantities of natural gas. As of June 2014, the Eagle Ford play ranked third and the Bakken play ranked seventh in terms of gas output from U.S. shale plays, as illustrated in Figure 3-101.[152] These plays are analyzed for future gas production below. Given that they are primarily oil plays, drilling rates and progression of drilling from sweet spots to lower quality areas will be governed by oil production—hence the analysis of these plays relies on the analysis in Part 2 of this report in order to determine likely future production.

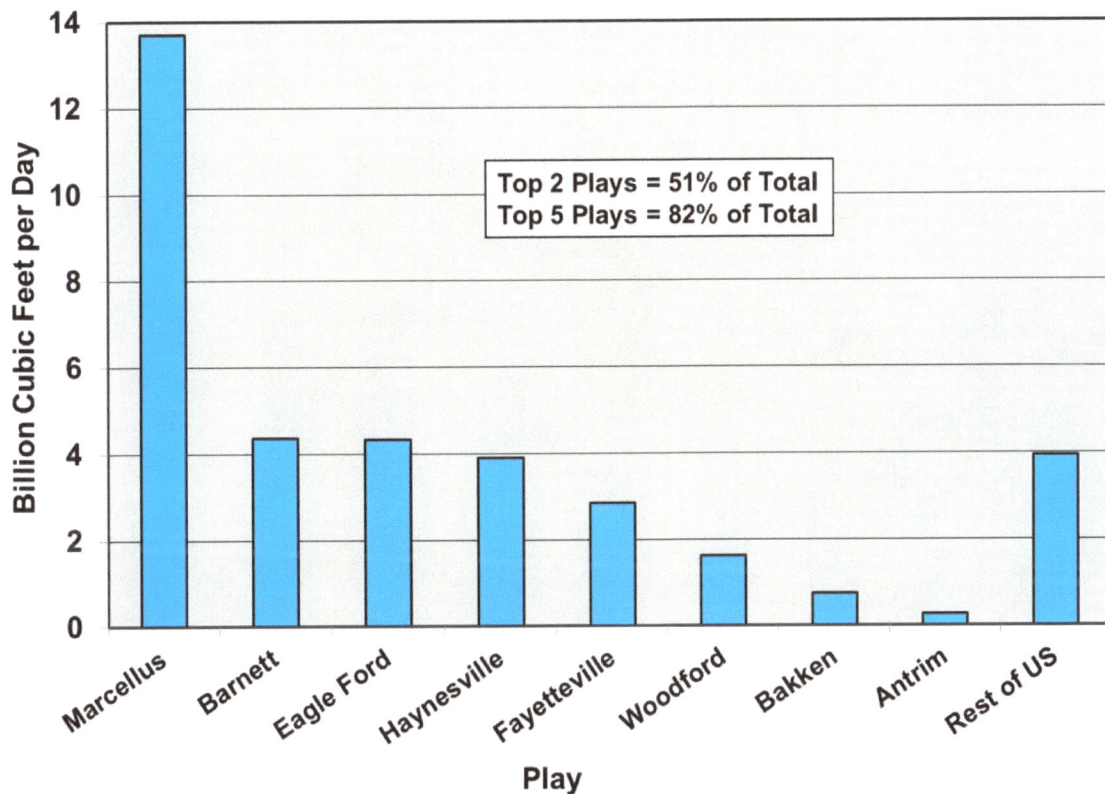

Figure 3-101. U.S. shale gas daily production by play as of June 2014.[153]

The Bakken tight oil play and especially the Eagle Ford tight oil play are also significant producers of shale gas.

---

[152] EIA, *Natural Gas Weekly Update*, retrieved July 2014, http://www.eia.gov/naturalgas/weekly/archive/2014/07_24/index.cfm. Note that the EIA in October 2014 published an estimate from the Utica of 1.174 Bcf/d, but this appears to be total gas production from Ohio, not specifically shale gas from the Utica Play; http://www.eia.gov/naturalgas/weekly retrieved October 9, 2014.

[153] EIA, *Natural Gas Weekly Update*, retrieved July 2014, http://www.eia.gov/naturalgas/weekly/archive/2014/07_24/index.cfm

3.4.1   Eagle Ford Play

The Eagle Ford play is divided into oil-, condensate- and gas-windows with increasing depth as discussed in Part 2 of this report. Therefore the best locations for oil production are not necessarily the same as the best locations for gas production. Figure 3-102 illustrates the distribution of well quality for gas production in the Eagle Ford as defined by highest one-month production (IP).

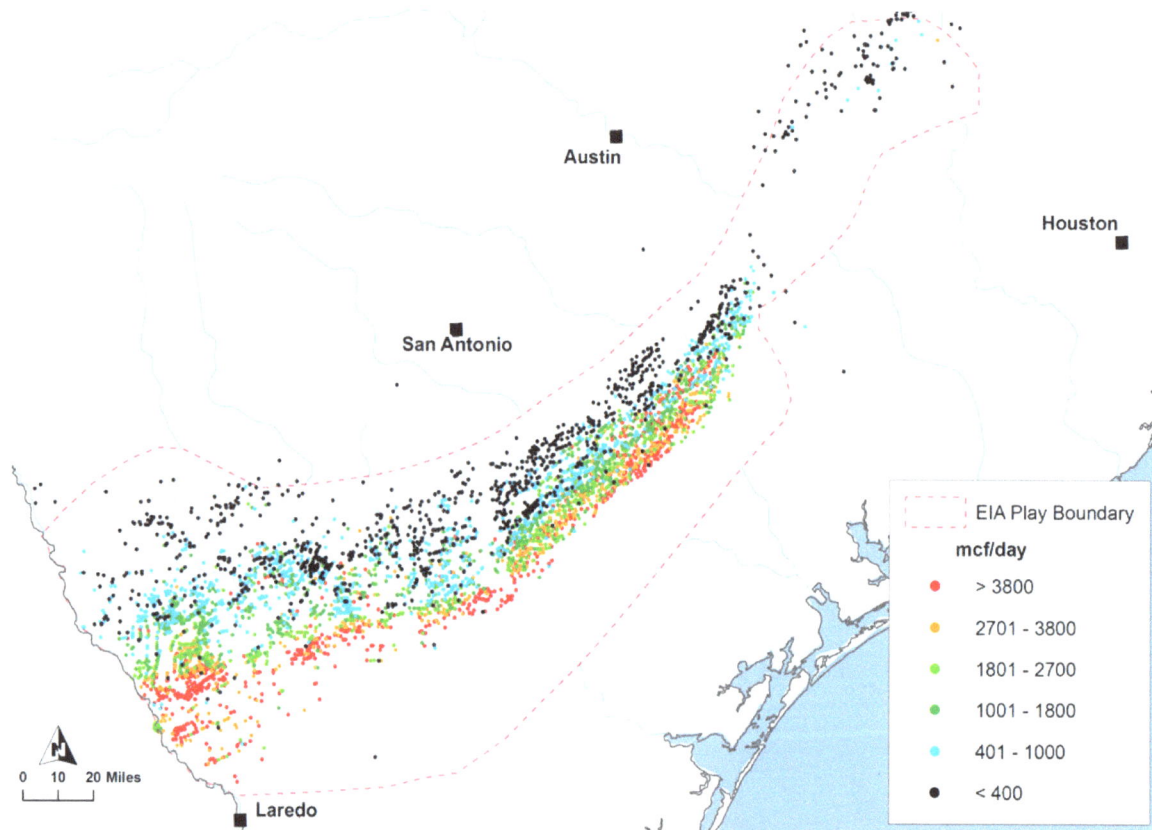

Figure 3-102. Distribution of wells in the Eagle Ford play as of mid- 2014, illustrating highest one-month gas production (initial productivity, IP).[154]

Well IPs are categorized approximately by percentile; see Appendix.

---

[154] Data from Drillinginfo retrieved August 2014.

Figure 3-103 provides a closer view of the main gas production area along with the counties utilized in the analysis.

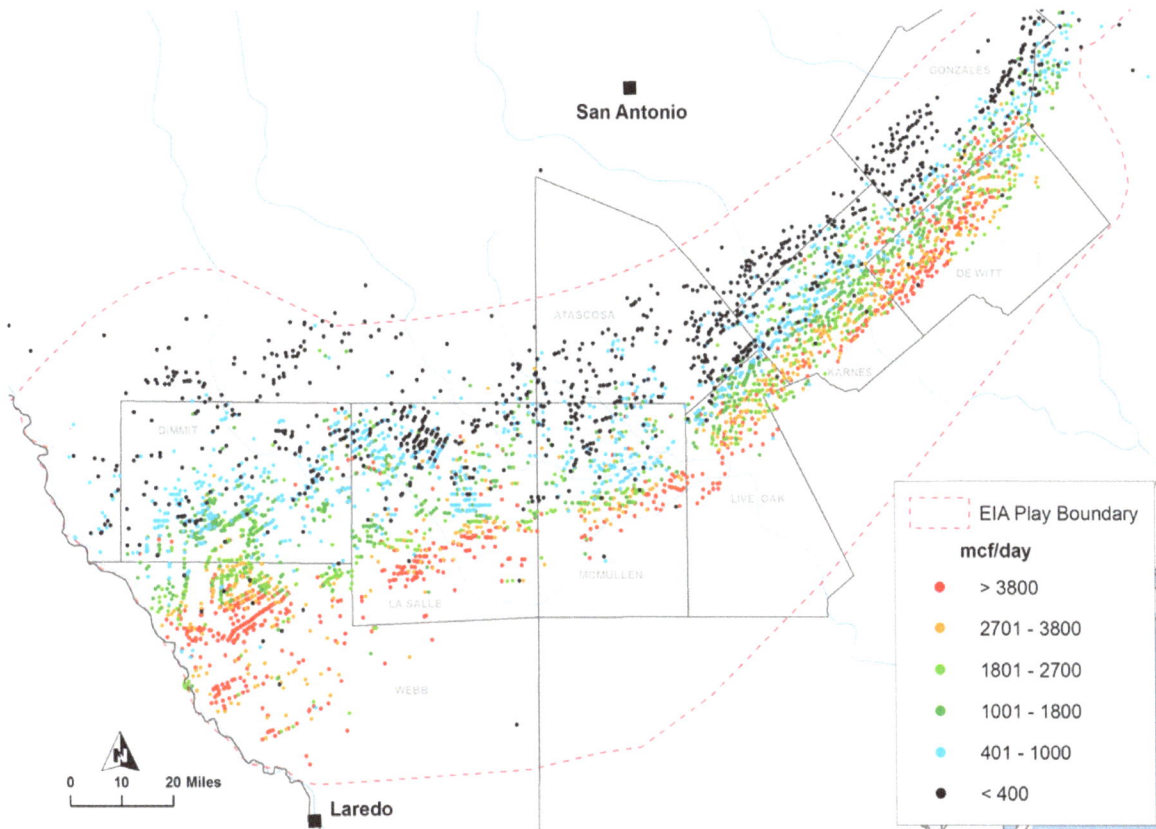

Figure 3-103. Detail of the Eagle Ford play showing distribution of wells as of mid-2014, illustrating highest one-month gas production (initial productivity, IP).[155]

Well IPs are categorized approximately by percentile; see Appendix.

---

[155] Data from Drillinginfo retrieved August 2014.

Figure 3-104 illustrates gas production in the Eagle Ford from 2007 through mid-2014. Production is nearing 5 Bcf/d from just over 10,000 producing wells.

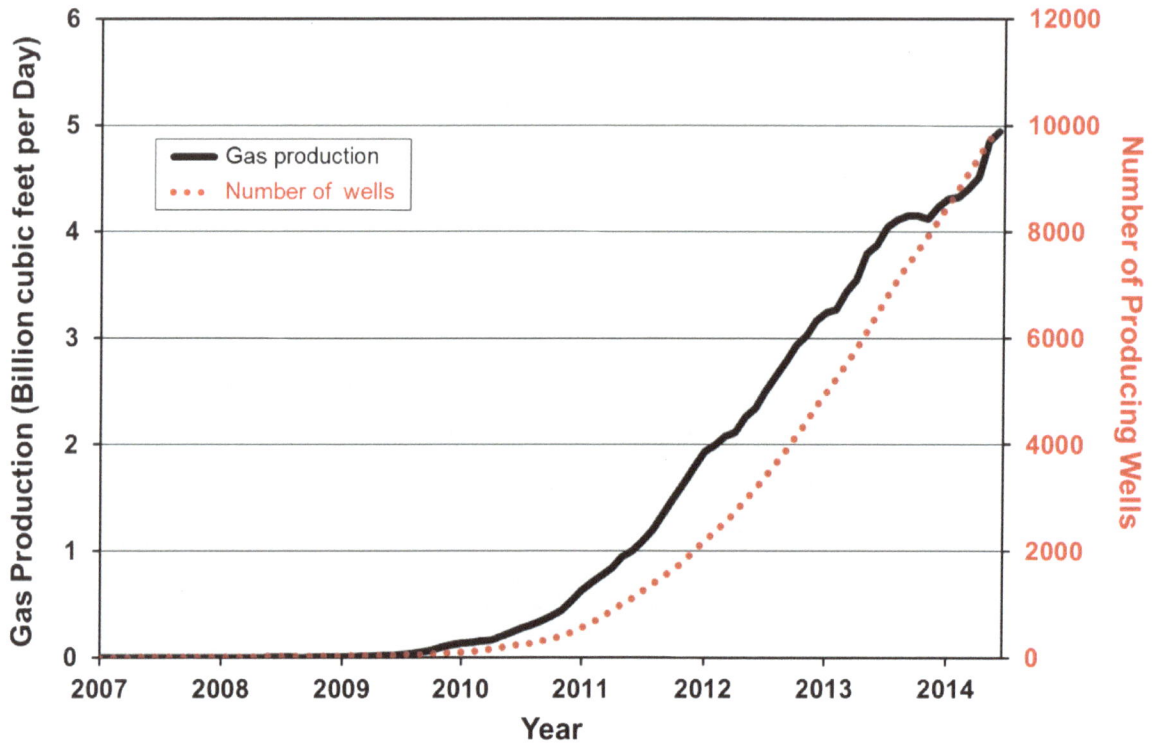

Figure 3-104. Eagle Ford play shale gas production and number of producing wells, 2007 through 2014.[156]

Gas production data are provided on a "raw gas" basis.

---

[156] Data from Drillinginfo retrieved September 2014. Three-month trailing moving average.

3.4.1.1   Critical Parameters

Other critical parameters include the average well decline, which is 80% over 3 years, and the average field decline, which is 47% for gas wells. The distribution of gas production by county is illustrated in Part 2 of this report, and the evolution of well quality over time is illustrated in Figure 3-105.

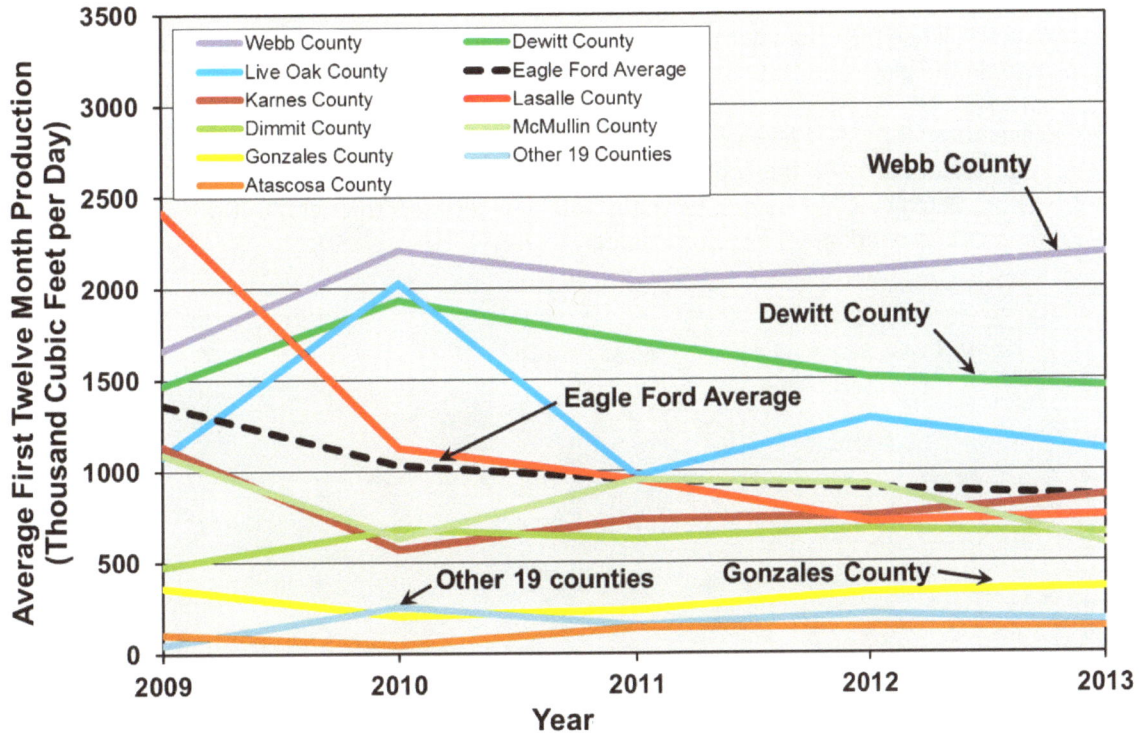

Figure 3-105. Average first-year gas production rates of wells by county in the Eagle Ford play, 2009 to 2013.[157]

Gas production is an important economic component of Eagle Ford wells as it comprises nearly 40% of the energy produced on average from the play (the distribution of energy production from the Eagle Ford on a "barrels of oil equivalent" basis is illustrated in Part 2 of this report). As can be seen in Figure 3-105, the average well quality from a gas production point of view has been declining. This is likely a result of drilling moving into areas more favorable for oil production and less favorable for gas production, and is not an indicator of what well quality for gas production will look like later on as sweet spots for oil production become saturated with wells. Webb County, for example, which is the best county for gas production but one of the worst for oil production, will see a lot more drilling in later stages of the play's development. Hence the average well quality, from a gas production point of view, is likely to increase in later stages of play development.

---

[157] Data from Drillinginfo retrieved September 2014.

3.4.1.2 Future Production Scenarios

Given that oil production is the driving force in the Eagle Ford at the current time, the "Most Likely Rate" scenario of the "Realistic Case" for oil production as outlined in Part 2 of this report is used for projection of future Eagle Ford gas production. This scenario assumes that more than 37,000 wells will be drilled in total (compared to just over 10,000 wells at present), and that drilling will continue at current rates of 3,550 wells per year and gradually fall to 2,000 wells per year as the play is drilled off. It also assumes that well quality for gas production will rise 50% from current levels as drilling moves from oil-prone areas back into gas-prone areas later in the play's development.

Figure 3-106 illustrates the "Most Likely Rate" projection for Eagle Ford production (see Part 2 of this report for other key parameters used for this projection). Production is forecast to rise considerably from current levels to nearly 6.5 Bcf/d by 2017 before declining. Total gas recovery through 2040 will be about 35.5 Tcf, or nearly 10 times the amount produced from the play so far.

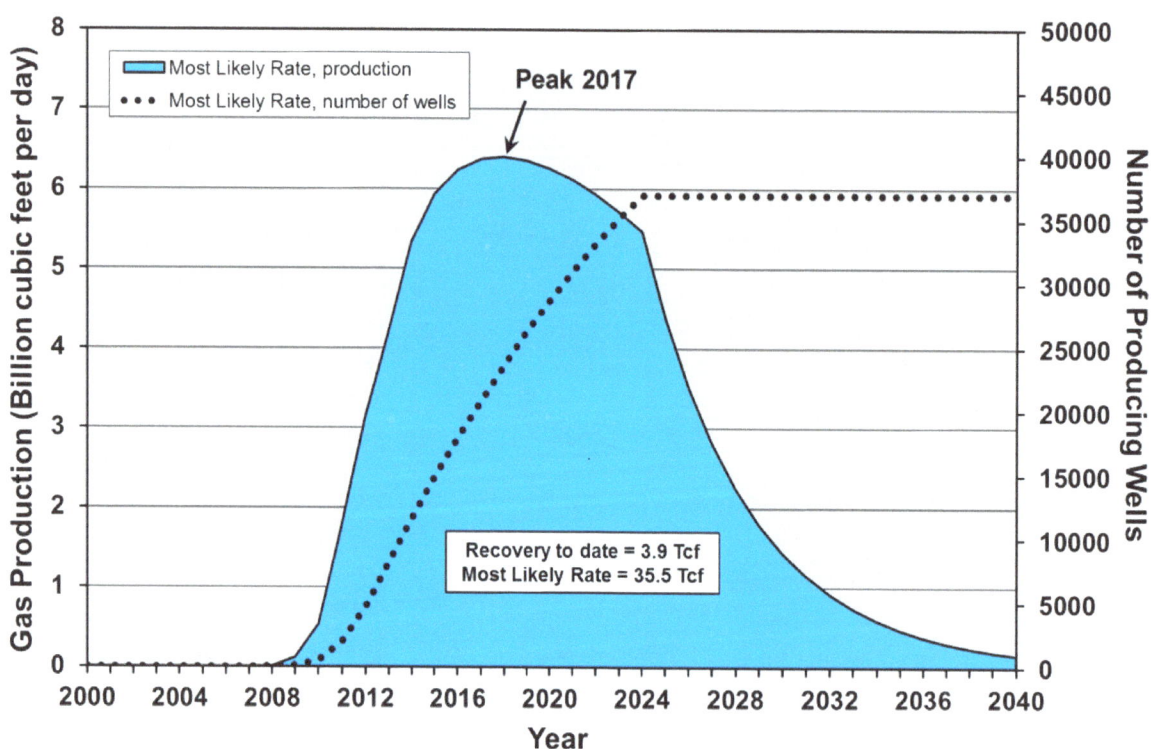

Figure 3-106. "Most Likely Rate" scenario of Eagle Ford production for gas in the "Realistic Case" (80% of the remaining area is drillable at six wells per square mile).

This projection assumes that well quality for gas production will rise in later stages of play development as drilling moves back into gas prone parts of the play.

3.4.1.3 Comparison to EIA Forecast

Figure 3-107 illustrates the comparison of the "Most Likely" drilling rate scenario to the EIA's reference case forecast. Several points are evident:

- The EIA is underestimating current production in the Eagle Ford in its forecast and highly overestimating production later on, after 2024. The EIA's near term forecast is invalidated by its own data as shown in Figure 3-107, which shows much higher current production.

- The EIA forecasts a recovery of 57.2 Tcf over the 2000-2040 period, or 21.7 Tcf more than the "Most Likely Rate" scenario over the same period.

- The EIA forecasts continuing growth in Eagle Ford gas production to an all-time high well over 7 Bcf/d in 2040. This is unrealistic given the data.

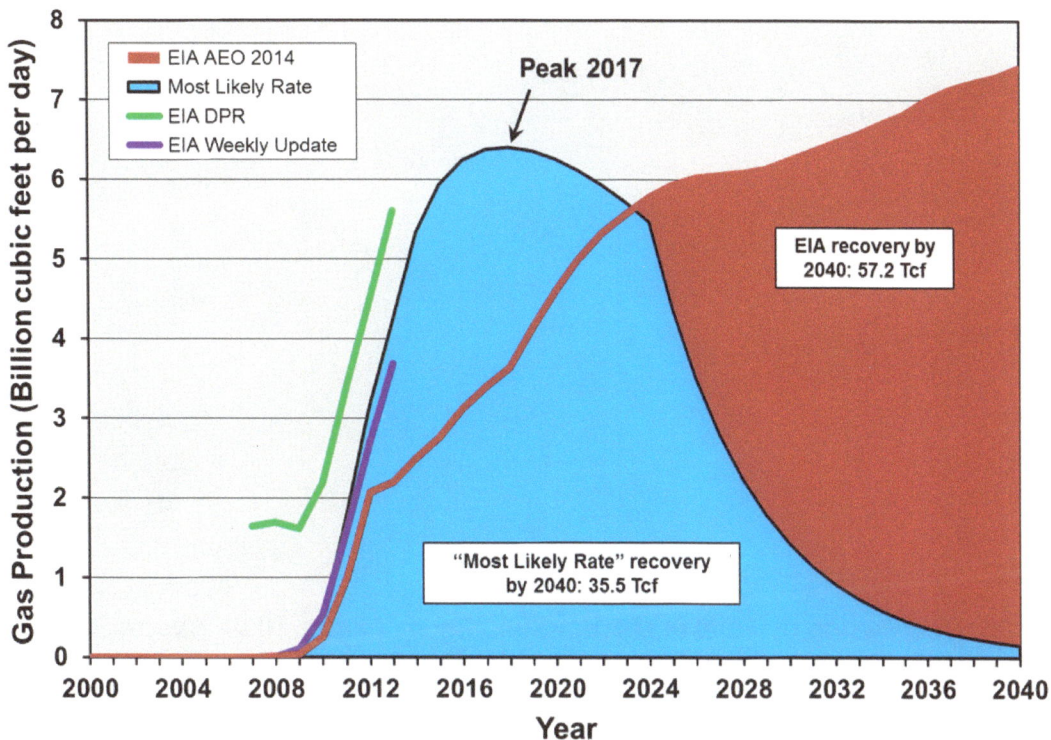

Figure 3-107. EIA reference case for Eagle Ford shale gas[158] vs. this report's "Most Likely Rate" scenario of the "Realistic Case," 2000 to 2040

Also shown are the EIA's Eagle Ford gas production statistics from its *Drilling Productivity Report* and its *Natural Gas Weekly Update*,[159] which contradict the early years of its AEO 2014 forecast. The EIA forecast is made on a "dry gas" basis, whereas the "Most Likely Rate" scenario forecast is made on a "raw gas" basis.

[158] EIA, *Annual Energy Outlook 2014*.
[159] EIA, *Drilling Productivity Report*, retrieved October 2014, http://www.eia.gov/petroleum/drilling. EIA, *Natural Gas Weekly Update*, retrieved October 2014, http://www.eia.gov/naturalgas/weekly.

3.4.2 Bakken Play

The Bakken play's areas of highest gas production per well are shifted a few miles west of the areas of highest oil production per well, but are generally in fairly close proximity. Figure 3-108 illustrates the distribution of well quality for gas production in the Bakken as defined by highest one-month production (IP)—see Part 2 of this report for a comparison to well quality for oil production.

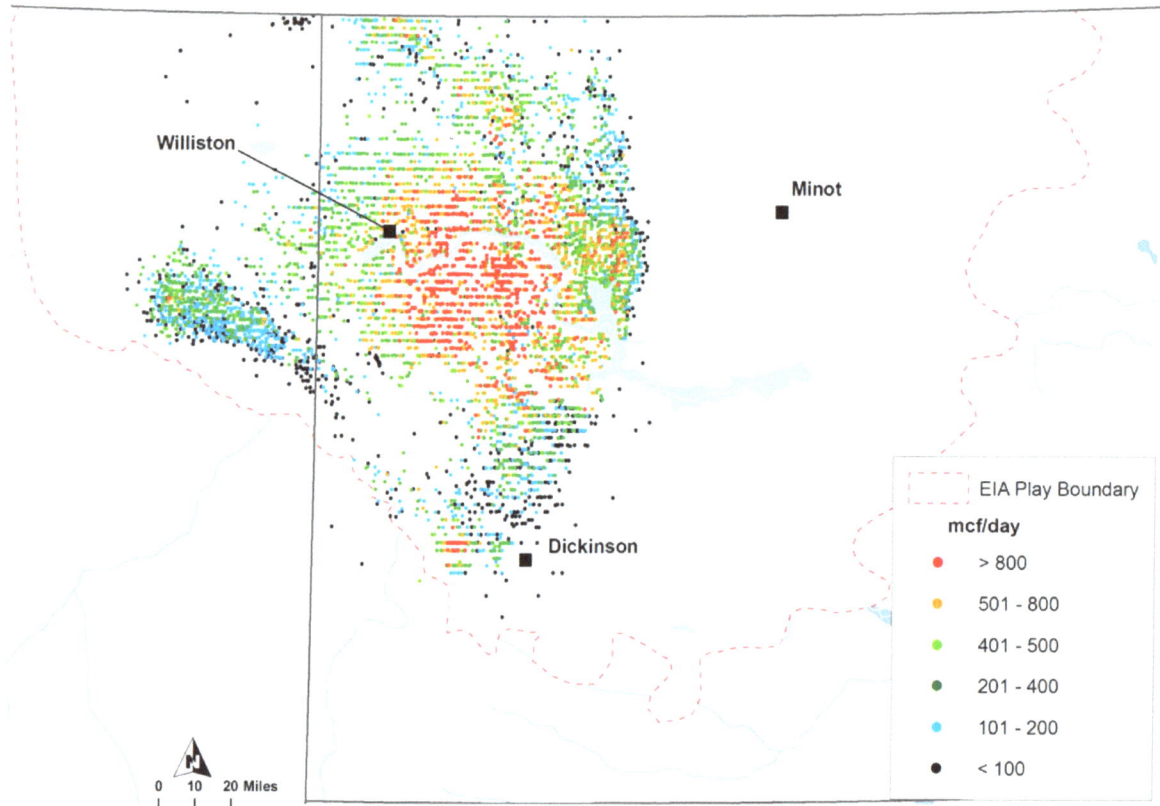

Figure 3-108. Distribution of wells in the Bakken play as of mid-2014 illustrating highest one-month gas production (initial productivity, IP).[160]

Well IPs are categorized approximately by percentile; see Appendix.

---

[160] Data from Drillinginfo retrieved August 2014.

Figure 3-109 provides a closer view of the main gas production area along with the counties utilized in the analysis.

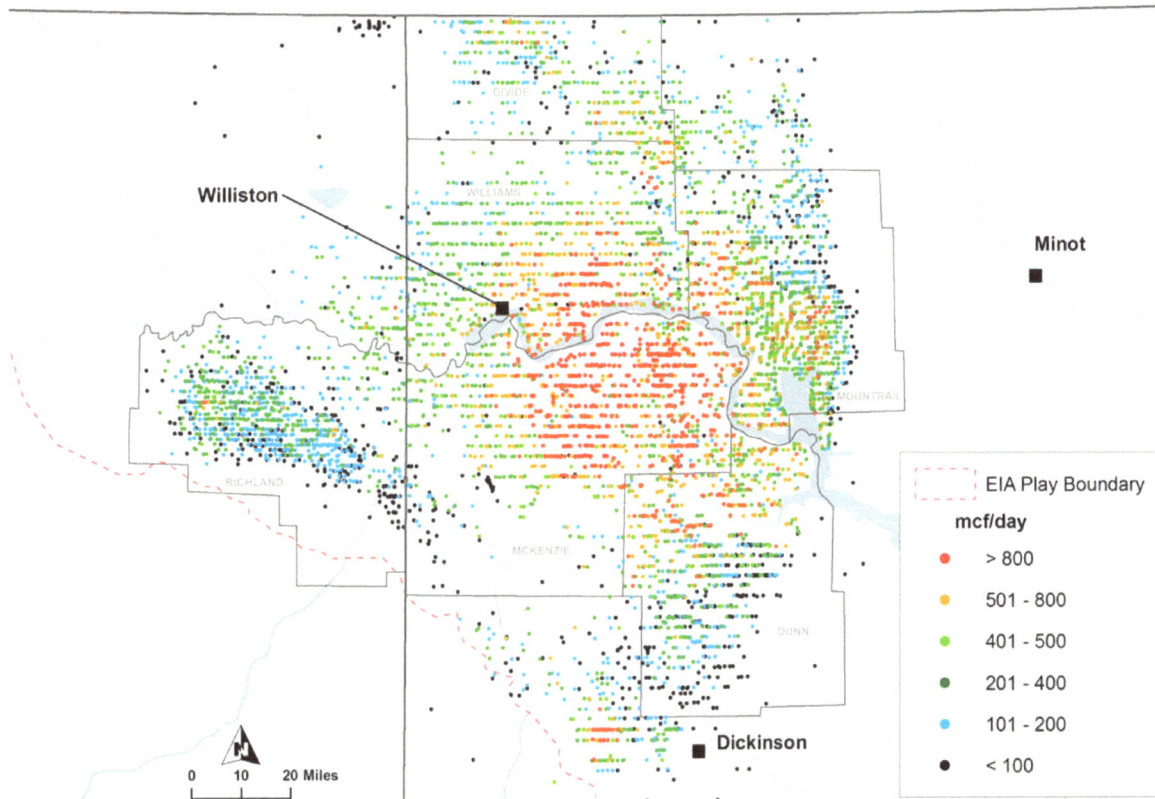

Figure 3-109. Detail of the Bakken play showing distribution of wells as of mid-2014, illustrating highest one-month gas production (initial productivity, IP).[161]

Well IPs are categorized approximately by percentile; see Appendix.

---

[161] Data from Drillinginfo retrieved August 2014.

Figure 3-110 illustrates gas production in the Bakken from 2003 through mid-2014. Production is about 1.1 Bcf/d from over 8,500 producing wells.

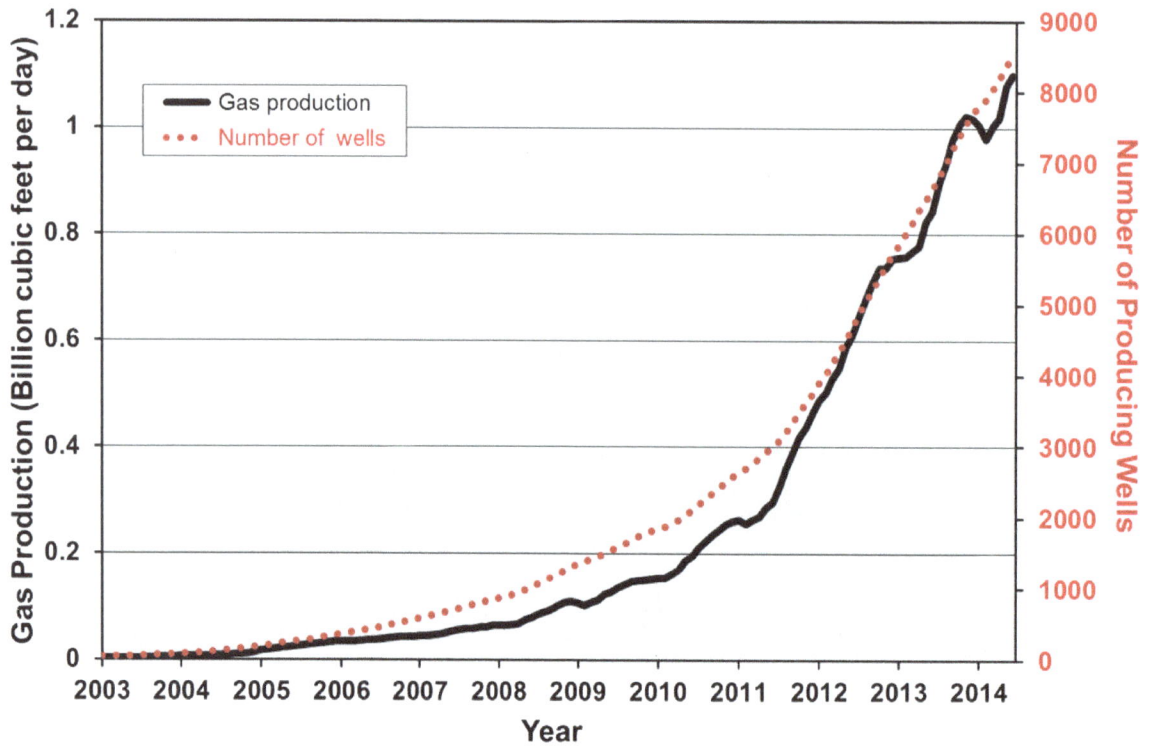

Figure 3-110. Bakken play shale gas production and number of producing wells, 2003 through 2014.[162]

Gas production data are provided on a "raw gas" basis.

---

[162] Data from Drillinginfo retrieved September 2014.

3.4.2.1 Critical Parameters

Other critical parameters include the average well decline, which is 81% over 3 years, and the average field decline, which is 41% for gas wells. The evolution of well quality over time for gas production is illustrated in Figure 3-111.

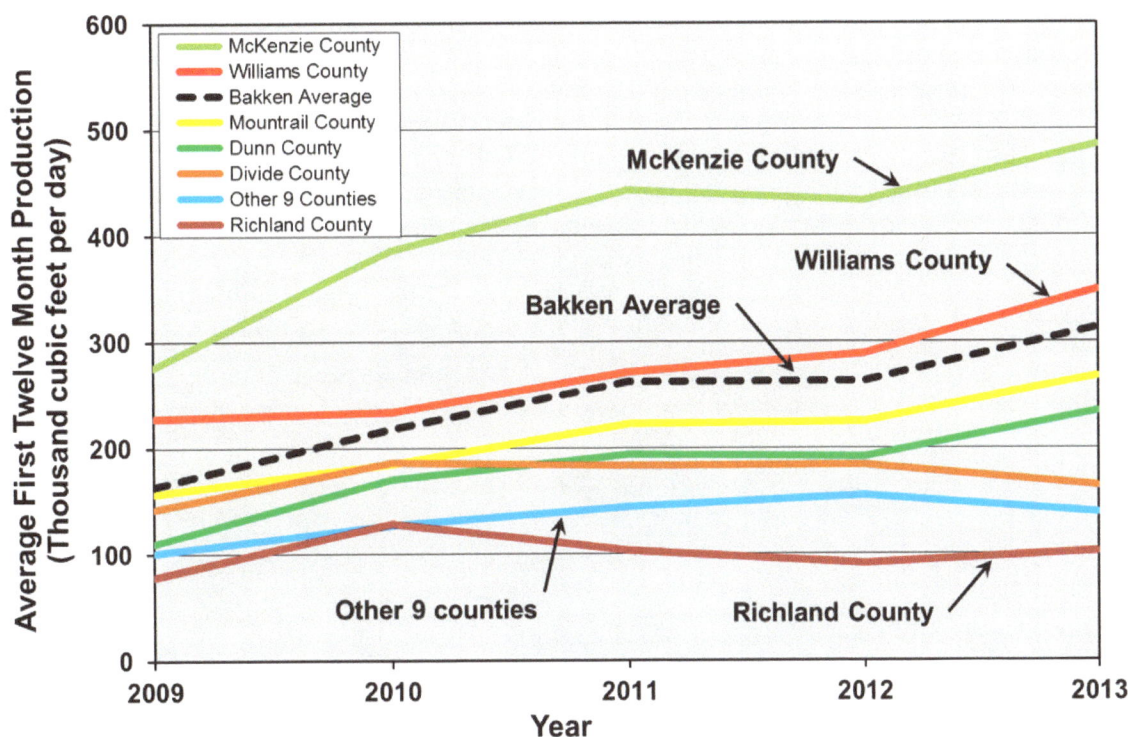

Figure 3-111. Average first-year gas production rates of wells by county in the Bakken play, 2009 to 2013.[163]

Gas production is a less important economic component of Bakken wells than for the Eagle Ford as only about 16% of the energy produced from the play is gas, and much of the gas is flared in areas remote from infrastructure (roughly 30% of gas production is flared).[164] New regulations on flaring will likely reduce the amount in future and divert more of this production to sales.[165] As can be seen in Figure 3-111, the average well quality from a gas production point of view has been increasing in the top four counties and declining or flat in the other 11 counties. Given the close proximity of high quality oil wells to high quality gas wells, the decline in well quality for gas as drilling moves to lower quality parts of the play is expected to parallel the decline in well quality for oil.

---

[163] Data from Drillinginfo retrieved September 2014.
[164] Styles, Geoffrey, August 2014, *The Energy Collective*, "Bakken shale gas flaring highlights global problem," http://theenergycollective.com/geoffrey-styles/449241/bakken-shale-gas-flaring-highlights-global-problem.
[165] Carroll, Joe, July 2014, *Bloomberg*, "Bakken Oil Explorers Told to Cut Flaring or Face Crude Caps," http://www.bloomberg.com/news/2014-07-01/bakken-oil-explorers-told-to-cut-flaring-or-face-crude-caps-1-.html.

3.4.2.2   Future Production Scenarios

Given that oil production is the driving force in the Bakken, and gas is a relatively small component of production, the "Most Likely" drilling rate scenario of the "Realistic Case" for oil production as outlined in Part 2 of this report is used for projection of future Bakken gas production. This scenario assumes that more than 32,000 wells will be drilled in total (compared to just over 8,500 wells at present), and that drilling will continue at current rates of 2,000 wells per year and gradually fall to 1,000 wells per year as the play is drilled off. It also assumes that well quality for gas production will decline from current levels as drilling moves from sweet spots for oil and gas into lower quality counties later in the play's development.

Figure 3-112 illustrates the "Most Likely Rate" scenario for Bakken production (see Part 2 of this report for other key parameters used for this projection). Production is forecast to rise considerably from current levels to roughly 1.3 Bcf/d by 2016 before declining. Total gas recovery through 2040 will be about 7.1 Tcf, or nearly 7 times the amount produced from the play so far.

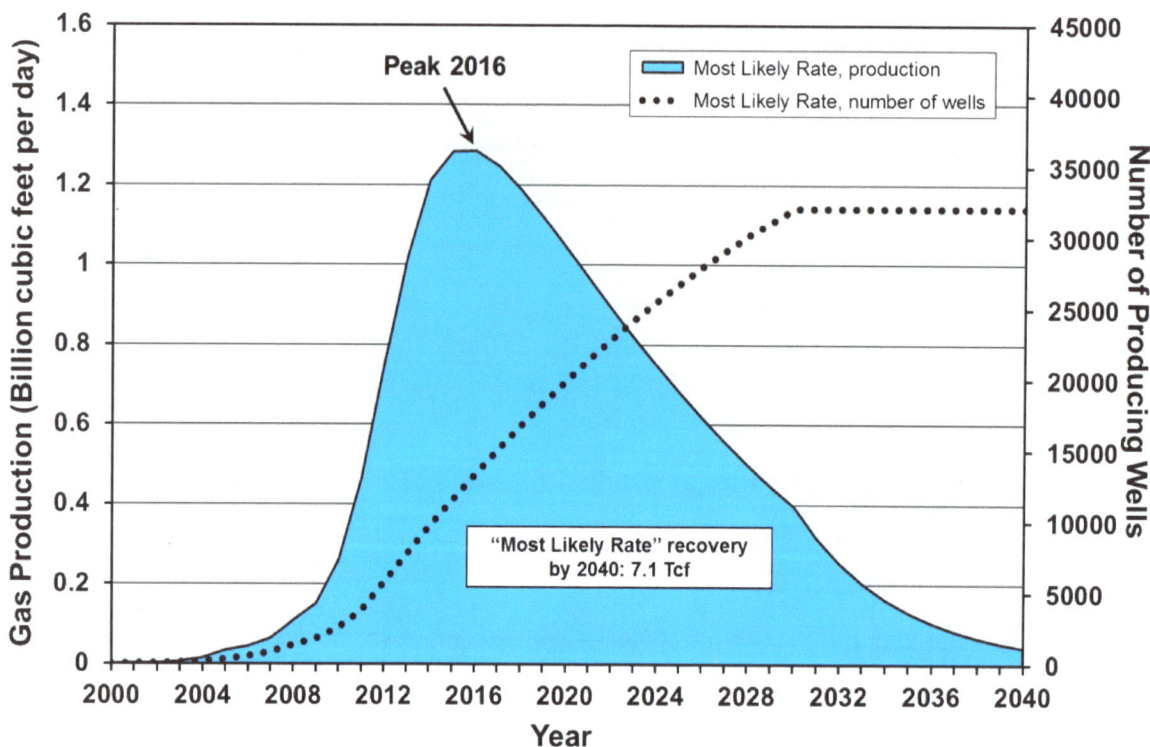

Figure 3-112. "Most Likely Rate" scenario of Bakken gas production in the "Realistic Case" (80% of the remaining area is drillable at three wells per square mile).

This projection assumes that well quality for gas production will parallel well quality trends for oil production as drilling moves into lower quality parts of the play.

3.4.2.3 Comparison to EIA Forecast

Figure 3-113 illustrates the comparison of the "Most Likely" drilling rate projection to the EIA's reference case forecast. Several points are evident:

- The EIA is highly underestimating current production in the Bakken in its forecast and overestimating production later on, after 2030. The EIA's near term forecast is invalidated by its own data as shown in Figure 3-113, which shows much higher current gas production.

- The EIA forecasts a recovery of just 5.1 Tcf over the 2000-2040 period, or 2.0 Tcf less than the "Most Likely Rate" scenario over the same period. However, it assumes production of 0.7 Tcf more gas after 2030. This is a result of the underestimates of current and short- to medium-term Bakken production.

- The EIA forecasts peak Bakken gas production at roughly the same time as this report (2016), albeit at production levels of less than half that of this report.

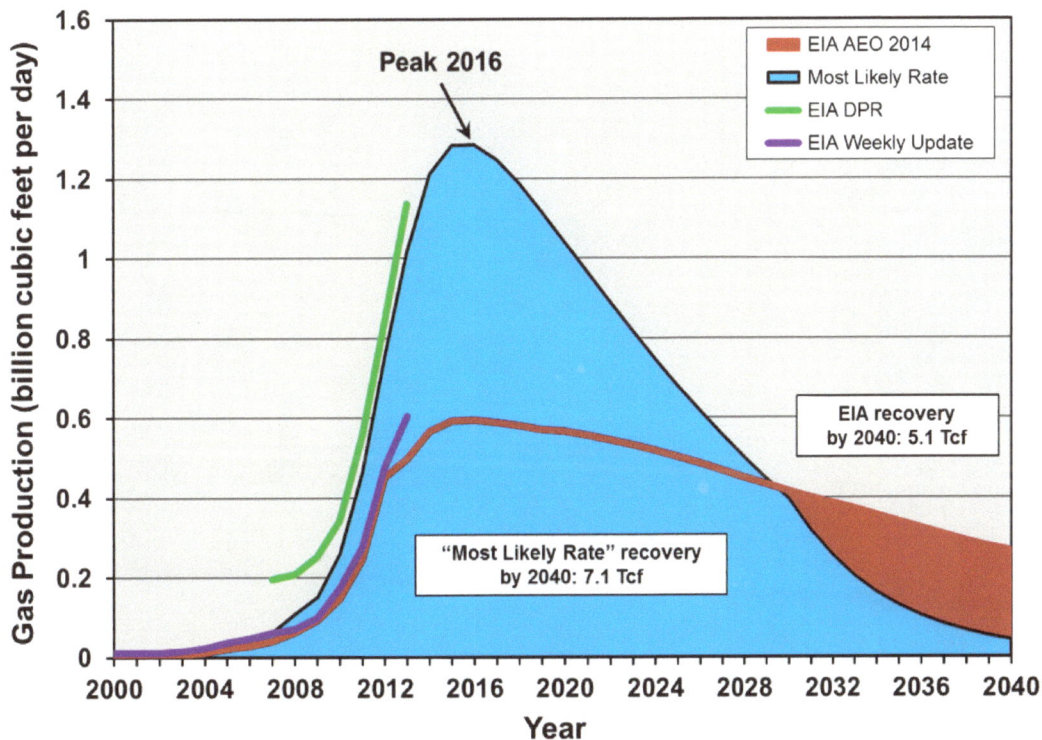

Figure 3-113. EIA reference case for Bakken shale gas[166] vs. this report's "Most Likely Rate" scenario of the "Realistic Case," 2000 to 2040

Also shown are the EIA's Bakken gas production statistics from its *Drilling Productivity Report* and its *Natural Gas Weekly Update*,[167] which contradict the early years of its AEO 2014 forecast. The EIA forecast is made on a "dry gas" basis, whereas the "Most Likely Rate" scenario forecast is made on a "raw gas" basis.

---

[166] EIA, *Annual Energy Outlook 2014*.
[167] EIA, *Drilling Productivity Report*, retrieved October 2014, http://www.eia.gov/petroleum/drilling. EIA, *Natural Gas Weekly Update*, retrieved October 2014, http://www.eia.gov/naturalgas/weekly.

3.5 ALL-PLAYS ANALYSIS

The foregoing analysis of shale gas plays has reviewed 88% of estimated June 2014, shale gas production[168] and 88% of the cumulative shale gas production that is forecast in the EIA's 2012-2040 reference case.[169] Although the EIA forecast for the Marcellus play is rated as "reasonable" and its forecast for the Bakken play is rated "conservative," the deficit left by being "very highly optimistic" on some of the other plays makes finding and developing the gas required to meet the overall forecast highly to very highly optimistic.

This section will further explore the outlook for overall U.S. shale gas production with a summary analysis of the plays' EIA forecasts, well quality, and production prospects to 2040.

3.5.1 Summary of EIA Forecasts

Table 3-6 summarizes the salient details of the EIA projections versus historical production and the EIA's estimates of "unproved technically recoverable resources" and "proved reserves."

Play	EIA Recovery 2012-2040 (Tcf)	Produc-tion to Date (Tcf)	EIA Unproved Resources as of January 1, 2012 (Tcf)	EIA Proved Reserves as of 2012 (Tcf)	Total Proved and Unproved Technically Recoverable (Tcf)	Percent of Unproved Resources and Proved Reserves Recovered by 2040 in EIA Forecast	Percent of Total Recovery in EIA Reference Case	EIA Production in 2040 (Tcf/year)	Optimism Bias
Barnett	44.4	15.60	20.3	23.7	44.0	101.0	10.1	2.15	Very High
Haynesville	97.2	9.41	70.9	17.7	88.6	109.8	22.0	3.37	Very High
Fayetteville	38.9	5.08	29.8	9.7	39.5	98.4	8.8	1.53	Very High
Woodford	22.8	3.14	16.8	11.1	27.9	81.6	5.2	0.82	High
Marcellus	127.2	9.70	118.9	42.8	161.7	78.7	28.8	4.57	Reasonable
Bakken	4.8	1.10	6.4	N/A	6.3	75.9	1.1	0.10	Conservative
Eagle Ford	56.7	3.90	60.3	16.2	76.5	74.2	12.8	2.70	Very High
Other	49.6	11.66	165.8	8.2	174.0	28.5	11.2	4.58	Unknown
Total	441.6	59.59	489.0	129.4	618.4	71.4	100.0	19.82	High to Very High

Table 3-6. Comparison of EIA reference case shale gas forecast assumptions[170] with unproved technically recoverable resources[171] and proved reserves[172] to cumulative production from shale gas plays.[173]

A determination of each play's "optimism bias" is included. Numbers may not add due to rounding.

---

[168] EIA, *Natural Gas Weekly Update*, retrieved July 2014, http://www.eia.gov/naturalgas/weekly/archive/2014/07_24/index.cfm.
[169] EIA, *Annual Energy Outlook 2014*, unpublished tables from AEO 2014 provided by the EIA. EIA, *Annual Energy Outlook 2014*, reference case forecast, Table 14, oil and gas supply, http://www.eia.gov/forecasts/aeo/excel/aeotab_14.xlsx.
[170] EIA, *Annual Energy Outlook 2014*, unpublished tables from AEO 2014 provided by the EIA.
[171] EIA, *Assumptions to the Annual Energy Outlook 2014*, http://www.eia.gov/forecasts/aeo/assumptions/pdf/oilgas.pdf.
[172] EIA, http://www.eia.gov/naturalgas/crudeoilreserves/index.cfm.
[173] Data from Drillinginfo retrieved August to September 2014.

3.5.2   Well Quality

A comparison of plays analyzed in this report reveals that they are highly variable in terms of well quality and that the Marcellus and Haynesville stand out as clearly superior. The estimated ultimate recovery (EUR) of wells has been reviewed in the discussion of each play in this report, with the caveat that these are merely estimates and subject to change as more data emerge on longer-term well productivity.

Another measure for comparison of plays is the average first-year production from wells. This metric builds in the current geology and the cumulative impact of all technological innovations in drilling and completions to date if the most recent year is used. Figure 3-114 illustrates the average first-year production of horizontal wells in the seven plays analyzed in this study for 2013 for both the average of all wells in the play and the average for wells in the best county. Although the best play from this comparison is clearly the Haynesville, the Haynesville has a much higher field decline rate than the Marcellus which will tend to equalize the two over time. It is clear, however, that high quality shale gas plays are not ubiquitous, and even within the top producers there is considerable variation in average well quality.

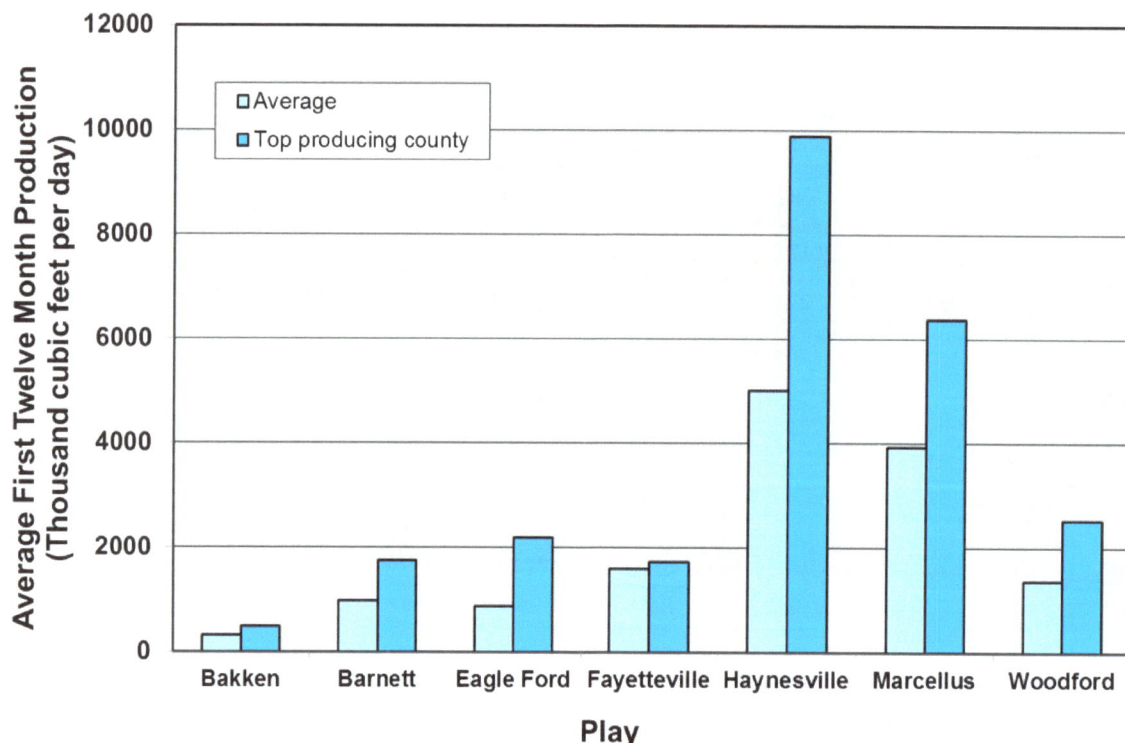

Figure 3-114. Average first-year gas production per well in 2013 from horizontal wells both play-wide and in the top-producing county for the plays analyzed in this report.[174]

---

[174] Data from Drillinginfo retrieved August to September 2014.

3.5.3   Production Through 2040

Figure 3-115 illustrates the sum of shale gas production from the plays analyzed in this report through 2040 in the "Most Likely" drilling rate scenario, along with the number of wells required to achieve it. Production from these plays peaks in 2016 at nearly 34 Bcf/d and declines to below 16 Bcf/d by 2040, or more than 50%. Total production over the 2000 to 2040 period is projected to be 291.7 trillion cubic feet. The Marcellus will make up 55% of production from these plays in 2040. Approximately 130,000 additional wells will need to be drilled by 2040 to meet the projections in Figure 3-115, on top of the 50,000 wells drilled in these plays through 2013. Assuming an average well cost of $7 million, this would require $910 billion of additional capital input by 2040, not including leasing, operating, and other ancillary costs.

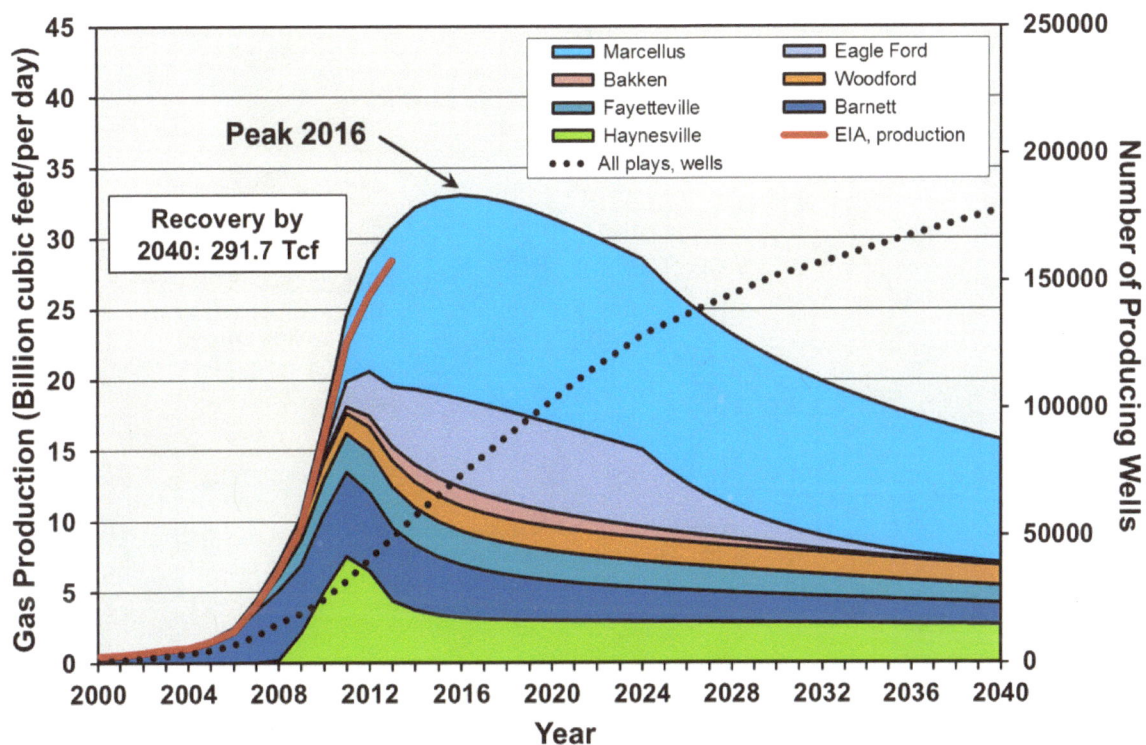

Figure 3-115. "Most Likely Rate" scenarios for the seven shale gas plays analyzed in this report and number of producing wells, through 2040.

The "Most Likely Rate" scenario projections here are made on a "raw gas" basis. 180,000 wells will be producing by 2040 in this scenario. Also shown is the EIA's production data for dry gas through August 2014 for these plays.[175]

---

[175] EIA, *Natural Gas Weekly Update*, retrieved October 2014, http://www.eia.gov/naturalgas/weekly.

Figure 3-116 illustrates the EIA's reference case forecast for shale gas compared to the projections in this report for the seven plays analyzed. This comparison is made on a "dry" basis, given that the EIA forecast is for dry gas.[176] As can be seen, actual production of shale gas from these plays is higher in the near term than the EIA forecast and higher yet for the EIA's own independent estimate (from its *Natural Gas Weekly Update*) of actual shale gas production through August 2014. In the longer term, however, the EIA forecast overestimates production from the plays in this report's "Most Likely Rate" scenario through 2040 by 147.4 Tcf, or 64%. The EIA further estimates that in 2040, production from the plays analyzed in this report with be 182% higher (nearly 3 times) than estimated herein, and that by 2040, another 49.6 Tcf will have been recovered from other plays not analyzed in this report. Indeed, if the analysis in this report is correct, in order to meet the EIA reference case forecast other plays will have to recover an additional 198.2 Tcf—nearly 4 times the EIA's own estimate for other plays.

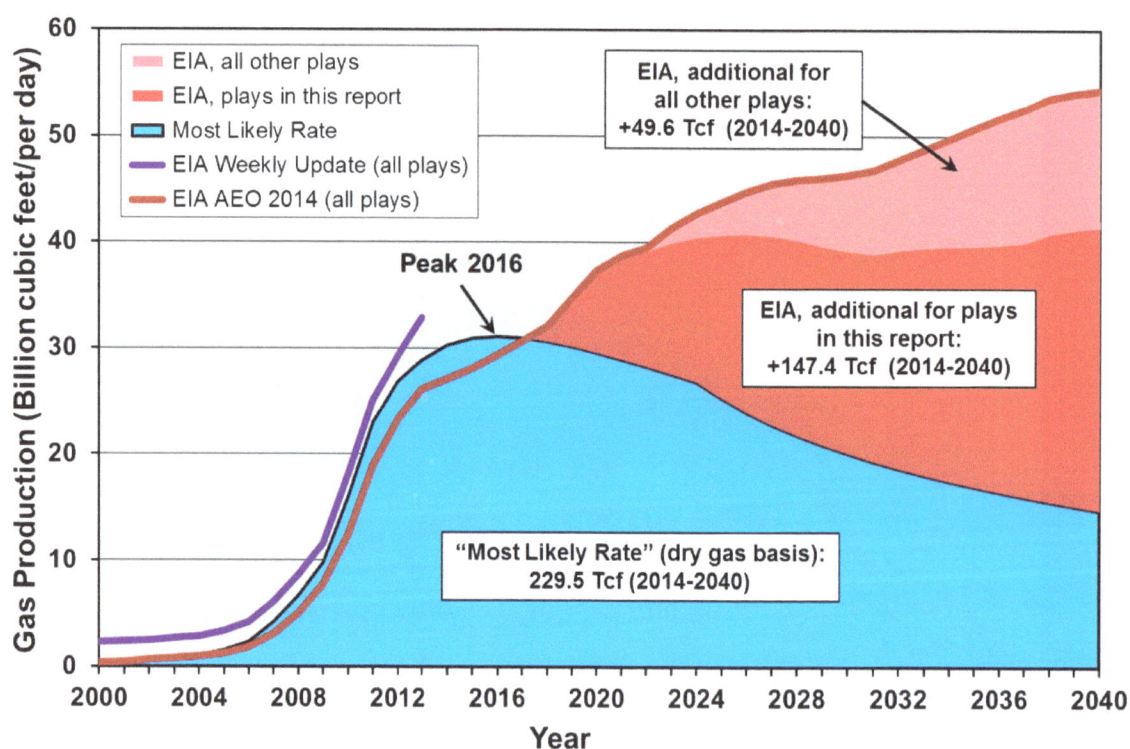

Figure 3-116. Totaled "Most Likely Rate" scenarios for the seven shale gas plays analyzed in this report, compared to the EIA's reference case forecast for these plays and for all plays.[177,178]

The "Most Likely Rate" scenario projections here are made on a "dry gas" basis. Also shown are the EIA's gas production statistics from its *Natural Gas Weekly Update*,[179] which contradict the early years of its AEO 2014 forecast.

---

[176] Dry gas has had liquids and other impurities removed and results in a shrinkage factor—in this case a shrinkage factor to dry basis is estimated at 6%, although the actual shrinkage factor varies by play and can be considerably higher for some plays—and lower for others.
[177] EIA, *Annual Energy Outlook 2014*, unpublished tables from AEO 2014 provided by the EIA.
[178] EIA, *Annual Energy Outlook 2014*, reference case forecast, Table 14, oil and gas supply, http://www.eia.gov/forecasts/aeo/excel/aeotab_14.xlsx.
[179] EIA, *Natural Gas Weekly Update*, retrieved October 2014, http://www.eia.gov/naturalgas/weekly.

3.6   SUMMARY AND IMPLICATIONS

The growth of U.S. shale gas production has been a game-changer in a natural gas supply picture that as recently as 2005 was thought to be in terminal decline. The assumption that natural gas will be cheap and abundant for the foreseeable future has prompted fuel switching from coal to gas, along with investment in new generation and gas distribution infrastructure, investment in new North American manufacturing infrastructure, and calls for exporting the shale gas bounty to higher-priced markets in Europe and Asia.

Given these assumptions—and the investments being made and planned because of them—it is important to understand the long-term supply limitations of U.S. shale gas. The analysis presented herein, which is based on one of the best commercial databases of well production information available.[180] finds that the continued growth in supply over the long term at low prices is highly questionable. Certainly production will rise in the short term, but with the likely collective peaking of the seven major plays analyzed in this report (which provide 88% of current and estimated long-term U.S. shale gas output) in the 2016-2017 timeframe, maintaining production or even stemming the decline will require maintenance of high drilling rates, along with the capital input to sustain them.

This report finds that major shale plays are variable in well quality, with some plays—like the Marcellus and Haynesville—being much more productive on average than the rest. Furthermore, the assessment of individual counties within plays reveals that well quality varies considerably, and that the best counties are attracting most of the drilling and investment—meaning that the poorer-quality counties, which account for most of the remaining drilling locations, will be drilled last. Given that field declines are steep, requiring 25-50% of production to be replaced each year, the levels of drilling and capital investment needed to maintain production will escalate going forward. Without the considerably higher prices needed to justify drilling in poorer quality rock, production will fall. The concept that high-quality shale gas plays are widespread is false, along with the concept that they are "manufacturing operations", where tens of thousands of wells can be drilled with the same productivity.

The EIA, which is viewed as perhaps the most authoritative source of U.S. energy production forecasts, has often overestimated future oil and gas production.[181] The analysis presented herein suggests that this is the case with respect to shale gas. A play-by-play analysis of the data with respect to the EIA forecasts reveals a high to very high "optimism bias" for most plays. The EIA assumes that 74% to 110% of its "unproved technically recoverable resources as of January 1, 2012" plus "proved reserves" will be recovered by 2040 for most plays. Unproved resources have no price constraints applied and are loosely constrained, compared to "reserves" which are proven to be recoverable with existing technology and economic conditions. Not only do the EIA's projections demonstrate a high or very high optimism bias, they also assume that the U.S. will exit 2040 with shale gas production significantly higher than today, at 54.3 Bcf/d. This is highly unlikely given a thorough analysis of the data.

The major shale plays analyzed in this report have produced just under 45 trillion cubic feet through 2013, and will certainly continue to produce more gas. This report projects that they will produce an additional 230 trillion cubic feet over the 2014-2040 period, with production of 14.8 Bcf/d in 2040, given unconstrained capital input and no restrictions in access to drilling locations. In contrast, the EIA forecasts 377 trillion cubic feet of gas will be recovered from the plays analyzed in this report over this period, and that production will be nearly three times as high in 2040 at 41.8 Bcf/d. Figure 3-117 illustrates the stark difference between the EIA's projections and this report's projections for the seven major shale gas plays analyzed.

---

[180] DI Desktop (formerly HDPI), produced by Drillinginfo.
[181] Hughes, J.D., 2013, *Drill Baby Drill: Can Unconventional Fuels Usher in a New Era of Energy Abundance?*, Post Carbon Institute, http://www.postcarbon.org/publications/drill-baby-drill.

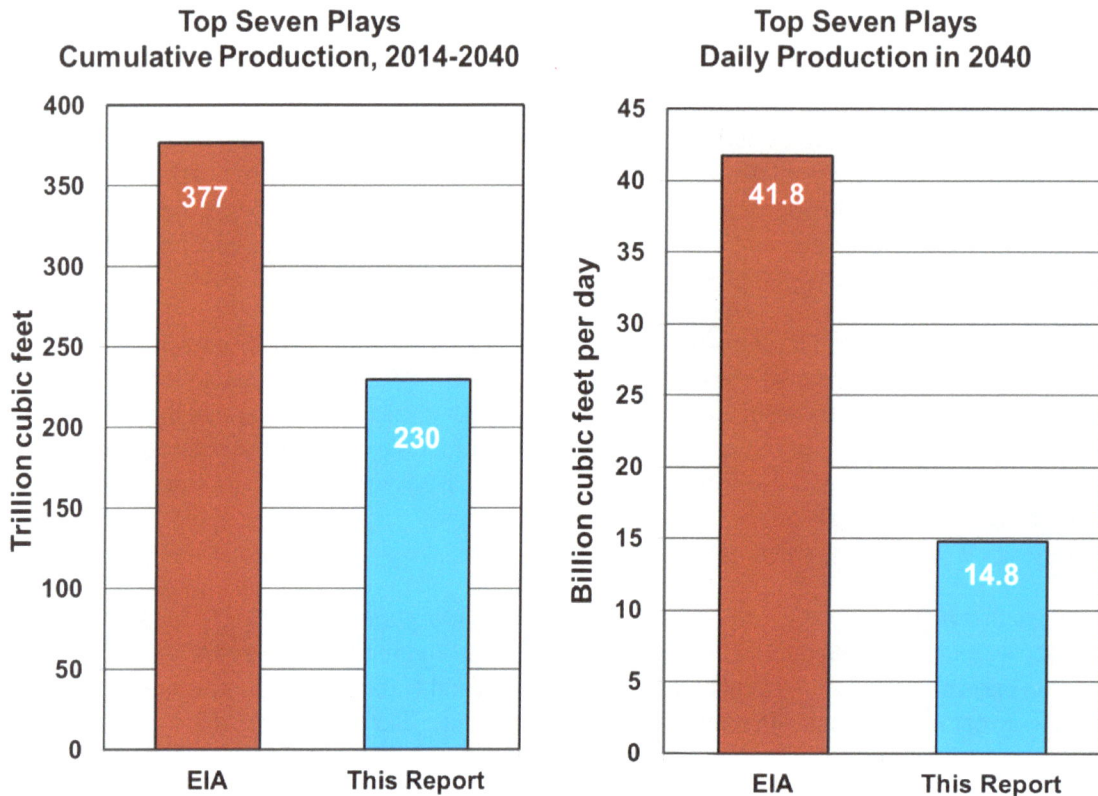

Figure 3-117. Projected cumulative gas production to 2040 and daily gas production in 2040, EIA projection[182] versus this report's projection.

The values given here are for the seven plays analyzed in this report. These plays constitute 88% of cumulative U.S. shale gas production from 2014 to 2040 in the EIA's reference case forecast.

The EIA's forecast strains credibility, given the known decline rates, well quality by area, available drilling locations, and the number of wells that would need to be drilled to make the forecast a reality. Given this report's "Most Likely" scenario estimate for the seven major plays analyzed, the remaining significant U.S. shale gas plays would need to produce 198.2 trillion cubic feet, or nearly 4 times the EIA's own estimate for "other" plays, by 2040. Failing to do this would jeopardize many current and future investments made on the assumption of a cheap, abundant, and long-term domestic gas supply. Most troubling from an energy security point of view is that much of the shale gas production will occur in the early years of this period, when decisions about long-term investment in exports and domestic infrastructure are being made—making any supply constraints later even more problematic.

The consequences of getting it wrong on future shale gas production are immense. The EIA projects that the U.S. will be a significant LNG exporter in 2040 (15% of total production—see Figure 3-2). Although the flush of shale gas production is likely to peak by 2020 and decline thereafter, there are 4 approved, 13 proposed, and 13 potential[183] LNG export facilities under consideration. The wisdom of liquidating as quickly as possible what will likely turn out to be a short-term bonanza should be questioned. A sensible energy policy would be based on this prospect.

---

[182] EIA, *Annual Energy Outlook 2014*, http://www.eia.gov/forecasts/aeo.

[183] FERC, September 30, 2014, "Approved LNG terminals," http://www.ferc.gov/industries/gas/indus-act/lng/lng-approved.pdf; "Proposed LNG terminals," http://www.ferc.gov/industries/gas/indus-act/lng/lng-export-proposed.pdf; "Potential LNG terminals," http://www.ferc.gov/industries/gas/indus-act/lng/lng-export-potential.pdf.

APPENDIX

WELL IP COLOR CODING

In certain Figures in this report, wells are displayed by initial productivity (IP) and color-coded by their approximate percentage rank for the wells for that play. These ranks are:

- Red, top 15% of wells (i.e., above the 85th percentile)

- Orange, next 15% of wells (i.e., between the 70th and 85th percentile)

- Light green, next 15% of wells (i.e., between the 55th and 70th percentile)

- Dark green, next 15% of wells (i.e., between the 40th and 55th percentile)

- Blue, next 20% of wells (i.e., between the 20th and 40th percentile)

- Black, bottom 20% of wells (i.e., below the 20th percentile)

The IP values of the respective categories have been rounded for simplicity. For example, if the lowest IP in top 15% of wells in the Barnett is 3,024 Mcf per day, the lower boundary of this category on the Barnett map will be rounded to 3,000. IP on these maps is defined as the highest one-month production.

ABBREVIATIONS

/d	per day
bbl	barrel
bbls	barrels
Bbbls	billion barrels
Bcf	billion cubic feet
Btu	British thermal unit (1,055 Joules)
CAPP	Canadian Association of Petroleum Producers
EIA	Energy Information Administration of the U.S. Department of Energy
ERCB	Alberta Energy Resources Conservation Board
EUR	estimated ultimate recovery
GDP	Gross Domestic Product
IEA	International Energy Agency, the energy watchdog of the Organization for Economic Cooperation and Development (OECD)
IP	initial productivity (i.e., of a well),typically the highest rate of production over well lifetime achieved in the first month of production
Kbbl	thousand barrels
LNG	liquefied natural gas
Mcf	thousand cubic feet
MMcf	million cubic feet
MMbbl	million barrels
MMBtu	million British thermal units
NEB	Canadian National Energy Board
SAGD	Steam-Assisted Gravity Drainage
Tcf	trillion cubic feet
TRR	technically recoverable resources
URR	ultimate recoverable resources
USGS	United States Geological Survey

GLOSSARY

Basin — A large depressed structural geological entity which is the loci of sedimentation over tens to hundreds of millions of years.

Bench — An informal term applied to discreet rock layers, assumed to be productive, of formations such as the Three Forks in the Bakken Field.

Crude oil — As used herein, conventional crude oil not including natural gas liquids, biofuels or refinery gains. Lease condensate is included in the EIA definition and has been differentiated in this report for plays like the Eagle Ford where it is a significant component.

Dry Gas — Natural gas that has had all impurities removed to end use specifications and is essentially nearly pure methane.

Formation — A formal name in stratigraphic nomenclature for a rock unit with recognizable attributes distributed over a wide area.

Horizontal well — A well typically started vertically which is curved to horizontal at depth to follow a particular rock stratum or reservoir.

Hydraulic fracturing ("fracking") — The process of inducing fractures in reservoir rocks through the injection of water and other fluids, chemicals and solids under very high pressure.

Multi-stage hydraulic-fracturing — Each individual hydraulic fracturing treatment is a "stage" localized to a portion of the well. There may be as many as 30 individual hydraulic fracturing stages in some wells.

Oil shale — Organic-rich rock that contains kerogen, a precursor of oil. Depending on organic content it can sometimes be burned directly with a calorific value equivalent to a very low grade coal. Can be "cooked" in situ at high temperatures for several years to produce oil or can be retorted in surface operations to produce petroleum liquids.

Petroleum liquids (also, "liquids") — All petroleum-like liquids used as liquid fuels including crude oil, lease condensates, natural gas liquids, refinery gains and biofuels.

Play — A prospective area for the production of oil, gas or both. Usually a relatively small contiguous geographic area focused on an individual reservoir.

Raw Gas — Gas as produced at the well head which often contains significant amounts of impurities such as carbon dioxide, hydrogen sulfide, nitrogen, and water vapor, as well as other contaminants and hydrocarbon liquids. Gas cleanup to a "dry basis" will result in shrinkage, which is variable depending on the reservoir, and may range from less than 3% to more than 12% by volume.

Reserve — A deposit of oil, gas or coal that can be recovered profitably within existing economic conditions using existing technologies. Has legal implications in terms of company valuations for the Securities and Exchange Commission. A detailed classification scheme is available from the SPE.[1]

---

[1] Society of Petroleum Engineers, *Guidelines for Application of the Petroleum Resources Management System*, November 2011, http://www.spe.org/industry/docs/PRMS_Guidelines_Nov2011.pdf.

Resource — Energy resources inferred to exist using probabilistic methods extrapolated from available exploration data and discovery histories. Usually designated with confidence levels. For example, P90 indicates a 90% chance of having a least the stated resource volume whereas a P10 estimate has only a 10% chance. Resources may be "in situ", which are all resources thought to exist in place, or "technically recoverable" but without any implied price needed for economic recovery. Shale gas and oil resources are referred to by the EIA as "unproved technically recoverable."

Risked scenario — The reduction of play area to account for the "risk" that all parts of a play will not be accessible for drilling (allowing for towns, parks etc.). This reduces the number of available drilling locations and therefore the ultimate production from a play.

Shale gas — Gas contained in shale with very low permeabilities in the micro- to nano-darcy range. Typically produced using horizontal wells with multi-stage hydraulic fracture treatments.

Shale oil—See "tight oil."

Stripper well—An oil or gas well that is nearing the end of its economically useful life. In the U.S., a "stripper" gas well is defined by the Interstate Oil and Gas Compact Commission as one that produces 60,000 cubic feet (1,700 m^3) or less of gas per day at its maximum flow rate. Oil wells are generally classified as stripper wells when they produce ten barrels per day or less for any 12-month period.

Tank-to-wheels emissions—Emissions generated from burning gasoline or diesel fuel not considering the emissions in the extraction and refining process.

Tight oil—Also referred to as shale oil. Oil contained in shale and associated clastic and carbonate rocks with very low permeabilities in the micro- to nano-darcy range. Typically produced using horizontal wells with multi-stage hydraulic fracture treatments.

Well decline profile—The average production declines for all wells in a given area or play from the first month on production. For most shale plays there are only four or five years of data given their relative youth, although operators routinely fit hyperbolic and/or exponential functions to this data and extrapolate well lives of 25 or more years. Also known as a well decline curve.

Well-to-wheels emissions—Full cycle emissions including those associated with extraction, refining and burning at point of use.